Clean Utilization of Coal

Coal Structure and Reactivity, Cleaning and Environmental Aspects

NATO ASI Series

Advanced Science Institutes Series

A Series presenting the results of activities sponsored by the NATO Science Committee, which aims at the dissemination of advanced scientific and technological knowledge, with a view to strengthening links between scientific communities.

The Series is published by an international board of publishers in conjunction with the NATO Scientific Affairs Division

A Life Sciences **B Physics**	Plenum Publishing Corporation London and New York
C Mathematical and Physical Sciences **D Behavioural and Social Sciences** **E Applied Sciences**	Kluwer Academic Publishers Dordrecht, Boston and London
F Computer and Systems Sciences **G Ecological Sciences** **H Cell Biology** **I Global Environmental Change**	Springer-Verlag Berlin, Heidelberg, New York, London, Paris and Tokyo

NATO-PCO-DATA BASE

The electronic index to the NATO ASI Series provides full bibliographical references (with keywords and/or abstracts) to more than 30000 contributions from international scientists published in all sections of the NATO ASI Series.
Access to the NATO-PCO-DATA BASE is possible in two ways:

– via online FILE 128 (NATO-PCO-DATA BASE) hosted by ESRIN, Via Galileo Galilei, I-00044 Frascati, Italy.

– via CD-ROM "NATO-PCO-DATA BASE" with user-friendly retrieval software in English, French and German (© WTV GmbH and DATAWARE Technologies Inc. 1989).

The CD-ROM can be ordered through any member of the Board of Publishers or through NATO-PCO, Overijse, Belgium.

Series C: Mathematical and Physical Sciences - Vol. 370

Clean Utilization of Coal
Coal Structure and Reactivity, Cleaning and Environmental Aspects

edited by

Yuda Yürüm
Department of Chemistry,
Faculty of Engineering,
Hacettepe University,
Ankara, Turkey

Kluwer Academic Publishers

Dordrecht / Boston / London

Published in cooperation with NATO Scientific Affairs Division

1992

Proceedings of the NATO Advanced Study Institute on Chemistry and Chemical Engineering of Catalytic Solid Fuel Conversion for the Production of Clean Synthetic Fuels
Akçay, Turkey
July 21–August 3, 1991

Library of Congress Cataloging-in-Publication Data

Clean utilization of coal : coal structure and reactivity, cleaning, and environmental aspects / edited by Yuda Yürüm.
p. cm. -- (NATO ASI series. Series C, Mathematical and physical sciences ; vol. 370)
Lectures presented at an institute held in Akçay, Edremit, Turkey, Jul. 21-Aug. 3, 1991.
Includes index.
ISBN 0-7923-1730-0 (alk. paper)
1. Coal--Cleaning--Congresses. 2. Coal gasification--Congresses. 3. Coal liquefaction--Congresses. I. Yürüm, Yuda, 1946- II. Series: NATO ASI series. Series C, Mathematical and physical sciences ; no. 370.
TP325.C488 1992
662.6'23--dc20 92-8533

ISBN 0-7923-1730-0

Published by Kluwer Academic Publishers,
P.O. Box 17, 3300 AA Dordrecht, The Netherlands.

Kluwer Academic Publishers incorporates the publishing programmes of D. Reidel, Martinus Nijhoff, Dr W. Junk and MTP Press.

Sold and distributed in the U.S.A. and Canada
by Kluwer Academic Publishers,
101 Philip Drive, Norwell, MA 02061, U.S.A.

In all other countries, sold and distributed
by Kluwer Academic Publishers Group,
P.O. Box 322, 3300 AH Dordrecht, The Netherlands.

Printed on acid-free paper

Printed in the Netherlands

CONTENTS

FOREWORD

This volume contains the lectures presented at the Advanced Study Institute on "Chemistry and Chemical Engineering of Catalytic Solid Fuel Conversion for the Production of Clean Synthetic Fuels" which was held at Akçay, Edremit, Türkiye, between July 21 and August 3, 1991. The book includes 23 chapters originally written for the meeting by distinguished scientists an technologists in the field. I would like to acknowledge the contribution of each of the authors in the book. Their efforts have shed light on our understanding in coal science research and better utilization of coal.

Three main subjects: structure and reactivity of coal; cleaning of coal and its products, and factors affecting environmental balance of energy usage and solutions for future, were discussed in the Institute and these are presented under six groups in the book. I hope that the book will be of great use to research workers from academic and industrial background.

Many people contributed to the success of the Institute on which this volume was based. I take this occasion to thank my colleagues who lectured in the Institute, both for their efforts during the two weeks and their expertly prepared lecture notes that reached to me in time.

The Institute was generously sponsored by the Scientific and Environmental Affairs Division of the NATO and their contribution is deeply acknowledged.

Finally, I acknowledge the people who contributed to the organizational aspects of the Institute. I include in this list for her endless encouragements my wife Perla Yürüm and for their efforts my graduate students Dr. Murat Azık, Abdülkerim Karabakan, Nurşen Altuntaş, Faize Akçaba and Demet Erçin. Special thanks are due to my student Demet Güler for typing of some of the manuscripts.

I should also convey my appreciation to the publishers themselves for their customary efficiency in bringing out this volume.

Ankara
February 1992

YUDA YÜRÜM
Director
NATO ASI on
Chemistry and Chemical Engineering
of Catalytic Solid Fuel Conversion for
the Production of Clean Synthetic Fuels

FOREWORD

LIST OF CONTRIBUTORS

Professor Larry Anderson
Department of Fuels Engineering
University of Utah
Salt Lake City, UT 84112-1183
U.S.A.

Professor Aysel Atımtay
Department of Environmental Engineering
Middle East Technical University
Ankara 06530
Türkiye

Professor Turgut Balkaş
Department of Environmental Engineering
Middle East Technical University
Ankara 06530
Türkiye

Dr. Frano Barbir
Clean Energy Research Institute
University of Miami
Coral Gables, FL 33124
U.S.A.

Professor Ekrem Ekinci
Department of Chemical Engineering
İstanbul Technical University
İstanbul 80626
Türkiye

Professor Alec F.Gaines
Institute of Marine Sciences
Middle East Technical University
P.O.Box 28
Erdemli, İçel 33731
Türkiye

Professor John W. Larsen
Department of Chemistry
Lehigh University
Bethlehem, Pennsylvania 18015
U.S.A.

Professor Jacob A.Moulijn
Department of Chemical Engineering
Delft University of Technology
Julianalaan 136
2628 BL Delft
Holland

Prof. Harold H.Schobert
Fuel Science Program
209 Academic Projects Building
The Pennsylvania State University
University Park, PA16802
U.S.A.

Professor Abdurrahman Tanyolaç
Department of Chemical Engineering
Hacettepe University
Beytepe
Ankara 06532
Türkiye

Professor Nejat Veziroğlu
Clean Energy Research Institute
University of Miami
Coral Gables, FL 33124
U.S.A.

Professor Yuda Yürüm
Department of Chemistry
Hacettepe University
Beytepe
Ankara 06532
Türkiye

PARTICIPANTS

F. Akçaba	Department of Chemistry, Hacettepe University, Ankara 06532, Türkiye
N. Akın	Department of Chemical Engineering, Boğaziçi University, İstanbul, Türkiye
N. Alpan	Department of Chemistry, Hacettepe University, Ankara 06532, Türkiye
N. Altuntaş	Department of Chemistry, Hacettepe University, Ankara 06532, Türkiye
H. Atakül	Department of Chemical Engineering, İstanbul Technical University, İstanbul 80626, Türkiye
M.Azık	Department of Chemistry, Hacettepe University, Ankara 06532, Türkiye
S. Besler	Department of Fuel and Energy, The University of Leeds, England
E. Bolat	Department of Chemical Engineering, Yıldız University, İstanbul, Türkiye
I. Cabrita	Ministerio da Industria a Energia, Laboratorio Nacional de Engenharia e Tecnologia Industrial, Lisbon, Portugal
A. Çalımlı	Department of Chemical Engineering, University of Ankara, Ankara, Türkiye
K. Ceylan	Department of Chemical Engineering, Inönü University, Malatya, Türkiye
T. Dağlı	Marine Sciences and Enviromental Research Group, TUBITAK, Ankara, Türkiye
B. Demirel	Department of Chemical Engineering, University of Ankara, Ankara, Türkiye
Z. Doğan	Department of Mining Engineering, İstanbul Technical University, İstanbul, Türkiye
G. Erbatur	Department of Chemistry, Çukurova University, Adana, Türkiye
O. Erbatur	Department of Chemistry, Çukurova University, Adana, Türkiye
D. Erçin	Department of Chemistry, Hacettepe University, Ankara 06532, Türkiye
M.I. Erol	Department of Chemical Engineering, University of Ankara, Ankara, Türkiye
J.L. Figueiredo	Departamento de Engenheria Quimica, Universidade do Porto, Porto, Portugal
D. Güler	Department of Chemistry, Hacettepe University, Ankara 06532, Türkiye
J.B. Hansen	Haldor Topsoe, Nymollevej 55, P.O.B 213, DK-2800 Lyngby, Copenhagen, Denmark
F. Kapteijn	Department of Chemical Engineering, University of Amsterdam, 1018 WV Amsterdam, The Netherlands
A.Karabakan	Department of Chemistry, Hacettepe University, Ankara 06532, Türkiye
A. Koyun	Department of Mechanical Engineering, Yıldız University, İstanbul, Türkiye
J.J.G. Martins	Departamento de Ciencias Engenheria, Universida do Minho, Guimaraes, Portugal
A.M. Mastral	CSIC, Inst. De Carboqumica, Apartado 589, 50080 Zaragoza, Spain
C. Mayoral	CSIC, Inst. De Carboqumica, Apartado 589, 50080 Zaragoza, Spain
P.E.H. Nielsen	Haldor Topsoe, Nymollevej 55, P.O.B 213, DK-2800 Lyngby, Copenhagen, Denmark
M. Paolucci	Dipartimento di Chimica, Universita Degli Studi di Roma, La Sapienza, Roma, Italy
J.J. Pornsakul	Department of Fuels Engineering, University of Utah, Salt Lake City, UT 84112, USA
E. Pütün	Department of Chemical Engineering, Anadolu University, Eskişehir, Türkiye
A. Pütün	Department of Chemistry, Anadolu University, Eskişehir, Türkiye
M. Seferinoğlu	Department of Chemistry, Hacettepe University, Ankara 06532, Türkiye

E.H. Şimşek	Department of Chemical Engineering, University of Ankara, Ankara, Türkiye
C. Snape	Department of Pure and Applied Chemistry, University of Strathclyde, Glasgow, Scotland
F. Söğüt	Department of Chemical Engineering, University of Ankara, Ankara, Türkiye
T. Sonino	Sonino and Associates, 16 Beeri Street, Rehovot 76352, Israel
B.Tantekin	Department of Chemical Engineering, İstanbul Technical University, İstanbul 80626, Türkiye
M. Tolay	Istituto Guido Donegani, Gruppo Enichem, Novara, Italy
P.J. Tromp	Department of Chemical Engineering, University of Amsterdam, 1018 WV Amsterdam, The Netherlands
S. Ünal	Department of Chemistry, Marmara University, İstanbul, Türkiye
T.S. Uyar	Marmara Research Center, TUBITAK, Gebze, Turkey
J.M. Vleeskens	SBN, POB 1, 1755 ZG Petten, The Netherlands
M. Weeda	Department of Chemical Engineering, University of Amsterdam, 1018 WV Amsterdam, The Netherlands
A. Williams	Department of Fuel and Energy, The University of Leeds, England
G. Yalın	Department of Chemical Engineering, Yıldız University, İstanbul, Türkiye
A. Yıldız	Department of Chemistry, Hacettepe University, Ankara 06532, Türkiye
W.H. Yuen	Department of Fuels Engineering, University of Utah, Salt Lake City, UT 84112, USA

THE PHYSICAL AND MACROMOLECULAR STRUCTURE OF COALS

J. W. LARSEN
Department of Chemistry
Lehigh University
Bethlehem, Pennsylvania 18015
U.S.A.

ABSTRACT. Coals are surely the most complex materials whose structure determination has ever been attempted by chemists. It is an endlessly fascinating material; each new level of understanding raising yet more questions and each new technique providing answers and new questions in equal measure. It is clear from this brief discussion that there are major areas of coal physical structure where we have essentially no experimental information. There are areas where our understanding is not limited by our observations, but by theory and our ability to understand our observations. In all of this, the interrelationship between coal reactivity and structure has been alluded to only briefly. One of the reasons for this is that, except under conversion conditions, there are a very limited number of reactions which have been performed on coals in a manner such that relative reactivities could be determined. Coals are fascinating materials. They are worth studying not only because such studies may have practical impact, but also because they provide a formidable challenge to both the experimental tools and the theoretical insight of chemists.

Other models of coal structure differing from this one in fundamental ways exist. The central feature of the differences is the importance and/or the spatial distribution of network active covalent bonds. The vitrinite structure discussed here has a uniform spatial distribution of such covalent bonds and a population of network active covalent bonds which continuously and smoothly decreases with increasing coal rank. Changes in either of these features lead to large changes in the structure model. One of the more interesting alternatives has been nicely summarized by Krichko and Gagarin. Their picture of cross-linked regions held together by non-covalent bonds is a modern, more quantitative version of the old "micelle model". A detailed comparison of the predictions of various models with experimental obsevations is beyond the scope of this article and will be the subject of another paper.

1. Introduction

Coals are extraordinarily complex materials, so complex that many kinds of data must be assembled and integrated to develop structure models. Their physical structure has not always been a popular research topic and there are many structural aspects for which relevant data are lacking or are contradictory. Research on coal physical structure has increased recently. New techniques are being applied and new kind of data are becoming available. Any structure must be tentative because of the material's complexity, because of the paucity of data and because of the new information becoming available will make possible more detailed and more correct models. The structural conclusions presented here are based on my interpretation of the available data data. I am confident of some features while other important features are uncertain. This model will change as more data become available and is best regarded as a "working model" for coal physical structure, not the final answer.

This being the situation, it may be necessary to justify producing a summary article on structure. One justification is that this activity is always worthwhile. Almost all scientific models are changing, either being elaborated or corrected. They differ in the rate of change. The rate of change of coal structure models is fast at present. A structure article will become obsolete quickly, but the fast pace of change demands the discipline of written summaries to

Y. Yürüm (ed.), Clean Utilization of Coal, 1–14.

test, to compare and to identify areas where more work is necessary. To know where one is going, it is necessary to know where one is. Summaries are useful to focus attention on inadequately explored areas and to draw attention to flaws by exposing different aspects of the models which do not fit well together. So there is value in setting down on paper an integrated structure model, even an imprecise one.

The comments in this article are limited to the vitrinite maceral. The structure will be presented in broad strokes with emphasis placed on major structural features. The arguments will be quite general and the supporting references will be illustrative, not exhaustive. The interplay of physical structure and chemical reactivity of coals is not well established, except possibly in the area of pyrolysis. A discussion of this relationship will not be part of this article.

This article is organized so that it moves from bulk properties down in scale to the atomic/molecular level. It begins with a discussion of the physical form of coals: Are they glassy or rubbery? How are these two forms interconverted and what are the different properties of the two forms? The next section deals with the macromolecular architecture and is an attempt to define the network structure of coals and some of the ways in which the network behaves. Finally, the last section deals with the atoms and molecules in coal and the nature of the interactions between them. A detailed catalog of the functional groups in coals and their populations will not be presented. The emphasis throughout will be on the physical structure, not chemical structure.

2. Solid Structure of Coal

2.1. GLASSES, RUBBERS AND THEIR INTERCONVERSION

Coals are glassy, strained, macromolecular solids. The first clear recognition that coals were glassy solids seems to have been made by Brenner (1). He associated the optical anisotropy of Illinois No. 6 coal viewed in thin section through a polarizing microscope with the existence of strain in a glassy system. He showed that this strain could be removed if the coal was swollen with a good swelling solvent such as pyridine. Furthermore, exerting pressure on the swollen coal deformed it and reestablihed optical evidence for strain. In his ground-breaking work, Brenner established that untreated coals were strained, that solvent swelling could lead to relaxation of strain and that the swollen coals were rubbery (1,2). Peppas first identified the thermal glass to rubber transition of coals in a careful differential calorimetry (DSC) study (3). He reported that a second order phase transition existed at around 325°C for all of the six (70% to 94%C dmmf) coals studied. This phase transition was assigned to a glass to rubber transition. Confirmation of this assignment was made two months later in the NMR study published by Lynch (4). In this work, the proton spin lattice relaxation times of many coals were measured as a function of temperature. It was demonstrated that a major increase in structural mobility occurred in most coals at similar temperatures to those of the second order phase transition reported by Peppas. Taken together, these two papers establish that a wide variety of coals are thermally transformed from glasses to rubbers at temperatures on the order of 300°C to 350°C.

The introduction of liquids into most glassy polymers results in adecrease in temperature of the glass to rubber transition (T_g). This behaviour was observed by Peppas who studied coals swollen with pyridine and observed a decrease in the temperature at which second order phase transition occurred (3). However, the temperature at which this transition occurred never dropped below 122°C. These observations seem to contradict those of Brenner, whose optical studies demanded that a glass to rubber transition occur below room temperature for Illinois No. 6 coal swollen with pyridine (1). This discrepancy seems to have been resolved very recently in a paper by Peter Hall which reported a second order phase transition for pyridine and N-methylpyrollidone (NMP) swollen coals at temperatures as low as 170 K (5). The existence of the high temperature phase transformation in swollen coals observed by

Peppas does not preclude the existence of a much lower temperature phase change. The very low temperature phase change could well be glass to rubber transition consistent with Brenner's observations. Its complete characterization awaits the results of spectral studies now underway. The maximum pyridine weight fraction in the swollen coals used by Peppas was about 0.45, while the coal studied by Hall was much more swollen (weight fraction pyridine 0.72). Peppas presented surprising data that the phase transition temperature was not strongly dependent on pyridine concentration above a weight fraction of 0.3. The data are scattered and the difference in the two results may be due to the amount of pyridine dissolved in the coals.

Glassy coals and rubbery coals are startlingly different materials (6). In the glassy state, molecular motion is very limited. Diffusion rates are very low because diffusion through the macromolecular solid involves moving portions of the macromolecule around to allow of the passage the diffusing molecule. Thermal bond cleavage in a glassy material will leave the fragments close to each other, trapped in a cage which presents them with a very limited set of potential reaction partners, those groups on the inside of their small cage. In a rubbery solid, molecular motion is similar to that in a non-cross-linked polymer solution of the same composition. Bulk diffusion is much more rapid, often by as much as 10^3. Reactive intermediates will have same opportunity to move about to seek the lowest energy reaction pathway. Cage effects, while certainly not absent, will be much smaller. In order to remove diffusional and steric constraints, it is best to run chemical reactions on rubbery coals rather than on glassy ones (7).

2.2. STRAIN AND STRUCTURAL ANISOTROPY

Coals have many anisotropic properties and this anisotropy is rank dependent. Only above about 85% carbon (dmmf) do easily detectable anisotropic properties regularly occur (Table 1). For low volatile bituminous coals and anthracites, molecular level anisotropy is quite apparent in X-ray diffraction results (9). The aromatic ring systems tend to be oriented parallel to the bedding plane. For lower rank coals, for example Illinois No. 6, there is little evidence for anisotropy in the bulk structural properties listed. X-ray results reveal only a small amount of ordering for coals of this rank.

TABLE 1. Some anisotropic properties of coals, [after van Krevelen (8)].

Property	Anisotropy appears above this rank, (%C, dmmf)
Vitrinite reflectance	85
Semi-conductance	94
Elasticity	94
Young's modulus	92
Thermal expansion	85

It was vey surprising that the initial solvent swelling of six low rank coals is highly anisotropic, even for a lignite (10). The experiment involved swelling thin sections of coals with three solvents: pyridine, THF, or chlorobenzene. Repeated swellings using pyridine were done. All coals swelled more perpendicular to the bedding plane than parallel to it in all three solvents. With pyridine, the linear expansion of the coal is approximately twice as great perpendicular to the bedding plane as parallel to it. Evaporation of the pyridine shrinks the coal thin section to a particle having different dimensions than the original particle. It is thicker (expanded to the perpendicular to the bedding plane) and less broad (contracted parallel to the bedding plane). Repeated swellings now return the coal to its original expanded rubbery state and removal of the solvent returns the coal to its new shape. The reversibility indicates that it is the new thicker shape which is the thermodynamic ground state and not the original shape (1), (see Figure 1).

This enormous anisotropy called for a re-examination of the orientation of organic groups in coal. Three different experimental approaches were used: IR linear dichroism (11); optical birefringence (12); and NMR (13). All three confirmed the essentially isotropic molecular structure of low rank coals. There is only a slight (perhaps 1.5^{o}) orientation of the aromatic ring systems parallel to the beding plane.

An explanation of this behavior must be consistent with the following facts. The organic structure of the coal is essentially isotropic on the molecular level. The initial swelling of coal to rubbery state followed by solvent removal leads to a change in the coal shape. Subsequent swellings are reversible and the swollen state is the same as that reached in the initial swelling. The explanation was originally proposed by Dr.Gary Carlson of Sandia National Laboratory (14). The coal as mined is strained. The coalification process taking place under great pressure and for long times deforms the coal, compacting it perpendicular to the bedding plane and expanding in the bedding plane. It is squeezed and flattened. Because it is a glassy solid, these deformations are locked in and not released when the coal is mined. It is locked into a deformed, high energy state. When the coal is swollen and becomes rubbery, the macromolecules can move about much more freely and the strain is relaxed. As the swelling solvent is removed, the chains still have enough freedom of motion in the slowly contracting material to find their lowest free energy packing arrangement. This involves a different set of interactions than the starting coal and results in a material of lower energy and different shape.

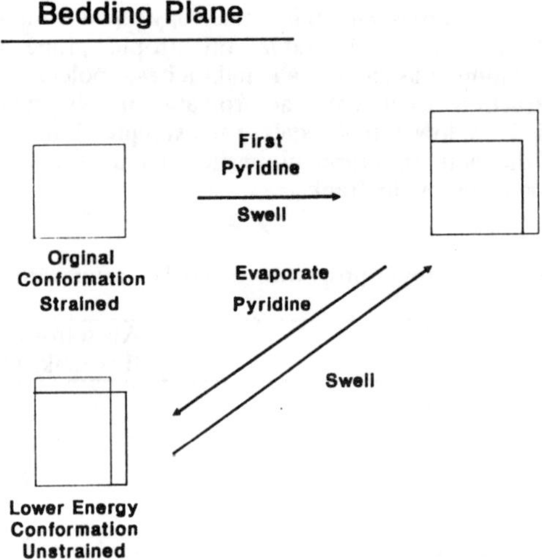

Figure 1. Diagram of size and shape changes during coal swelling.

The great anisotropy in coal swelling is not due to anisotropic organization of organic functional groups. These groups are arranged essentially randomly in bituminous and sub-bituminous coals (except for low-volatile bituminous). The anisotropy is the result of the locked-in deformation of glassy coals. The locked-in strain is certain to have a large influence on coal properties. There are as yet very few measurements of the properties of relaxed (unstrained) coals. These few are discussed below.

Because native coals are strained, a driving force for their conformational rearrangement to a less strained state exists. A coal with limited internal mobility should slowly alter its conformation, moving from its initial higher free energy strained state to a lower free energy, more highly interacting, unstrained state. Limited structural mobility can be conferred by partial solvent swelling. Such swelling should result in slow rearrangement of coals. Evidence for this has recently been published (15,16). The pyridine extraction of coals results in a conformational isomerization to a lower free energy state as evidenced by solvent swelling and extractability changes. If Illinois No. 6 coal is warmed in the poorly swelling solvent chlorobenzene, it undergoes a transformation taking 2 weeks as it slowly relaxes to lower free energy, less strained state. During this process, both extractability and solvent swelling decrease. A very limited amount of data has been collected on the reactivity of rearranged coals. Rearranged Illinois No. 6 coal is significantly less reactive in low temperature conversion to pyridine solubles by reaction with tetralin (17). The rearrangement of Pittsburgh No. 6 coal in chlorobenzene gives a coal whose heat capacity is half that of the starting strained coal (18). The hysteresis in solvent uptake observed by Duda and Hsieh is best explained by slow rearrangement of the coal during solvent uptake (19). Not have been given enough time to reach its ground state, each exposure to solvent yielded a different structure on its removal. Reports of reactivity increases caused by solvent swelling followed by solvent removal have recently appeared (20,21). These reactions were carried out under H_2 pressure, but were otherwise similar to those reported by Larsen (17) who reported a reactivity decrease. Solvent swelling presumably gives rise to a rearranged, more highly associative structure. It is difficult to understand how such a rearrangement would increase conversion in some cases and decrease it in others. The relationship between coal conformation and reactivity is not understood.

2.3. PORE STRUCTURE AND DIFFUSION

This view of coal as a glassy macromolecular solid can be used together with an analysis of gas adsorption data to yield a model for coal pore systems which is importantly different from the accepted view. The accepted view of coal pore structure was proposed some time ago by Bond and holds that coals contain an interconnected network of slit-like pores involving numerous bottlenecks (22-24). In this model, a molecule could diffuse through the pore system and if it could squeeze pass the bottlenecks, would eventually reach the entire internal surface area of coals. This internal surface area was thought to be accurately measured by CO_2 adsorption and usually to be between 100 and 400 m^2/g. It is known to be rank dependent (25). If N_2 at liquid nitrogen temperatures was used to measure the surface area by adsorption, much lower values were found (25). These discrepancies were reconciled by asserting correctly that pore diffusion is an activated process and N_2 at liquid nitrogen temperatures does not have sufficient thermal energy to overcome the activation barrier to pass the bottlenecks in the pore network. It thus reports only the external surface area. I believe this picture is incorrect. The correct model is isolated coal pores, like bubbles in a solid.

A recent BET study of the adsorption small hydrocarbon gases, CO_2 and N_2 by Illinois No. 6 coal contains data which are inconsistent with the interconnected pore structure (26). The data are presented in Figure 2. Molecules of very similar size, for example CO_2 and ethane, give startlingly different surface areas. If diffusion through a bottleneck pore system was involved, molecules of similar size should report very similar surface areas. Also, molecules of similar size but very different shape often report similar surface areas. One would expect molecules having different dimensions to have different accessibilities to different sections of the pore system and thus report different surface areas. Similar data exist for all of the Argonne coals (27). A startling observation contained in this report is the fact that the surface areas reported measured using BET adsorption on Illinois No. 6 coal varied with the 11th power of the molecular diameter.

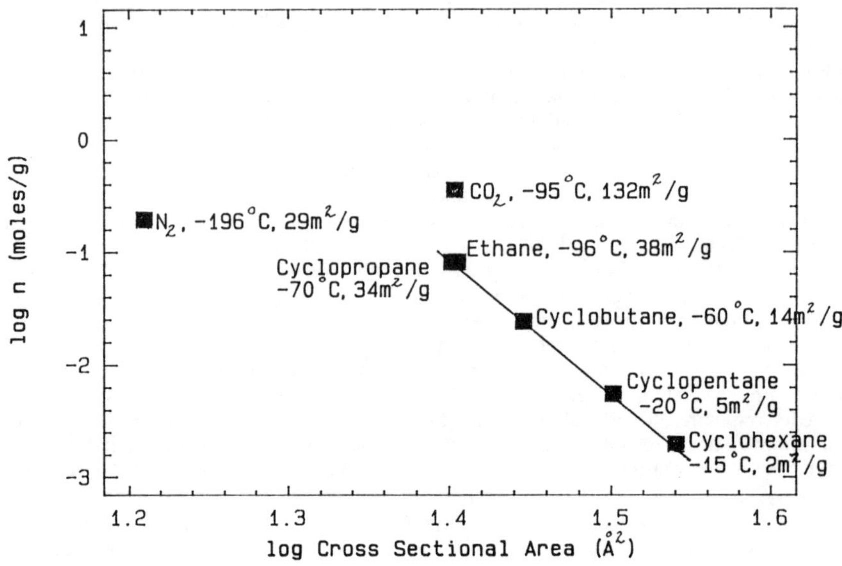

Figure 2. BET tiling experiment on Argonne Illinois No. 6 coal, 77.0% C, 5.7% H, 10.0% O, 5.4% S, 16.0% ash (daf).

These data can be understood much more easily if the coals do not contain an interconnected system of pores, but contain isolated pores. The pores are like a discrete series of bubbles, independent cavities occurring within the solid coal. They have a range of sizes and their surface is fractal with fractal dimensionality varying with rank. Accessibility to the pores involves diffusion through the solid, glassy coal. Small molecules such as hydrogen and helium can readily penetrate the material, so helium densities give an accurate measure of pore volume. Carbon dioxide is known to swell coals slightly and interact with it (28,29). It has rapid diffusion pathways enabling it to reach all of the pore surface by diffusion through the solid. Surface areas measured using CO_2 adsorption are similar to those obtained using small angle X-ray scattering (30). As molecular size increases, diffusion rates into the coal drop off sharply. This is well known with glassy polymers (31,32). Because of their slow diffusion rates, larger molecules will reach only a small fraction of the coal's pore system during the time required for a BET measurement. The diffusion rates are usually so slow that the BET data appear good, that is appear to be equilibrium data, but in fact are not. As molecular size increases, diffusion rates decrease sharply and the surface area reported by the BET measurement also drops. As probe molecule size increases, smaller fractions of the total pore network are being probed. For non-interactive hydrocarbons containing six or more carbons, the diffusion rates are so slow that even at room temperature only the external surface is sampled. The coal "surface" depends strongly on the nature of the molecule which is interacting with coal. Strongly inreacting molecules such as pyridine, THF, small alcohols and I_2 diffuse rapidly throughout the coal (19,33,34). For rapidly diffusing molecules which dissolve in and swell coals, it makes little sense to think of coals as having discrete surface areas. Small molecules such as hydrogen and helium diffuse rapidly because of their small size. Interactive (H-bonding) molecules can usually diffuse through coals failry rapidly. Non-interactive molecules diffuse through coals quite slowly. Unless there is specific chemical interaction with the adsorbate, coals are best regarded as glassy, low-surface-area solids.

3. Macromolecular Architecture

3.1. DEFINITION OF A CROSS-LINKED NETWORK

Before discussing in detail the macromolecular architecture of coals, I wish to define clearly the nomenclature which I will be using. The chief building blocks of coal vitrinites are aromatic and hydro-aromatic systems which I will refer to as clusters. Two possibly imaginary examples of such systems are given below. Each cluster can be linked to one, two, three or more clusters. I will refer to the number of links of one cluster to other clusters as its

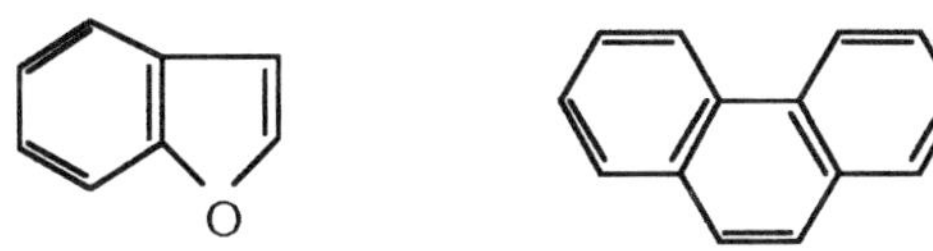

valence. For example, if two clusters shown were linked to each other by an ether group, both would be monovalent. If the clusters in a piece of vitrinite were all monovalent or divalent, the vitrinite would consist of a group of linear macromolecules whose average chain length (number of clusters in a chain) depends upon the ratio of divalent to monovalent clusters. The incorporation of trivalent clusters into such a system results in the formation of a macromolecular network. It is by definition cross-linked and insoluble. In its limiting form, all the clusters are linked together to form a single very large molecule comprising the piece of vitrinite. A cluster which is trivalent or higher valent will be called a branch point. If only a few trivalent clusters are present in the system, one will have a very loose network with high value for the number average molecular weight between cross links (M_c). This is perhaps the most useful way to describe the network. The term cross-link density is commonly used, but I wish to avoid it in discussing coals because it implies that there is a special kind of bond called the cross link. Selective cross link cleavage in coals is often discussed as if there were some special cross links to cleave. Any bond which connects one cluster to another can be thought of as a cross link. Breaking any bond which reduces the valence of a cluster makes the network looser; increases M_c. For this reason. I prefer to discuss the network in terms of its M_c value or the population of branch points. In addition to covalent linkages, the clusters will be associated by all the other known types of non-covalent interactions. These will be discussed in detail in the next section.

3.2. NETWORK BEHAVIOR DURING DEPOLYMERIZATION

There is significant difference between the products produced by bond cleavage in a network system and by cleavage in a system of linear macromolecules. This difference is important to coal chemistry. Consider first a linear macromolecule. If a bond linking any pair of the clusters is broken, there now are two molecules whose total mass is the same as the original molecule. The molecular weight will have been cut in half. Continued random cleavage of this linear macromolecule will result in a steady decrease in the average molecular weight. The resulting molecular weight distribution is calculable using statistical mechanics. As random bond clevage continues, the average molecular weight drops if one is dealing with a system of linear macromolecule.

If one is dealing with a three-dimensionally cross linked macromolecular network, random bond cleavage will yield soluble products whose molecular weight will at first increase with increasing bond cleavage and only later will decrease with increasing bond cleavage. Consider random bond cleavage in a macromolecular network. First, cleavage need not free a soluble fragment from the insoluble network. It may simply break a bond between two clusters and effectively increase M_c without liberating a fragment. On average, low molecular weighy portions of the network will be bonded to the rest of the network by few bonds, while very high molecular weight arrays will be bonded to the rest of the network by many bonds. As bonds are broken at random, the first fragments which are liberated will be of low molecular weight because breaking only very few bonds will free them. As the network slowly comes apart under the impact of more and more bond breaking, larger and larger pieces will be freed. The mathematics describing this procedure have been well worked out and in a very interesting example, applied to lignin depolymerization (35). In the early stages of coal depolymerization, the molecular weight of the products can be expected to increase as more bonds are cleaved and more material is freed from the network.

3.3. NATURE OF THE EXTRACTABLES

In addition to the insoluble portion of coals, most coals contain significant amounts of material which dissolve in good solvents and which can be separated from the insolubles. The question of relationship of this material to the insoluble network is an active one (36). In my view, the coal network-extractable system is best treated as a sol-gel system. In this model, the extracts (the sol) are essentially chemically identical to the insoluble portion of the coal (the gel). The only difference between sol and gel is that the extractable molecules have not been cross linked into the network or, more likely, are fragments which have broken off of the network during the coalification process. There is one caveat which must be applied to this model. It results from the great heterogeneity of the material from which coal is formed. There will exist in the coal the products from terpines, resins and all manner of other molecules which are not appropriately functionalized to be incorporated into the coal macromolecular network and which do not start out as part of the lignin, which is the primary precursor of vitrinite. Lignins are cross-linked networks. Some of the unincorporatable materials are part of the extractable portion of coals. I wish to exclude them from the discussion that follows. The extractables consist of a mixture comprised primarily of true sol (the stuff of the network) and a small amount of material which is not lignin derived. This non-sol material tends to be aliphatic in nature and is responsible for the observation that the H/C ratio of extracts is often higher than that of the parent coal (8).

The close resemblance in elemental analysis and functional group distribution between extracts and their parent coals has long been noted (8). A recent NMR study demonstrated that the functional group distribution of the extracts is essentially identical to that of coals from which they are derived (37). This provides strong support for the view that the network and extractables constitute a sol-gel system.

Considering only bituminous coals, for which pyridine is an excellent solvent, the amount of extractable material increases up to about 87% C (dmmf), then drops off rapidly (8). Over this same rank range, a number of groups of investigators have demonstrated that the average molecular weights of those extracts also increase (8,38). This is exactly what is expected of a depolymerizing macromolecular network (39). The coalification of bituminous coals is a net depolymerization. Depolymerization is not the only chemical process going on. More aromatic rings are being formed, there are small increases in the average size if the PNA systems and there are numerous changes in functional groups. But on a macromolecular level, the overall result is an increase in M_c and a decrease in the concentration of branch points in the coals (38,39).

One of the principle chemical changes which is occurring during coalification is deoxygenation. Both phenolic hydroxyl and ether groups are being lost (40). These ethers are important linkages between clusters and their loss should result in an increase in Mc. As these bonds are cleaved, more and more material will be broken off from the network and will become soluble in good solvents such as pyridine. If coals are covalently cross-linked networks, the average molecular weight of this material will be increasing. This is exactly what is observed.

In the next section, I will deal with the reason for the sharp fall off in extractability above 88%. The role of ether linkages in the coalification process will be considered in somewhat more detail.

4. Bonding and Interactions

4.1. IMPORTANT ASSOCIATIVE INTERACTIONS

In addition to the covalent bonds holding atoms together in the coal, there are a variety of non-covalent interactions which are responsible for attractive forces existing between groups of atoms in the coal. In this section, these interactions will be briefly individually described and a guess made as to their rank dependence.

Ionic bonds exist in lignite and sub-bituminous coals (4). These involve primarily the catboxylate anion and a variety of cations, including alkali metals and alkaline earths. These ionic bonds may serve to link two clusters together and thus form part of the macromolecular network. This can be visualized most easily as a calcium ion associated with two different carboxylate anions, one from each of two different clusters. This would form an ionic linkage between the clusters and the ionic bond would be part of the network structure.

Another effect of the ionic groups is to decrease the organic solubility of the coal molecules containing the ionic groups. Most organic solvents used for exhaustive coal extraction (e.g. THF or pyridine) are not good solvents for carboxylate salts. Otherwise soluble molecules which happen to contain one or more carboxylate salts will be insoluble in the extracting solvent due to the presence of the ionic group. One study of the solvent swelling and extraction of low rank coals before and after demineralization led to the conclusion that the effect of ionic groups on pyridine extractability was due more to decreases in solubilty greater than that its role as a linkage in the macromolecular network (41). For numerous good reasons, this was a tentative conclusion and the structural role of ionic linkages in coal remains a major area deserving careful investigation.

Phenolic hydroxyl is one of the major oxygen functional groups in coals and most of the hydroxyls are hydrogen bonded (42,43). It has amply been demonstrated that hydrogen bonds are an important associative interaction holding the network together, that there are a range of hydrogen bonds structures and energies and that coal hydrogen bonds can be broken and replaced by a coal solvent hydrogen bonds to varying extent depending upon the basicity of the solvent (42-44). It is also worth pointing out that O-H""O hydrogen bonds are long interactions. The density of ice, which is more hydrogen bonded than water, is less than the density of water because of the long O-H""O hydrogen bond distance. Extensively hydrogen bonded structures in coals will tend to be more open structure with groups held at long distances by the hydrogen bonds.

Dipole-dipole interactions will be important in coals. A consideration of any of the published coal structures reveals a high population of heteroatom sustituted aromatics and hydroaoromatics, most of which have substantial dipole moments. These will have a tendency to associate with each other to maximize the interaction energy.

In a similar way the interaction of a permanent dipole with a readily polarizable large aromatic system can be significant (45). This is another short range ($1/r^6$) interaction which will be an important associative force in all coals. Again. it only takes some consideration of the type of structures thought to be present in coals to make it clear that such interactions will be important.

Finally, the induced dipole-induced dipole, or London interactions, which occur between all atoms and groups of atoms, will be present in coals. These will be especially strong in materials which have π systems so extensive that they are colored (46). It is only recently that the collective role of these interactions in coal structure has been explored (15,47).

It has recently been asserted that charge transfer interactions between aromatic clusters will be an important associative interaction in coals (47,48). Charge transfer between neutral aromatic donors and acceptors is certainly not an important associative force. Both theory and experiment agree that the associative interactions in the ground state of a charge transfer complex is that collection of noncovalent interactions which would normally occur between the two partners (49,50). The existence of charge transfer from one of the partners to the other on excitation does not stabilize the ground state. There is nothing extra contributed to the ground state interaction by the existence of charge transfer in an excited state. The assignment of a special role to simple pairwise charge transfer interactions in coal chemistry is without merit (51).

The matter becomes significantly more complicated if cooperative interactions between π systems are considered. It has been known for years that π systems in high rank coals were sufficiently delocalized for electronic transitions to lie in the infrared range (8). The π systems are so extensive that the HOMO-LUMO gap has shrunk to IR energies. Recent studies have demonstrated the occurence of this IR electronic excitation in coals containing as little as 86% carbon (12). The extensive overlap of π systems to form extended arrays is possible and occurs. The energetic consequences of this associative interaction in coals have not been considered in detail. If one considers the strength of these interactions in crystalline polynuclear aromatic molecules, it seems likely that these highly extended cooperative interactions involve strong forces. Coronene melts at 438°C and the solubility of naphthalene and other small PNA's is very low in most solvents. The interactions due to the extended overlap of π systems probably confer significant stability.

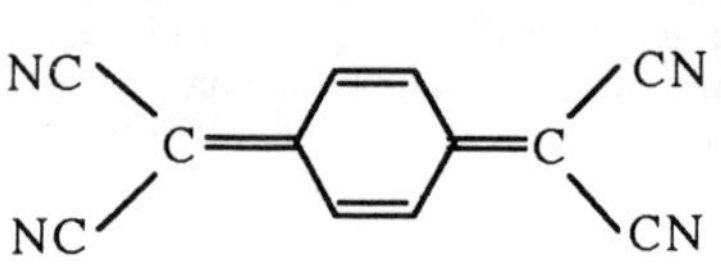

The addition of good single electron acceptors such as tetracyanoquinodimethane (TCNQ) and tetracyanoethlene (TCNE) results in the formation of large scale electronic valence bands caused by the presence of the electron acceptors which form part of the extended π systems (54). The bright blue or bronze colored coals which result are most interesting materials. It seems likely that the cooperative overlap of π systems of many individual clusters to form very extended π systems is an important interaction in coals.

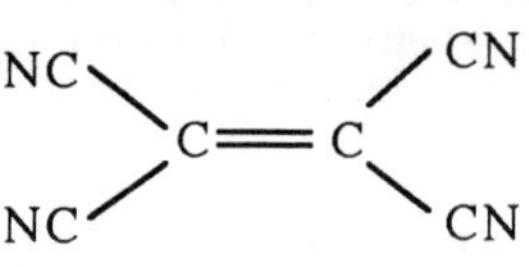

Cooperative aromatic-aromatic interactions should be especially important in high rank coals. Several papers have demonstrated the importance of non-covalent associations in such coals, Stock and Mallya compared the extractability of methylated and buthylated coals (52). For coals containing more than 86% C, there are striking extractability increases caused by butylation, but a negligible increase caused by methylation. The larger butyl group sterically disrupts aromatic-aromatic interactions and enhances extractability by destabilizing the coals. The smaller methyl group causes little steric disruption and so does not enhance extractability. The reaction of maleic anhydride with coals has a similar effect and an identical explanation has been offered (15,53). Finally, heating coals with solvents are themselves quite polarizable also results in enhanced extraction yield (15). All three papers illustrate the importance of non-covalent associative interactions in high rank coals.

4.2. THE COALIFICATION PROCESS

In this section I shall attempt to trace in a qualitative fashion the changes in all of these interactions as the rank of the coal increases. This will be done in a highly qualitative manner with only the major trends identified. Quantitative estimates of the changes in the populations of the interactions or even of their relative importance cannot be made at this time. Again, these comments are limited to vitrinite.

Vitrinite is formed from lignin and the changes which occur as lignin is transformed to lignite and sub-bituminous coals have recently been discussed by Pat Hatcher (55). Lignin is itself a macromolecular network through the initial coalification processes. It does not depolymerize to a collection of monomers which then repolymerize. What it does do is lose methane from methoxy groups and then undergo a series of rearrangement and condensation reactions which are responsible for the initial formation of furans, phenols and some cross linking. Internal redox process involve the conversion of some benzylic alcohols to carboxylic acids. These are capable of ion exchange with cations from the surrounding waters. Most of the rings bear more than one hydroxyl group. As coalification continues more condensation reactions occur and larger ring systems begin to be formed. In lignites and sub-bituminous coals, ionic interactions are important, but decrease rapidly and disappear together with the carboxyl group at about 80% C (dmmf). Hydrogen bonding interactions are very important in low rank coals and lignites because of high population of hydroxyl groups. It has been estimated that there are 6 times as many hydrogen-bond branch points as covalent branch points in Illinois No. 6 coal (44). Dehydroxylation occurs slowly over the whole rank range and with it the importance of hydrogen bonding drops off. I assume the decreases in hydroxyl population and hydrogen bonding are parallel. The complex set of dipolar interactions almost certainly changes strongly in character as the coalification process proceeds. The lignites are highly polar materials and not highly polarizable compared to high rank coals. Accordingly, dipole-dipole interactions will predominate in the low rank materials. As deoxygenation continues and ring sizes grow, the dipole moments will decrease while the polarizability increases. Thus the dipole-dipole interactions while London forces are growing steadily in importance.

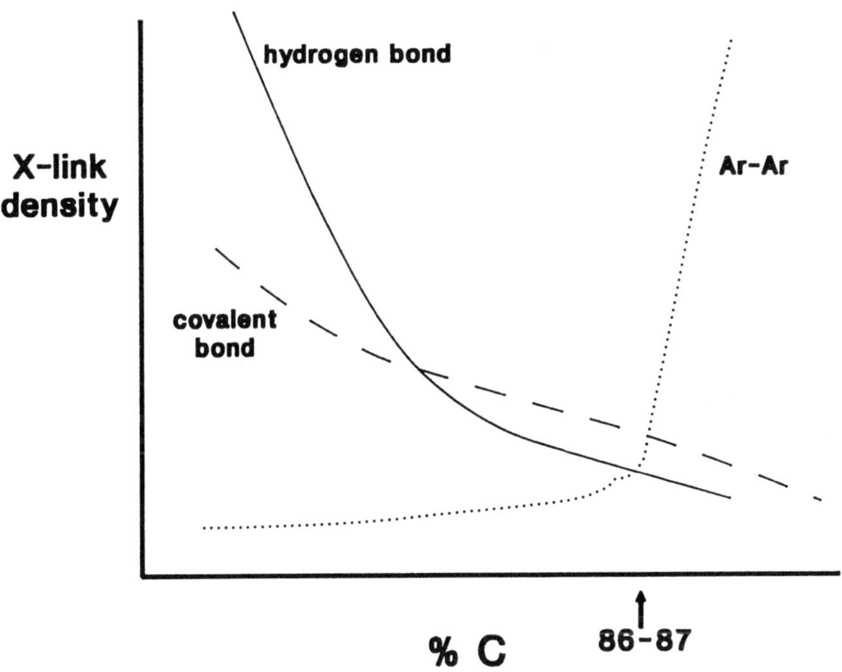

Figure 3. Qualitative changes in cross-link density during coalification.

It is not yet possible to what is happening to the covalent M_c in the lignin to sub-bituminous coal range. The macromolecular structure of these materials has not been studied in detail. Using very strong bases (tetraalkylammonium hydroxides), some high swellings of lignites have been observed suggesting low covalent cross link densities (56). However, these very strong bases could also be cleaving bonds. In general, swelling values in good low-rank-coal solvents like THF are not very different from the swelling values of higher rank coals in pyridine. The molecular weights of materials solubilized from low rank coals by vigorous treatment with bases run into the millions (57). There is no evidence for extensive depolymerization during the coalification of these low rank materials. The arguments for the coalification of bituminous being a net covalent depolymerization have already been presented and will not be repeated here. Figure 3 shows a very qualitative view of the relative importance of the various interactions in coals. There is much speculation in the figure, especially concerning the relative populations of the various interactions.

It is clear that there is a major structural alteration in coals occurring at about 87% carbon. Changes in reactivity and physical structure occur, extractability decreases sharply and the system behaves as if it were becoming much more highly cross linked. If these "more highly cross linked" coals are treated with reagents which will destroy aromatic-aromatic interactions by stericly separating the aromatic groups or interrupting the π systems while separating the aromatic groups, the extractability of the coal increases greatly (52,53). Likewise, if one uses a thermal treatment severe enough to overcome strong noncovalent interactions and not severe enough to break covalent bond, significant extractabilities can be achieved (15). There is no sudden shift in the elemental composition of the coals at about 87% carbon and no reason to expect that the depolymerization process associated with coalification should suddenly and become a polymerization process.

During the coalification process across the range of bituminous coals the size of the PNA systems remain nearly constant (58,59). Both NMR and X-ray suggest that PNA size grows slightly, but that the increase in the fraction of aromatic carbons in coals is due primarily to an increase in the number of aromatic systems. It is this increase in concentration of aromatic systems which I believe is responsible for the sudden shift in coal properties at about 87% carbon. I believe the process is analogous to that of a condensation polymerization. In a condensation polymerization the soluble molecules react with each other to form dimers, trimers and oligomers of increasing size, all of which remain soluble until the gel point is reached. At the gel point, there is a discontinuity in the system and the first insoluble network precipitates. Once the gel point is reached, the amount of insoluble network increases very rapidly. In coals, the concentration of aromatic systems is increasing steadily across the bituminous range. At 87% carbon the concentration of aromatics is sufficiently high so that very long range cooperative π-π interactions can occur, similar to those which occur in crystals of polynuclear aromatics. These interactions are occurring in an amorphous system, not a crystalline one, but if the concentration of aromatics gets high enough, there is no reason this cannot happen. These highly cooperative, extended interactions will be highly stabilizing. I postulate that "cross linking process" which is responsible for the sudden shift in coal behaviors is the formation of cooperative extended π-π interactions. The coal behaves as if the system had suddenly begun to polymerize. In one way it has, though it is not covalently polymerizing; it is noncovalently polymerizing.

5. References

1. Brenner, D. (1985), Fuel 64, 167-173.
2. Brenner, D. (1985), Am.Chem.Soc.Fuel Div. 30.
3. Lucht, L.M., Larsen, J.W. and Peppas, N.A. (1987), Energy and Fuels 1, 56-58.
4. Sakurovs, R., Lynch, L.J, Maher, T.P. and Banerjee, R.W. (1987), Energy and Fuels 1, 167-172.
5. Hall, P.J. and Larsen, J.W. (1991), Energy and Fuels 5, 228-229.
6. van Krevelen, D.W. and Hoffyzer, P.J. (1976), Properties of Polymers, Elsevier, New York.
7. See Table II, Larsen J.W., Green, T.K. Choudry, P. and Kuemmerle, E.W. (1981), in Gorbaty, M.L. and Ouchi, K. (editors), Adv. in Chem. Ser. 192, 277-291.

8. Van Krevelen, D.W. (1981), Coal, Elsevier, New York.
9. Hirsch, P.B. (1954), Proc.Roy.Soc. (London) a226, 143-169.
10. Cody, C.D.Jr., Larsen, J.W. and Siskin, M. (1988), Energy and Fuels 2, 340-344.
11. Cody, C.D.Jr., Larsen, J.W. and Siskin, M. (1989), Energy and Fuels 3, 544-551.
12. Cody, C.D.Jr., Larsen, J.W., Siskin, M. and Cody, C.D.Sr. (1989), Energy and Fuels 3, 551-556.
13. Roberts, J.E., Coleman, K.M., Alaimo, M.H. and Larsen, J.W. (1991), in press.
14. Carlson, G., personal communication.
15. Nishioka, M and Larsen J.W. (1990), Energy and Fuels 4, 100-106.
16. Larsen, J.W. and Mohammadi, M. (1990), Energy and Fuels 4, 107-110.
17. Larsen, J.W., Azik, M. and Korda, A. (1992), Energy and Fuels, in press.
18. Hall, P.J. and Larsen, J.W. unpublished observations.
19. Hsieh, S.T. and Duda, J.L. (1987), Fuel 66, 170-178.
20. Joseph, J.T. (1991), Fuel 70, 139-144.
21. Baldwin, R.M., Kennar, D.R., Nguanprasert, O. and Miller, R.L. (1991), Fuel 70, 429-433.
22. Bond, R.L. (1956), Nature 178, 104-105.
23. Mahajan, O.P. and Walker, P.L.Jr. (1976), in Karr, C.Jr. (editor), Analytical Methods for Coal and Coal Products, Vol. 1, Academic Press, New York.
24. Marsh, H. (1987), Carbon 25, 49-58.
25. Gan, H., Nandi, S.P. and Walker, P.L.Jr. (1972), Fuel 51, 272.
26. Larsen, J.W. and Wernett, P. (1988), Energy and Fuels 2, 719-720.
27. Wernett, P. (1991), Ph.D.Thesis, Lehigh University.
28. Reucroft, P.M. and Patel, H. (1986), Fuel 65, 816-820.
Reucroft, P.M. and Sethuraman, A.R. (1987), Energy and Fuels 1, 72-75.
29. Walker, P.L.Jr., Verma, S.K., Rivera-Utrolla, J. and Khan, M.R. (1988), Fuel 67, 719-726.
30. Spitzer, Z. and Uliccky, L. (1977), Fuel 55, 212-224.
31. van Krevelen, D.W. and Hoffyzer, P.J. (1976), Properties of Polymers, Elsevier, New York.
32. Berens, A.R. and Hopfenberg, H.B. 91982), J.Mem.Sci. 10, 283-303.
33. Barr-Howell, B.D. and Peppas, N.A. (1986), Chem.Eng.Commun. 43, 301-315.
34. Otake, Y. and Suuberg, E.M. (1989), Fuel 68, 1609-1612.
35. Yan, J.F. (1981), Macromolecules 14, 1438-1445.
Yan, J.F. and Johnson, D.C. (1981), J.Appl.Polym.Sci. 26, 1623-1635.
36. Given, P.H., Marzec, A., Barton, W.A., Lynch, L.J. and Gerstein, B.C. (1986), Fuel 65, 155-163.
37. Davis, M.F., Quinting, G.R., Bronnimann, C.E. and Maciel, G.E. (1989), Fuel 68, 763-770.
38. Larsen, J.W., Mohammadi, M., Yiginsu, I. and Kovac, J. (1984), J.Geochemica et Cosmochimica Acta 48.
39. Larsen, J.W., and Wei, Y-C. (1988), Energy and Fuels, 2, 344-350.
40. Whitehurst, D.D., Mitchell, T.O. and Farcasiu, M. (1980), Coal Liquefaction, Academic Press, New York.
41. Larsen, J.W., Pan, C-S. and Shawver, S. (1989), Energy and Fuels 3, 557-561.
42. Larsen, J.W. and Baskar, A.J. (1987), Energy and Fuels 1, 230-232.
43. Painter, P.C., Sobkowiak, M. and Youtcheff, J. (1987), Fuel 66, 973-978.
44. Larsen, J.W., Green, T.K. and Kovac, J. (9185), J.Org.Chem. 50, 4729-4735.
45. Kauzmann, W. (1959), Quantum Chemistry, Cp.13, Academic Press, New York.
46. Grunwald, E, and Price, E. (1964), J.Am.Chem.Soc. 86, 4517-4525.
47. Nishioka, M. (1991), Energy and Fuels 5, 487-491.
48. Nishioka, M. (1991), Energy and Fuels 5, 523-525.
49. Dewar, M.J.S. and Thompson, C.C.Jr. (1966), Tetrahedron Suppl. 7, 97.
Bentley, M.D. and Dewar, M.J.S. (1967), Tetrahedron Lett. 5043.

50. Le Ferre, R.J.W., Radford, D.F., Ritchic, G.L.D. and Stiles, P.J. (1967), Chem.Comun. 1221.
Houper, H.O. (1964), J.Chem.Phys. 41, 599.
51. It has been suggested that the reaction of maleic anhydride with coals is due to charge transfer compexation (48). Our unpublished ^{13}C CPIMAS studies of Illinois No. 6 coal reacted with maleic anhydride added to the coal (~11wt%); does so in chemical reactions which alter the hybridization the non-carbonyl carbons from sp^2 to sp^3. A chemical reaction, not charge transfer complexation is occurring.
52. Mallya, N. and Stock, L.M. (1986), Fuel 65, 736-738.
53. Quinga, E.M.Y. and Larsen, J.W. (1987), Energy and Fuels 1, 300-304.
54. Larsen, J.W., Flowers, R.A.II., Hall, P., Silbernagel, B.G. and Gebhard, L.A. (1991), Proc.Intl.Conf.on Coal Science 1.
55. Hatcher, P.G. (1990), Org.Geochem. 16, 959-968.
56. Matturro, M.G., Liotta, R. and Isaacs, J.J. (1985), J.Org.Chem. 50, 5560.
57. Olson, E.S., Diehl, J.W. and Froelich, M.L. (1987), Fuel 66, 992-995.
58. Cartz, L. and Hirsch, P.B. (1960), Phil.Trans.Roy.Soc.(London) A252, 557-604.
59. Solum, M.S., Pugmire, R.J. and Grant, D.M. (1989), Energy and Fuels 3, 187-193.
60. Krichko, A.A. and Gagarin, S.G. (1990), Fuel 69, 885-891.

FUNDAMENTALS OF COAL PYROLYSIS AND LIQUEFACTION

A. F. GAINES
Marine Institute
Middle East Technical University
Erdemli
İçel 33731
Türkiye

ABSTRACT. This chapter describes what is known of the dependence of the pyrolysis and liquefaction of solid fuels on the chemical structure of their constituent maceral groups, on the design of the reactors and on the parameter defining their operation. The role of catalysts is not considered.

1. Introduction

In the progression from the mining of coal to processing a clean synthetic fuel, this chapter considers the basic chemistry and chemical engineering of pyrolysis and liquefaction. In particular, it will be shown how starting from the chemical structure of coal, the yields of liquid products are determined by the parameters regulating pyrolysis and liquefaction and how this regulation arises naturally from the mechanism of the process. It will also be shown how reactor design influences both the yields and characteristics of the liquid products, emphasis being given to the distinction between primary and retrogressive reactions. In this chapter, the effects of solid catalysts -save as they are inherent in coal- will not be considered either as they influence the primary conversion or as they are used to refine the technological properties of the liquid products. Much of the material discussed will be taken from the author's own research and some familiarity with the subject will be assumed. It is hoped that the understanding generated by the chapter will lead to better and even novel syntheses of clean fuels. The need for such processes is urgent. The motivation for using solid fuels, despite the environmental and health hazards pointed out in other chapters, is the human desire for increased comfort. Tables 1 and 2 show the annual consumption of energy and electricity per head of population. It is the difference in these figures between different parts of the world which distinguishes developed from developing societies.

TABLE 1. Energy consumption per head of population in 1981 (tons of coal equivalent)

USA	UK	USSR	CHINA	INDIA
10.2	4.6	5.7	0.6	0.2

TABLE 2. Electricity consumption per head of population in 1984 (kw hrs.)

ASIA	EUROPE	N.AMERICA	S.AMERICA	AFRICA
550	14800	7341	1186	393

(Russia is not included in this table. The statistics are from United Nations Year Books.)

Y. Yürüm (ed.), Clean Utilization of Coal, 15–31.

It is clear that India, Africa and especially China, with its large reserves of coal, will wish to increase their energy consumption ten-fold. Given the size of their populations, the global environmental consequences could be grotesque. One is faced either with having to persuade developing societies to forego the use of solid fuels or with transferring the best engineering design of which we are capable.

2. The Starting Material

In this chapter coal will be considered as consisting of three groups of organic mineral (maceral):

a. Vitrinite, composed of aromatic, often phenolic, nuclei associated through hydroaromatic linkages (1).

b. Liptinities, formerly termed exinites, composed of aliphatic chains substituted by alkyl hydroxy aromatics (1).

c. Inertinites which, in the absence of detailed knowledge, may be considered to have similar structures to high rank vitrinites.

Chemical structures of fragments of solubilised material are listed in reference (2). It will sometimes be useful to regard coals and their constituent maceral groups as organic matrices containing extractable material. Further details of coal structure will be introduced as they are needed.

3. Pyrolysis and Hydropyrolysis at Slow Rates of Heating

The archetypal apparatus for the study of coal pyrolysis at low rates of heating, prompted by the need to understand the formation of metallurgical coke, was the Gray-King furnace (3). Experiments in such apparatus established that coal pyrolysis starts at about 350°C and that whilst the total yield of volatile material always increases with temperature, prolonged heating at 600°C is sufficient to give the maximum yield of tar. The pyrolysis of vitrinites and of vitrinite rich material in the Gray-King furnace may be represented by the following sequence of consecutive and parallel reactions (4):

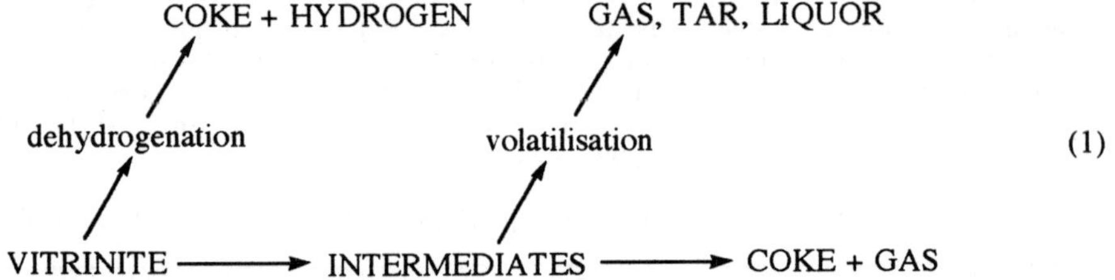

Hydroaromatic structures dehydrogenate to give an extended aromatic network which becomes transformed into coke or char. This is not only consistent with the pyrolysis of such model compounds as tetralin (5) and of partially deuterated coals (6), it has been observed in lignites by infra-red spectroscopy (7). The hydrogen released by the dehydrogenation diffuses through the pyrolysing solid and causes rupture of aliphatic bonds. This reaction, which may be represented by

$$R\text{-}X + H \longrightarrow XH + R \qquad (2)$$

requires little activation energy and results in the elimination of alkyl groups to form gas, the formation of water from hydroxyl groups and the scission of aliphatic linkages between aromatic nuclei and it is not until such reactions are nearly complete that significant quantities of hydrogen gas are evolved from the pyrolysis. (In another chapter, Schobert describes a different - but with our present experimental knowledge, perhaps equally valid - mechanism in which dehydrogenation of hydroaromatic material and bond scission occur simultaneously, the hydrogen being transferred by radicals. The consequences of such a radical hydrogen transfer would be essentially the same as those of the description of pyrolysis given here). The free radicals formed by reaction 2 have been observed (8). Some radicals recombine to yield extended structures which eventually form coke or char together with gas. The presence of mobile intermediate material (metaplast) of moderate molecular mass confers fluidity (9), relatively small changes in the concentration of intermediates making large changes to the fluidity (10). If vitrinites are pyrolysed slowly in a static bed, reaction 1 is sufficient to model both the rate of development of fluidity and also the loss of weight measured by thermogravimetric analysis (9). More complex schemes, reviewed by Gavalas, Howard and Solomon (12) have been developed to model the formation of individual compounds during the pyrolysis. Mass transfer processes play a prominent role in determining the amount and composition of volatiles obtained during pyrolysis of coals, particularly of softening coals (11). The development of fluidity plays a key role in the regulation of mass transfer (12), especially in controlling the transfer of hydrogen into, within and out of the reaction zone. Table 3 shows results from the pyrolysis of a coal which was but weakly caking and gave little fluidity in nitrogen but which gave much fluidity in hydrogen atmospheres.

When pyrolysis produced little fluidity (in nitrogen) volatilisation unlocked the pore structure and led to an increase in char microporosity. Pyrolysis in hydrogen generated fluidity which slowed mass transfer and led to a decrease in coke microporosity.

TABLE 3. Microporosities and surface areas of cokes formed under pressure at 500°C (4).

Sample	Initial H/coal ratio, KPa/g	Final pressure, MPa	Surface area, m^2/g	Microporosity volume, ml/g
Coal	—	—	162.3	0.0614
Nitrogen coke	—	1.90	181.5	0.0629
Nitrogen coke	—	2.95	206.2	0.0742
Hydrogen coke	20	4.45	96.6	0.0334
Hydrogen coke	80	5.75	57.5	0.0199
Hydrogen coke	140	6.66	61.0	0.0214
Hydrogen coke	190	10.03	33.7	0.0115
Hydrogen coke	140	2.50	102.8	0.0376
Hydrogen coke	140	4.50	68.5	0.0239
Hydrogen coke	140	6.66	61.0	0.0214
Hydrogen coke	140	13.20	33.0	0.0115

The original coal containing 83.3%C daf., was heated to 500°C at 4.2K/minute and was held at 500°C for two hours.

Of course, pyrolysis in hydrogen caused increased tar and gas yields. Maximum yields of gas appear to be generated by high ratios of hydrogen to coal and high pressures but maximum yields of tar require an optimum ratio of hydrogen to coal but not excessive pressure since further increase in the pressure and the hydrogen to coal ratio generates gas from the tar (4).

The tars obtained from slow pyrolysis in a static bed such as a Gray-King furnace are complex mixtures of alkanes, alkyl phenols, alkyl aromatics and heterocyclic material. They are even more complex than coke oven tars and the "high temperature" Gray-King assay simulates the process in a coke oven by the addition of a secondary cracking furnace arranged such that tars issuing from the primary furnace at 600°C pass through a static bed of carbon at 900°C. This results in a diminution of the tar yield and a simplification of the chemical structures present in the tar. The high temperature tars, like coke oven tars, contain polynuclear aromatics, the low temperature tars having become dealkylated and alkanes having cracked, as have such reactive compounds as polyhydric phenols. The role of the second "cracking" Gray-King furnace introduces the concept of secondary reactions which are "retrogressive in the sense that they may counter the primary reactions of pyrolysis and reduce yields of liquid products. These retrogressive reactions are not necessarily confined to the second Gray-King furnace; investigation of the primary static bed shows that retrogressive reactions occur below 600°C as the products of pyrolysis diffuse through the bed, especially when mass transfer is limited by the development of fluidity (4).

4. Rapid Pyrolysis

As interest in steel making was overtaken by curiosity about combustion and electricity generation, and as chemical engineering developed fluidised and entrained beds, coal scientists forsook slow rates of heating and considered the design of the reactors. In a Gray-King furnace bituminous coals, rich in vitrinite, pyrolyse to yield about 70% of coke or char and about 10% of each of tar, liquor (water) and gas. It is clear that retrogressive reactions reduce the yield of tar and one realises that, in order to maximise the tar yield one requires a reactor in which liquid products are removed from the hot zone as rapidly as they are formed. This implies

a. fine particles of coal- so that liquid products may escape readily and so that problems of heat transfer are minimised.

b. a bed in which particle contact is avoided- for the same reasons as in (a).

c. rapid heating- to encourage liquid products to 'explode' out of the coal and to minimise their residence time in the bed.

d. a "carrier" to remove pyrolysis products from the hot reaction zone.

Table 4 shows pyrolysis yields from a reactor- a "wire mesh" reactor- which embraces these principles (13). The design of the reactor consists of a stainless steel sample holder comprised of a sandwich of two layers of wire-mesh stretched between two electrodes. The mesh acts as an electrical resistances heater. About 7 mg of particles of 100-150 microns diameter form a monolayer of nontouching particles between the wire-mesh sandwich. Gas sweeps through the wire-mesh at 0.1 m/s to carry pyrolysis products from the bed. Linear heating rates of 1000 K/s (still orders of magnitude slower than those achieved in combustion) take the temperature to its final value which may then be held for specified lengths of time.

There can be no guarantee that no retrogressive reactions at all occur in the wire-mesh reactor though this is believed to approximate to the truth. Figures 1 and 2 compare pyrolysis yields from the wire-mesh reactor with those obtained, using the same coal, from two other laboratory reactors designed for rapid heating, one, a fluidised bed and the other, a "hot rod" (14). In the hot rod reactor a shallow plug of coal is maintained within a metal cylinder. The cylinder is used as the electric resistance heater -hence the name "hot rod"- and gas flows through the plug of coal.

Table 5 compares the parameters of the three designs of reactor. Table 4 and Figures 1 and 2 show that tar yields from vitrinite rich material at rapid rates of heating may exceed those from a Gray-King furnace by a factor of two or three and that total yields of volatile material may easily exceed those obtained from a standard volatile test. The maximum yields of tar from the wire-mesh and fluidised bed reactors are very similar -within experimental error- but the tar yields from the fluidised bed reactor generally exhibit a maximum as a function of peak temperature which suggests the occurrence of retrogressive reactions, at least at the higher temperatures. Even under optimum conditions tar yields from the hot rod reactor remain lower than those obtained from the wire mesh reactor, indication that in the hot rod reactor significant quantities of tar are removed by retrogressive reactions. A detailed discussion of the experimental parameters controlling the extent of retrogressive reactions, given in reference (14), convinces one that the kinetics of pyrolysis depend on the design of the reactor. Figures 1 and 2 confirm that, as perhaps one would expect, fast rates of heating eg. 1000K/s give similar results to Gray-King pyrolyses in that tar formation appears to start near to 350°C and is substantially complete around 600°C. This is comforting, it implies that the chemistry used to interpret Gray-King pyrolyses, such as the reactions 1, can also be applied to rapid pyrolyses.

TABLE 4. Pyrolysis yields in a wire-mesh reactor (13).

Fuel	%C	Maceral group analysis			T1	V1	T1000	V1000
		%V	%L	%I				
POINT OF AYR								
whole coal	85.2	83.8	6.0	10.2	20.7	33.6	26.1	42.4
clarain	82.6	93.0	5.0	2.0	18.1	34.7	24.4	40.8
durain	84.9	30.0	27.0	43.0	26.8	40.5	30.1	44.2
vitrinite	84.8	91.0	5.0	4.0	20.5	33.9	24.6	40.1
liptinite 1	85.7	30.0	61.0	9.0	43.4	56.8	47.1	62.0
liptinite 2	84.8	29.5	52.3	18.2			47.9	62.5
inertinite	84.2	17.0	3.0	80.0	15.4	30.4	16.1	31.3
LINBY								
whole coal	82.3	73.0	15.0	12.0	26.7	37.6	29.8	40.9
vitrinite	77.6	85.0	6.0	9.0	20.3	37.8	29.5	45.2
liptinite	79.1	16.0	70.0	14.0	45.5	59.9	48.9	64.9
inertinite	78.2	35.0	4.0	61.0	19.6	35.8	26.3	42.4
CORTONWOOD								
whole coal	86.5	63.2	15.4	21.4	26.7	37.6	29.8	40.9
vitrinite	85.9	97.0	3.0	0	24.7	39.2	26.5	42.1
liptinite	85.1	0	90.0	10.0	54.9	65.5	53.7	70.7
inertinite	85.7	7.0	0	93.0	21.0	33.4	22.5	35.8
FREYMING								
whole coal	82.3	93.0	3.5	3.7	20.3	36.7	28.4	44.2
vitrinite	83.5	99.2	0.4	0.4	18.2	34.4	26.3	43.0
DINNINGTON								
vitrinite	81.4	96.0			14.3	31.0	21.2	34.2
liptinite	84.3	32.0	68.0	0	47.3	58.9	48.8	63.1

%C: % carbon w/w daf

Maceral group analysis, %V, %L, %I = % vol/vol dmmf vitrinite, liptinite, inertinite.

T1: Tar yield, %w/w daf coal; heated to 700°C at 1K/s

V1: Total volatile yield, %w/w daf coal; heated to 700°C at 1K/s

T1000: Tar yield, %w/w daf coal; heated to 700°C at 1000K/s

V1000: Total volatile yield, %w/w daf coal; heated to 700°C at 1000K/s

All pyrolyses were conducted in a flow of helium and the peak temperature of 700°C was always held for 30 seconds.

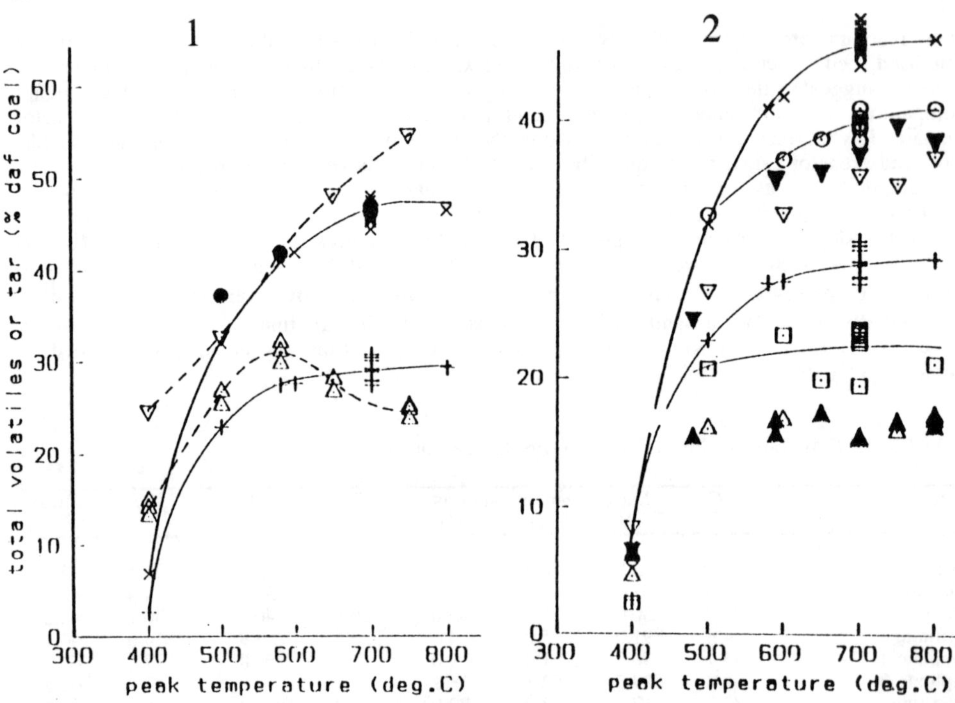

Figure 1. Pyrolysis of Linby coal in fluidised-bed and wire mesh reactors (14):
Fluidised-bed: Δ: tar yields, ∇: total volatile yields.
Wire mesh: (1000K/s, 30s hold, helium flow 0.1m/s), +: tar yields, X: total volatiles
(1000K/s, 1000s hold, helium flow 0.1m/s), ●: total volatiles.

Figure 2. Pyrolysis of Linby coal in wire mesh and hot rod reactors (14):
Wire mesh: (1000K/s, 30s hold, helium flow 0.1m/s), +: tar yields, X: total volatiles,
(1K/s, 30s hold, helium flow 0.1m/s), ⊡: tar yields, ʘ: total volatiles.
Hot rod: (5K/s, 200s hold, helium flow 0.1m/s), Δ: tar yields, ∇: total volatiles
(5K/s, 200s hold, hydrogen flow 0.1m/s) solid symbols as above.

TABLE 5. Comparison of the parameters of three reactors (14)

Reactor type	Fluidised Bed	Wire-Mesh	Hot Rod
Size of sample	1-15 g	5-25 mg	50 mg-1 g
Maximum temperature (^{o}C)	900	1200	900
Pressure (bar)	1	0-160	1-150
Rate of heating	slow to flash	slow to 5000 K/s	slow to 20 K/s
Flow of carrier gas	3-5(fluidisation)	slow to 0.3 m/s	slow to 10 m/s
Retrogressive reactions	in bed and in freeboard	minimal but non-zero	may be minimised at very high flows
Accuracy/repeatability	tar + 2-3% char + 5%	tar + 2-3% char +1-2%	tar + 2-3% char + 3-4%

Table 4 and Figures 1 and 2 show the effect of the rate of heating and the peak temperature on the yields of tar and volatile matter in the wire-mesh reactor. These results are amongst those which demonstrate that the tar yields quoted at 1000K/s are the maximum obtainable at atmospheric pressure. It is quite possible that larger yields might be obtained under reduced pressure but it is already abundantly clear that minimisation of retrogressive reactions leads to greatly increased tar yields. Table 4 shows that the tar yields of maceral group concentrates increase in the order inertinite < vitrinite< liptinite. Table 6 shows that, if one assumes that the tar and the volatile yields from a solid fuel are the additive sums of the tar and volatile yields of its maceral group components, then one can deduce a consistent set of yields for 'pure' maceral groups (13). Obviously these may vary with rank. The additivity of the yields of the maceral groups needs careful interpretation. Evidently, the maceral groups are not pyrolysing independently of each other. There is a sense in which the pyrolysis of inertinite, for example, is different if it is pyrolysing as an inertinite concentrate (when it is pyrolysing as a non-coherent solid powder) to when it is pyrolysing as part of a coking coal (when it is pyrolysing as part of a fluid or plastic mass in which mass transfer is circumscribed). What Table 6 is indicating is that the chemical skeleton of the organic matrix of each maceral group disrupts in a characteristic manner independently of the presence of other maceral groups. Thus, in liptinites, a crosslinked aliphatic chain substituted by alkyl hydroxy aromatics (1), disrupts by breakage of the aliphatic chain (possibly by a random scission process) to give relatively high yields of liquid products and volatile material. Liptinite concentrates may be observed to become fluid and relatively limpid throughout most of the pyrolysis (15). Vitrinites disrupt differently; the pyrolysis of their aromatic skeleton obeys reactions 1. The formation of tar by the scission of the skeleton competes with the formation of coke or char following dehydrogenation of the hydroaromatic linkages. As a result the tar yield is less than that from liptinite, the proportions of tar and coke/char being determined by the experimental conditions. In the absence of a hydrogen donor, the reduction of tar yields because of the simultaneous formation of coke/char is inevitable. The scission of the skeleton occurs because of attack on it by hydrogen and this hydrogen is generated by the coke/char forming reaction 4. In consequence, about as much aromatic material forms tar as forms coke by the dehydrogenation of the hydroaromatic network (16). Competition between simultaneous reactions forming tar and coke respectively provides a straightforward explanation of the variation of tar yields with heating rate observed in vitrinite rich material but other explanations are possible and Tables 4 and 6 show some examples of tar yields from liptinites and inertinites which also varied with the rate of heating.

5. The Liquefaction of Solid Fuels in Tetralin

The liquefaction of solid fuels may be divided into two; the primary solubilisation, usually by hydrogenation and the refining of the soluble products by further hydrogenation. Little will be said about the second stage though it is of great importance to the production of a clean synthetic fuel.

The criteria for minimising retrogressive reactions during the pyrolysis of solid fuels may also be applied to their solubilisation. Most of the recent laboratory developments of the original Bergius and Pott Broche prosesses have used closed reactors in which liquefaction products have been constrained to remain in the hot reaction zone for appreciable periods of time (17). Such circumstances clearly favour the modification both of conversions and of the nature of the liquid products by retrogressive reactions, especially since most liquefactions are performed at moderate pressure. Retrogressive reactions may be minimised by carrying out liquefactions in a flowing solvent reactor (18) in which a small, fixed bed of powdered coal is permeated by a hot hydrogen donor solvent which flows through the bed and carries products out of the reaction zone.

The results to be discussed in this chapter were obtained from a cylindrical, metallic, tubular reactor in which a plug of powdered (100-150 micron) coal, diluted tenfold with sand of the same size in order to minimise contact between the coal particles and prevent the development of fluidity, was swept by hot tetralin (18). Liquefactions were carried out with tetralin flowing at 0.9 ml/second at a pressure of 70 bar; the coal was heated at 5 K/second to a selected temperature between 300 and 450°C for a pre-set period of time. Minor variations in the tetralin flow rate, the pressure or the rate of heating caused little change in conversions. The tetralin flow rate of 0.9ml/second was always sufficient to remove solubilised liquefaction products continuously from the reaction zone and into a heat exchanger within 10 seconds.

TABLE 6. Calculated and experimental pyrolysis yields (13).

Fuel		T1	V1	T1000	V1000
POINT OF AYR					
100% liptinite	calculated yield	59.7	73.3	64.2	78.4
100% vitrinite	calculated yield	19.4	31.8	22.5	38.8
100% inertinite	calculated yield	12.9	28.5	12.9	27.9
whole coal	calculated yield	21.2	34.0	24.0	40.1
	experimental yield	20.7	33.6	26.1	42.4
clarain	calculated yield	21.3	33.8	24.4	40.6
	experimental yield	18.1	34.8	24.4	40.9
durain	calculated yield	27.5	41.6	29.6	44.8
	experimental yield	26.8	40.5	30.1	44.2
LINBY					
100% liptinite	calculated yield	58.2	71.1	59.1	75.1
100% vitrinite	calculated yield	17.9	35.9	28.2	44.0
100% inertinite	calculated yield	17.7	33.6	23.1	39.3
whole coal	calculated yield	23.9	40.9	32.2	48.1
	experimental yield	24.2	40.2	30.7	46.6

The pyrolysis conditions, the fuels and the symbols are those used in Table 4. Calculated yields have been deduced assuming that yields are additive sums of the yields of the component maceral groups.

Detailed comparison of the liquefaction of the same sample of Point of Ayr coal in the flowing solvent reactor and in a minibomb reactor operated at the same temperature for the same duration of time (19) has demonstrated the success of the design of the flowing solvent reactor in minimising retrogressive reactions. When tetralin was used as the hydrogen donor solvent conversions in the flowing solvent and the minibomb reactors were found to be quite similar. When, however, a less powerful hydrogen donor solvent, 1-methyl naphthalene, was used secondary char formation was observable in the minibomb but not in the flowing solvent reactor. As Figures 3 and 4 illustrate, size exclusion chromatography suggested that the relative proportions of higher molecular mass material in the liquefaction products from the flowing solvent reactor increased as a function of increasing reaction temperatures and hold times. This is what one expects if liquefaction is the disruption and subsequent solubilisation of a cross-linked network. The opposite effect was observed with samples from the minibomb reactor, the products becoming relatively lighter with increasing temperature and hold time and this was clearly the consequence of retrogressive reactions. Thus, one may write

COAL ⟶ SOLUBLE PRODUCTS (3)

PRODUCT MOLECULE +
PRODUCT MOLECULE ⟶ CHAR PRECURSOR + LIGHTER PRODUCT (4)
or PARTICLE

though this is an over simplification. The major retrogressive char forming reactions, depicted as reaction 4 and observable in the minibomb but not the flowing solvent reactor, were slow. In the minibomb reactor at 450°C these reactions had a characteristic time of 10-100 seconds which increased to over 1600 seconds at 400°C. Not only was there insufficient time for these retrogressive reactions to develop in the flowing solvent reactor but, as reaction 4 predicts, the enhanced dilution of the products inherent in the design of the reactor decreases the rate of the retrogressive reactions even further (19).

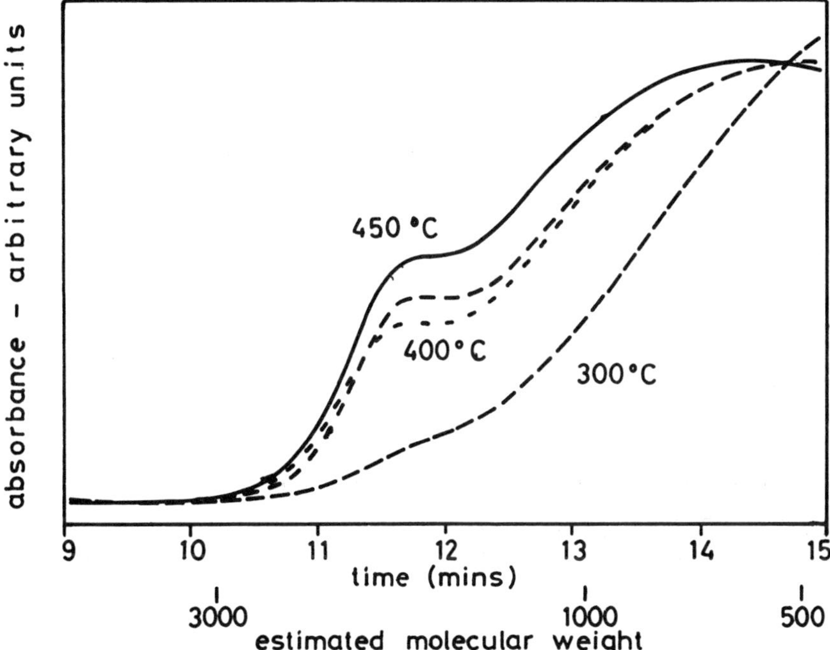

Figure 3: Size exclusion chromatograms of tetralin/cold THF soluble products from Point of Ayr coal liquefied in a flowing solvent reactor under standard - Table 9 - conditions with various peak temperatures (19).

Tables 7 and 8 illustrate how yields of liquefaction products from maceral group concentrates vary with the temperature and the hold time in the flowing solvent reactor (20). Table 9 shows yields of liquefaction products under standard conditions which may be compared with the tar and volatile yields obtained from the wire-mesh pyrolyser and displayed in Table 4.

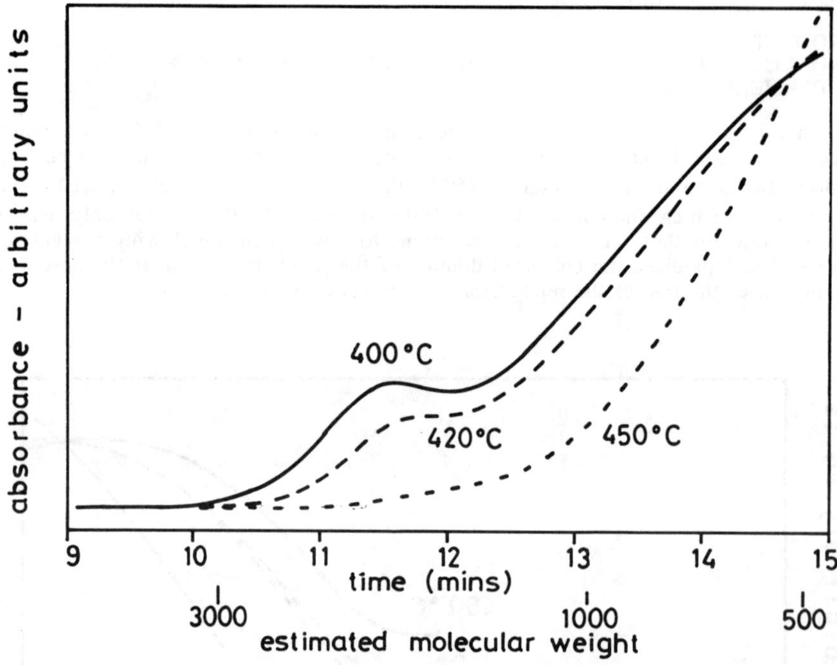

Figure 4: Size exclusion chromatograms of THF soluble products from Point of Ayr coal liquefied with tetralin in a minibomb for 1600 s at various temperatures (19).

TABLE 7. Total liquefaction yields (% daf) from Point of Ayr maceral concentrates (20).

Peak Temperature	Hold time at peak temperature (seconds)							
	0		100		400		1600	
	V	L	V	L	V	L	V	L
300°C	15.9	.	19.2	.	19.0	.	.	.
350°C	18.2	.	29.2	.	29.0	24.7	.	.
400°C	25.3	.	42.6	22.4	54.6*	34.7	.	65.9
450°C	37.2	.	73.7	48.1	84.1	85.7	.	.
450CH	.	.	17.1	35.6	.	.	.	.

V = Vitrinite concentrate used in Table 4.
L = Liptinite 2 concentrate used in Table 4.
Standard heating rate of 5 K/s to the peak temperature, standard pressure of 70 bar and standard tetralin flow of 0.9 ml/s.
* = 500 seconds.
450CH = Hexadecane used as solvent.

TABLE 8. Total liquefaction yields (% daf) from Linby coal and maceral concentrates (20).

	Peak Temperature (^{o}C)			
	350	400	420	450
whole coal	.	.	.	75.1
vitrinite	22.3	33.2	61.9	70.4
liptinite	29.4	.	.	70.1
inertinite	.	27.4	46.3	53.4

Vitrinite, liptinite and inertinite are the concentrates used in Table 4. Standard conditions in the flowing solvent reactor as in Table 7. Hold time at peak temperature = 100 seconds.

TABLE 9. Total liquefaction yields (% daf) under standard conditions in flowing solvent reactor (20).

	Whole coal	Vitrinite	Liptinite	Inertinite
Point of Ayr	69.4	73.7	48.1	38.7+
Linby	75.1	70.4	70.1	53.4
Cortonwood	68.7	79.6	82.7	65.8
Freyming	60.8	75.1	52.7	17.1
Dinnington	79.4	70.1	63.5	unexamined

The liquefaction conditions are as in Table 8. The samples are those in Table 4.
+: single result requiring confirmation.

Table 7 shows that at 300°C the total liquefaction yield (that is, the total loss in weight experienced by the sample as a result of liquefaction) was barely larger than the yield of extractable material (pyridine extracts some 18% of Point of Ayr Vitrinite). Conversions were still low at 350°C and at these temperature there was little increase in the conversion when the duration of liquefaction was prolonged. Little pyrolysis occurs below 350°C but analyses by gas chromatography/mass spectrometry discussed later in the chapter show that the liquids formed at these comparatively low temperatures consisted of both extractable material and material released by hydrogenation. Tables 7 and 8 show that at 400 and 450°C liquefaction yields became substantial and increased significantly as the duration of the experiment was increased from 100 to 400 seconds. It is often observed that, under mild conditions, liptinites give lower conversions than vitrinites, whilst at 450°C and hold times of 400 seconds conversions are comparable. In other words liptinites do not appear to solubilise as rapidly as vitrinites and their activation energy for liquefaction appears to be higher (20). Table 7 includes the yields obtained when pyrolysis, and not liquefaction, was carried out at 450°C and the products were swept away in a stream of hexadecane. As one expects, the yields were much lower than those obtained under the same conditions in tetralin. Pursuing this further, comparison of Tables 4 and 9 demonstrates that liquefaction yields, even when the liquefaction was prolonged for the comparatively short time of 100 seconds, were generally higher than the total volatile yields generated by pyrolysis in a wire-mesh reactor.

It is apparent from the results illustrated in Tables 7, 8 and 9 that liquefaction yields vary from one coal to another- and clearly with rank- and that much work will be needed before one can unravel all the chemistry involved. It is already obvious, however, that, not suprisingly, inertinites give low liquefaction yields and that the effects of liquefaction - when compared with pyrolysis - are more dramatic for vitrinites than for liptinites. The difference between the liquefaction and pyrolysis yields shown in Tables 4 and 9 is always greater for a vitrinite than for its associated liptinite. Partly this is simply because liptinite pyrolysis gives rather high yields of volatile products. It is more interesting to consider the effect of liquefaction on the vitrinites. It has already been shown that tar and volatile yields from the pyrolysis of vitrinites are necessarily low since char formation is essential to provide the hydrogen needed for the generation of liquid products by reaction 2. When, in liquefaction, hydrogen is provided by tetralin, the rates of the subsequent bond scission reactions are greatly increased and are no longer dependent on simultaneous char formation. Indeed, dehydrogenation reactions, precursors of char formation, might be expected to be inhibited by the pressures employed in liquefaction. This is the chemical explanation of how much larger yields of volatile material are given by vitrinites on liquefaction than on pyrolysis.

6. Liquefaction Products and the Mechanism of Liquefaction

Those liquefaction products obtained from the flowing solvent reactor which were soluble in hot tetralin but not in pentane, LPIP, comprised some 25-50% of the original fuel. Since retrogressive reactions have been minimised, the chemical structure of this material should be similar to that of the original fuel, as the infrared spectrum shown in Figure 5 illustrates, any differences being indicative of the chemistry of the liquefaction reactions. LPIPs being pitch-like solids, detailed information of the course of liquefaction may be obtained from their ^{13}C CPMAS nmr spectra. Illustrative ^{13}C CPMAS nmr spectra are shown in Figure 6. The spectra are similar to those of the starting material, the differences being quantitative but not qualitative. Figure 6 reminds one, however, that the differences in the chemical structures of vitrinites and liptinites is reflected in their ^{13}C nmr spectra. In particular, liptinites have a large, often spiky, peak between 29 and 30 ppm, characteristic of long methylene chains, which is not present in vitrinites spectra. Nevertheless, it is convenient to describe the spectra by dividing them into eight chemical shift regions; four - (a), (b), (c) and (d) - in the aromatic region, and four - (w), (x), (y) and (z) - in the aliphatic region (Figure 6). The regions (b), (c) and (d) embrace the chemical shifts of aromatic carbon atoms attached to hydrogen, carbon and oxygen atoms respectively (1, 21). Region (a) includes substituted phenols and certain heterocyclic structures including indoles (1, 21). Region (w), 0 to 15 ppm, includes methyl groups on the ends of alkyl chains (liptinites) and alpha to aromatic rings but shielded by two adjacent chemical groups (vitrinites) (21). Region (x), 15 to 24 ppm, contains methyl groups alpha to an aromatic ring (21) which are comparatively unshielded. Region (y), 24 to 40 ppm, contains methylene groups in long chains (liptinites) and also in hydroaromatic structures alpha to an aromatic ring (vitrinites) (21). Region (z), from 40 ppm onwards, contains methines, diarylmethanes and methylenes both alpha to an aromatic ring and adjacent to a methine group.

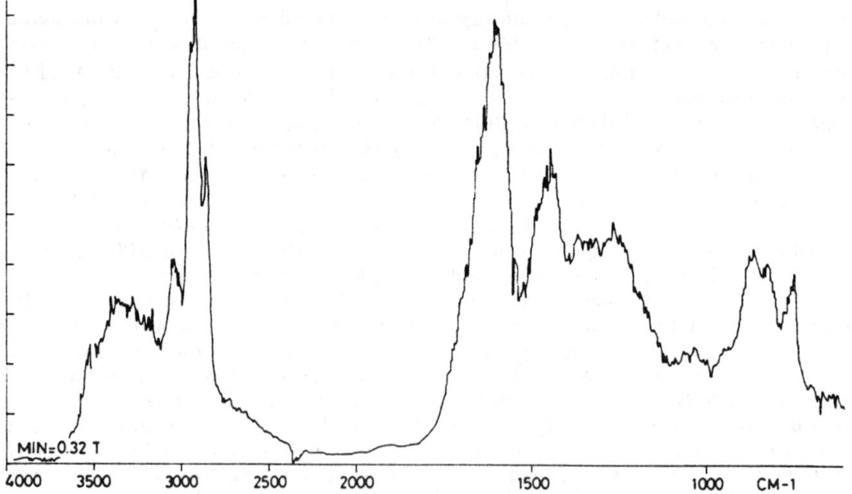

Figure 5: Infrared spectrum of LPIP from Point of Ayr vitrinite liquefied in the flowing solvent reactor at 450°C under standard -Table 9- conditions.

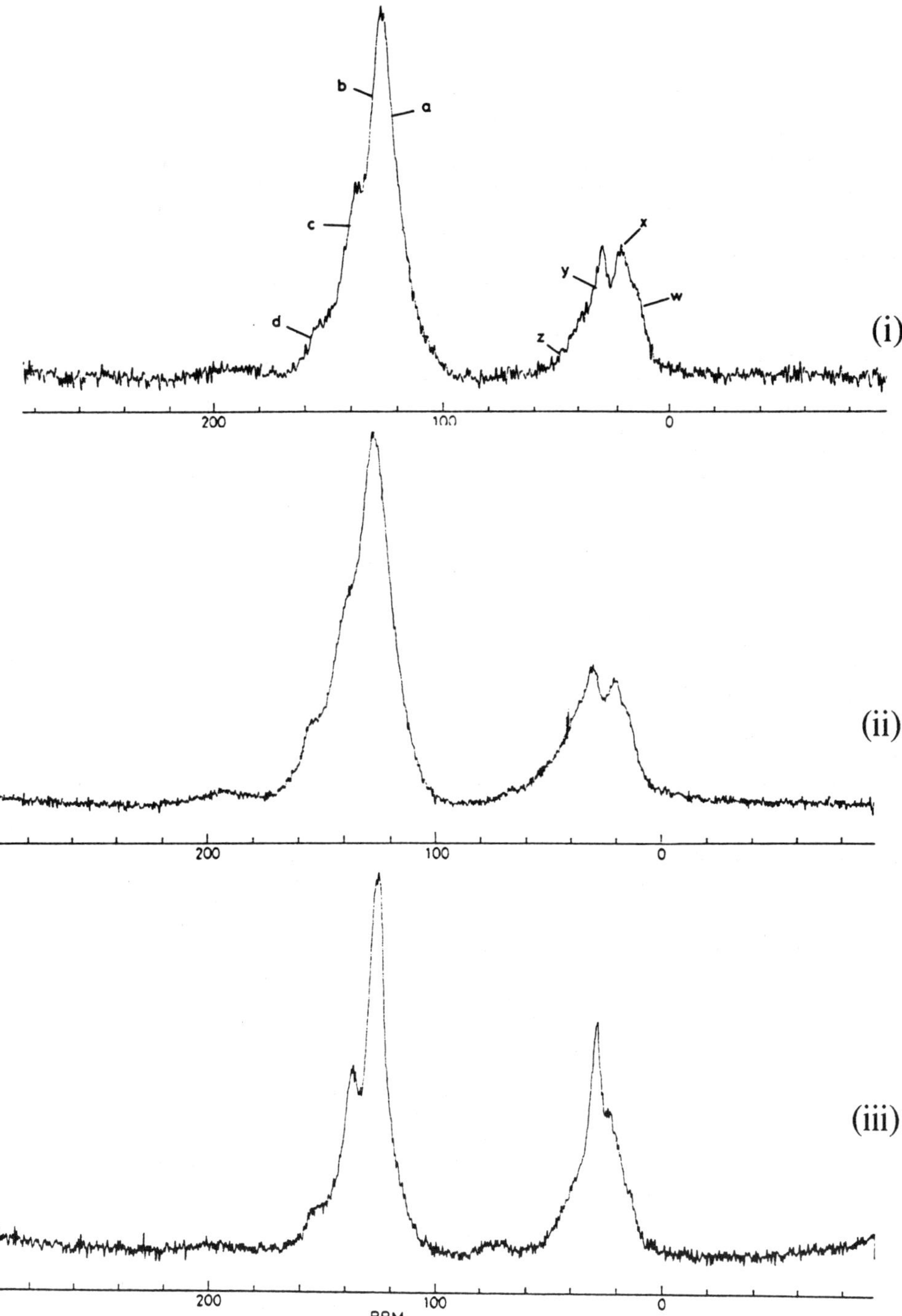

Figure 6: ^{13}C CPMAS nmr spectra: (i) LPIP from Point of Ayr vitrinite liquefied in the flowing solvent reactor at 450°C under standard conditions. (ii) Point of Ayr vitrinite. (iii) LPIP from Dinnington liptinite liquefied in the flowing solvent reactor at 450°C under standard conditions.

The liquefaction of liptinites in the flowing solvent reactor was accompanied not only by a small loss in the atomic H/C ratio but also by a consequent small increase in the values of the aromaticity, f_a. Davis et al., (1), observed the aromatic structures present in four different, pyridine-extracted liptinites (sporinites) to be virtually identical. Accordingly, the nmr spectra show the aromatic structures present in LPIPs obtained from liptinites to have been little affected by liquefaction. Liptinites contain much long chain aliphatic material, (1) and much of the 0-15 ppm region (w) of aliphatic absorption will be due to the methyl groups on the ends of the chains and the dominant peak near 29 ppm, in region (y) which distinguishes the shapes of vitrinite and liptinite nmr spectra, consists of absorptions by methylene groups delta and further from an aromatic nucleus. Comparison of the spectrum of the LPIP from Cortonwood liptinite with that of the original liptinite suggests that the increase in the value of f_a caused by liquefaction was due to the removal of long aliphatic chains: that is, absorption by methylene groups in the 24-40 ppm region and by chain terminating methyl groups in the 0-15 ppm region diminished. Gain in region (x) of the Cortonwood spectrum suggests that liquefaction created methyl groups alpha to an aromatic ring. The proportion of absorption in the region from 40 ppm onwards also increased. If one uses the spectra obtained from liptinite residues by Davis et al. (1) as a reference, comparison of the aliphatic portions of the spectra from LPIPs from Dinnington, Linby and Point of Ayr liptinites also suggests that liquefaction was accompanied by removal of methylene chains and the formation of methyls alpha to aromatic rings. Absorption due to methines and diaryl methanes also appears to have increased in these LPIPs but it seems that Dinnington and Linby liptinites gave increased absorption in the 0-15 ppm region.

Liptinites (sporinites) consists of an essentially aliphatic network with pendant substituted aromatic nuclei. Pyrolysis occasions the disruption of the aliphatic network and this results in the observed high yields of volatiles and the development of marked fluidity. Liquefaction and pyrolysis appear to follow a similar course. The hydrogenation of alkyl substituents on the aromatic rings connotes the removal of methylene and the formation of alpha methyl groups as we have observed. The hydrogenation of the aliphatic network would cause its disruption and the formation of liquid products. The hydrogenation would occur at points of structural weakness (in the bond energetic sense) such as alpha methine groups (as in polystyrene). This would lead to the formation of additional alpha methyl groups but, when hydrogenation is less extensive or when it occurs at branch points in the aliphatic network, it will result in the formation of methyl groups on the ends of shortened chains and absorbing in the 0-15 ppm range. The net loss or gain of methyl groups absorbing in the 0-15 ppm region clearly depends on the balance between the loss of alkyl substituents and the formation of terminal methyl by hydrogenation of the aliphatic network. Such a balance will vary from liptinite to liptinite. As one has seen, the removal of alkyl chains provides the mechanism for the decrease in the atomic H/C ratio and the small increase in aromaticity which accompanies the formation of LPIPs from liptinites even though it is transfer of hydrogen atoms from tetralin which is causing the disruption of the aliphatic network. The only nmr feature remaining unexplained is the creation of structures absorbing in the region where methines and diarylmethanes usually absorb and this demands further study. It suggests that there may be some feature of sporinite structure which still eludes our grasp. Everything which has been written so far implies that the details of the chemistry of liquefaction will apply also to pyrolysis. Since no hydrogen is supplied by a donor solvent, one expects the pyrolysis of liptinites to be accompanied by the formation of some olefines.

Liquefaction of vitrinites caused little change in the features of the aromatic part of their nmr spectra. There appeared to have been an interchange of carbon atoms between regions (a) and (b) of the nmr spectra but the number of aromatic carbon atoms attached not to hydrogen but to carbon and oxygen atoms - regions (c) and (d) of the spectra - remained unchanged in LPIPs. When oxygen groups are removed during the liquefaction of a solid fuel this must therefore occur through retrogressive reactions or during the refining stage of the hydrogenation. The apparent interchange of carbon atoms between regions (a) and (b) may be explicable by the transfer of hydrogen atoms between different aromatic systems during the liquefaction process eg. between diaromatic and polyaromatic rings.

Vitrinites are essentially hydroaromatic, containing little long chain alkyl material. One therefore supposes that, in LPIPs, the methyl groups in the 0-15 ppm range were alpha to an aromatic ring and shielded by two adjacent chemical groups and the methyl groups in the 15-25 ppm range were alpha to an aromatic ring and comparatively unshielded; many, probably most, of the methylene groups in the 24-40 ppm range will be in hydroaromatic structures and alpha to an aromatic ring whilst the methine groups in the 40 - end region will be close, often alpha, to an aromatic ring though diarylmethanes and methylenes which are both alpha to an aromatic ring and adjacent to a methine group can also absorb in the region (21). Inspection of the aliphatic regions indicated that the formation of LPIPs from vitrinite concentrates was accompanied by marked increases in region (y) and decreases in region (z) of the nmr spectra and thus liquefaction appears to have led to the formation of methyl groups alpha to an aromatic ring and the concommitant destruction of methine groups (or of certain methylenes alpha to an aromatic ring). Comparison of the Point of Ayr LPIP spectrum with that from a supercritical pyridine extract of the same vitrinite concentrate clearly indicated that the changes in the aliphatic region of the ^{13}C nmr spectra which accompanied liquefaction were due, not to selective solubilisation, but to chemical reaction. The formation of alpha methyls by the hydrogenation of methines (or of certain alpha methylenes) appears to have been the major chemical reaction during the liquefaction of vitrinites in the flowing solvent reactor. Thus LPIPs were formed by the rupture (hydrogenation) of hydroaromatic material containing methine and alpha methylene groups to give fragments in which previous linkages had been replaced by methyl groups (20). Negligible numbers of aliphatic carbon atoms were lost as simple volatile molecules during this liquefaction, aromaticities remained constant and atomic H/C ratios increased. The process is symbolised in reaction 5:

$$\mathrm{R\,Ar\,CH}\begin{cases}\mathrm{CH_2\,Ar'\,R^*}\\ \mathrm{CH_2\,Ar^*\,R'}\end{cases} \longrightarrow \mathrm{R\,Ar\,CH_3 + CH_3\,Ar'\,R^* + CH_3\,Ar^*\,R'} \qquad (5)$$

This is presumably the major pathway during which radicals and intermediates are formed in reaction 1. The significance of methines in vitrinite structures has not received much previous discussion though Doğru et al. (2) noted the preponderance of branching at the alpha position to an aromatic ring during their investigations of reductively ethylated solid fuels.

The nmr spectrum from LPIP obtained by liquefying the Linby inertinite concentrate indicated that solubilisation had followed the same (or a very similar) mechanism as that by which LPIP material had been formed from vitrinites. There was an increase in the formation of methyl groups and a marked decrease in the presence of methine (or alpha methylene) linkages. The increase in methyl groups in the 0-15 ppm region was noteworthy, suggesting that at least some methyl groups formed by liquefaction were alpha to an aromatic ring but shielded by adjacent substituents (21). Liquefaction appeared to cause little change in the proportion of aromatic carbons substituted by carbon or oxygen atoms but there appeared to have been redistribution of aromatic carbons between regions (a) and (b). Unfortunately, the Linby concentrate contained but 61% of inertinite and spectra from LPIP material obtained from Point of Ayr and Cortonwood inertinite concentrates containing 80 and 90% of inertinite repectively, showed inconsistent patterns. The spectrum of the LPIP from Point of Ayr inertinite was consistent with the scission (hydrogenation) of methine linkages and the formation of methyl groups alpha to aromatic rings but to a much lesser extent than the corresponding vitrinite LPIP. The methine content of the LPIP from Cortonwood inertinite was also reduced but the aliphatic part of the spectrum was dominated by a peak close to 30 ppm. Although this is close to the chemical shift where long alkyl chains absorb, since the parent concentrate consisted of 90% inertinite it seems more reasonable to suppose that the peak arose from methylene groups in tetralin structures with substituents at the alpha position (21). Such structures could obviously have been formed by hydrogenation of aromatic - or, more probably, hydroaromatic - material. There is little in these spectra to indicate why inertinites give low liquefaction yields and, clearly, it is the unexamined residues which resisted hydrogenation.

Gas chromatography/mass spectrometry of the volatile components of liquefaction products from the flowing solvent reactor showed them to consist of a complex mixture containing significant concentrations of alkanes and of alkylated/hydroxylated/aromatics. These compounds apparently embraced a variety of branched and cyclic structures and certainly at high conversions, simple sequences of n alkanes, alkyl benzenes, alkyl phenols or alkyl naphthalenes etc. were not, in fact, distinguishable by their intensity. Though probably present such compounds were submerged by the large numbers of more complicated structures. Products obtained by liquefaction of Point of Ayr coal in tetralin - or in 1- methyl naphthalene - in a minibomb reactor at the same temperatures as liquefactions in the flowing solvent reactor produced not only lower molecular mass (lighter) products, as discussed earlier in this section, but volatile material which was less diverse than that from the flowing solvent reactor. Thus, branched chains were less apparent and n alkanes dominated gas chromatography/mass spectrometry spectra. Similarly, the aromatic material from minibomb liquefactions was less varied in structure. These mass spectral results are supported by many similar product analyses obtained by previous workers, (17), whose reaction conditions one now realises to have been retrogressive. It is apparent that the relatively lower molecular mass, but doubtless hydroaromatic, primary material produced by the liquefaction of vitrinites is readily degraded to leave simpler structures as liquefaction products. More extensive retrogression - caused by even longer reaction times and lower concentrations of available hydrogen - would be expected to simplify product structures even more. Analogy with pyrolysis would suggest the further loss of alkanes, of complex phenols and of side chains from aromatic structures (3). Fine tuning of retrogression could be significant in refining the technological properties of a clean synthetic fuel. By this stage of retrogression, it may no longer be possible to discern dominant reaction pathways. One is generating a multitude of radicals and polarisable molecules from the original hydroaromatic material and virtually all possible combinations of these radicals and molecules take place, eventually resulting in randomised 'scrambling' of all the hydrogen atoms (22).

7. Acknowledgments

Although I am guilty of providing a restricted list of references, I hope that the extent to which this chapter is indebted to the many pioneers of coal science will be perfectly obvious. I am personally grateful to my colleagues and students who participated in the work. My especial thanks are due to Drs. J. R. Gibbins and R. Kandiyoti who taught me all I know about reactor design. The work has been funded by the National Coal Board, the British Gas Corporation, Shell, British Petroleum, the scientific research councils of Britain and Turkey (SERC and TUBITAK) and, most recently, [Research Contract No. EN3V.0052:UK(H) directed by Dr. R. Kandiyoti] by the European Community.

8. References

1. Davis, M. R., et al. (1988), Fuel 67, 960.
2. Doğru, R., Gaines, A.F., Olcay, A. and Tuğrul, T. (1979), Fuel 58, 823.
3. Lowry, H. H. (1945), Chemistry of Coal Utilisation, Vols 1 and 2, J. Wiley and Sons, New York.
4. Noor, N. S., Gaines, A. F. and Abbott, J. M. (1985), Fuel 64, 1274.

5. Mc Pherson, W. P., et al. (1985), Fuel 64, 457 and references therein.
Mushrush, G. W., et al. (1988) J. Anal. and Appl. Pyrolysis, 14, 17.
6. Gaines, A. F. and Yürüm, Y. (1976), Fuel 55, 129.
7. Burawas, S., Gaines, A. F., Hasadsri, T., Prasertwitayakij, A. and Sucharitakul, N. (1970), Fuel 49, 180.
8. Uebersfeld, J., Etiene, A. and Combrisson, J. (1954), Nature 174, 614.
Ingram, D. J., et al. (1954), Nature 174, 797.
9. van Krevelen, D. W., Huntjens, F. J., and Dormans, H. N. M. (1956), Fuel 35, 42.
10. Gaines, A. F. (1958), Residential Conference on Science in the Use of Coal, C-58, published by the Institute of Fuel.
11. Suuberg, E.M. (1985), in R.H.Schlosberg (ed.), 'Chemistry of Coal Conversion', Plenum Press, New York, p. 67.
12. Howard, J. B. (1981), in M. A. Elliott (ed.), 'Chemistry of Coal Utilisation', 2nd. Suppl. Vol., Wiley-Interscience, p. 665.
Gavalas, G. R. (1982), Coal Pyrolysis, in 'Coal Science and Technology', Elsevier, 111.
Solomon, P. R., et al. (1988), Energy and Fuels 2, 405.
13. Chun-Zhu Li, et al. (1991), International Coal Science Conference, Newcastle.
14. Gönenç, Z. S., Gibbins, J. R., Katheklakis, I. E., and Kandiyoti,R. (1990), Fuel 69, 383.
15. Davis, M. R., Abbott, J. M. and Gaines, A. F., (1985), Fuel 64, 1362.
16. Dicker, P. H., Gaines, A. F. and Stanley, L. (1963), J. Appl. Chem 13, 455.
17. Makabe, M. and Ouchi, K. (1985) Fuel, 64, 1112.
Collin, P.J., Gilbert, T. D., and Wilson, M. A. (1983), Fuel 62, 450.
Given, P. H., et al. (1982), Report to the US Department of Energy, DOE/ET/10587-T.
18. Gibbins, J. R., and Kandiyoti, R., (1990), Fuel Proc. Tech., 24, 237.
19. Gibbins, J. R., Kimber, G. M., Gaines, A. F. and Kandiyoti, R. (1991), Fuel 70, 380.
20. Gaines, A. F., et al. (1991), International Coal Science Conference, Newcastle.
21. Snape, C. E., Ladner, W. R. and Bartle, K. D., (1979), Anal. Chem, 51, 2189.
22. Noor, S.N., Gaines, A.F. and Abbott, J.M. (1986), Fuel 65, 67.

SOLUBILIZATION OF COAL

L. L. ANDERSON and W. H. YUEN
Department of Fuels Engineering
University of Utah
Salt Lake City, UT 84112
U.S.A.

ABSTRACT. Coal solubilization, or essentially complete dissolution, entails a series of complex processes which involve reactions of the solvent as well as coal. Effective solubilization requires hydrogen from the solvent or other source outside of the coal since its hydrogen content is so low. Solubilization conceptually consists of using a solvent (and sometimes a catalyst) to break up the coal structure into fragments that can be dissolved by the solvent and stabilization of those fragments in a solution. Solubilization can be used to facilitate chemical structure studies or it may be the first stage of a coal liquefaction process . Despite the long history of studies on coal dissolution there are many aspects of solubilization that need further study to achieve a complete understanding of the phenomena, mechanisms, and conditions which control coal dissolution reactions.

1. Introduction

Solvents have been used to study the behaviour of coal under various conditions to determine the character of occluded material in pores to facilitate characterization of coal, and as a first step in coal liquefaction processes. Many coal liquefaction processes are considered to be composed of two main stages, with the first being solubilization. Several of the most promising direct liquefaction processes rely on the solubilization step to make liquefaction yields acceptable. These include the Exxon Donor Solvent Process, Solvent Refined Coal (II), and the catalytic-catalytic close-coupled integrated two-stage liquefaction process (CC-ITSL) .

The actions of solvents upon coal have been studied by many investigators and terms relating to such actions have evolved which make precise explanations difficult. Solvent extraction, solvolysis, extractive disintegration, activated extraction, and solubilization are some of the terms used to describe solvent treatments of coal. In addition, solvents are described as specific, nonspecific, novel, hydrogen-donor, etc. For this topical discussion on coal solubilization, some definitions will first be given so that the later discussion will be framed in a proper perspective.

2. Solvents, Extraction and Solubilization

Oele et al. made the unfortunate choice of describing low-yield extraction below 100°C as nonspecific (1) when, in fact, this type of extraction is the most specific. This "nonspecific extraction" usually involves extraction of resins, waxes and other occluded material but not the main skeletal structure of coal . These authors did not define terms for solvents but described "groups of operations" consisting of:

Y. Yürüm (ed.), Clean Utilization of Coal, 33–38.

a. Nonspecific extraction.
b. Specific extraction with effective solvents below 200°C.
c. Extractive disintegration at 200-350°C where the solvent is not appreciably changed.
d. Extractive chemical disintegration at temperatures exceeding 400°C where the solvent becomes a reactant in the process.

Berkowitz (2) has differentiated between various extraction regimes as:
a. Extraction at or below the boiling point.
b. Extraction under solvolytic conditions (200-350°C at autogenic solvent pressures)
c. Extraction near 400°C when thermal decomposition, solvolysis, and hydrogenation reactions are combined.

Here solvolysis is defined as any coal-solvent interaction at "conditions that are more severe than conventional Soxhlet extraction, but do not simultaneously promote incipient 'active' thermal decomposition of the coal". These extraction processes are those which only approach complete "solubilization" or the conversion of coal's organic material into a solution or into soluble species. Solubility here is somewhat arbitrary and depends on the solvent used to determine whether a conversion product is soluble. At different times benzene, pyridine, tetrahydrofuran, cresol, or other solvents have been used to determine whether the products from extraction or solubilization were liquid (soluble) or unreacted coal (insoluble). Yields as determined by THF or pyridine are often over 90 percent of maf coal when a hydrogen donor and/or hydogen gas under pressure are present during the solubilization.

Solubilization has been studied extrensively as a first step or stage of coal liquefaction processing. In this context solubilization is designed to produce a preponderance of low molecular weight species at conditions that maximize formation and capping of free radicals and minimize processes which lead to condensation or polymerization products (residue, char, or coke). These reactions require that hydrogen from a solvent, the gas phase or some other source be available. Many hydrogen donor solvents have been used and hydrogen donor ability (HDA) measured in terms of the relative amount of dihydroanthracene (DHA) produced by a solvent after reaction with anthracene (3). However, Tagaya et al. (3) found that other characteristics may be more important than HDA in determining a solvent's ability to solubilize coal. In their work n-butylamine (HDA=0.05) gave higher yields of THF-extractable (TE) material than did tetralin (HDA=0.24); see Figure 1.

Their data showed conversion of Taiheiyo coal in n-butylamine to THF extractables (TE) to be about 93 percent after 24 hours whereas TE yields for tetralin were approximately 20 percent for the same reaction time. At all temperatures above 200°C n-butylamine gave higher yields than tetralin or n-butylamine/tetralin mixtures. These authors attribute the high yields to both physical factors (strong physical solvent ability) and high donor ability. To optimize the solubilization yields a binary solvent system consisting of n-butylamine, and hydrogen donors (tetralin and 9,10-dihydoanthracene was used). No synergistic effect on conversion was observed (Figure 2).

In batch autoclave experiments Joseph (4) found that preswelling of low rank coals as well as bituminous coal increased both the yield of soluble products and the quality of those products. Table 1 shows this effect as well as the influence of swelling agent. This work showed that tetrabutyl ammonium hydroxide (TBAH) was an effective swelling agent that improved subsequent liquefaction results. THF and methanol were effective for subbituminous and bituminous coals but not for lignite. Joseph also showed that preswelling of Wyodak subbituminous coal eliminated the need for fine grinding to get high liquefaction conversion (Table 2).

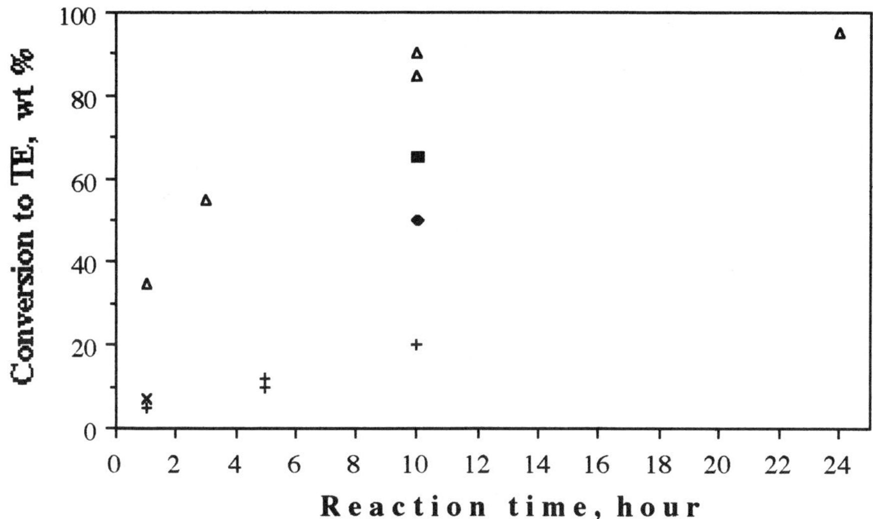

Figure 1. Liquefaction of Taiheiyo coal at 300°C using tetralin (+), n-butylamine (Δ) and di-n-butylamine (x) and of O-alkylated Taiheiyo coal using tetralin (♦) and n-butylamine (■). TE: tetrahydrofuran extractable material, [after Tagaya et al. (3)].

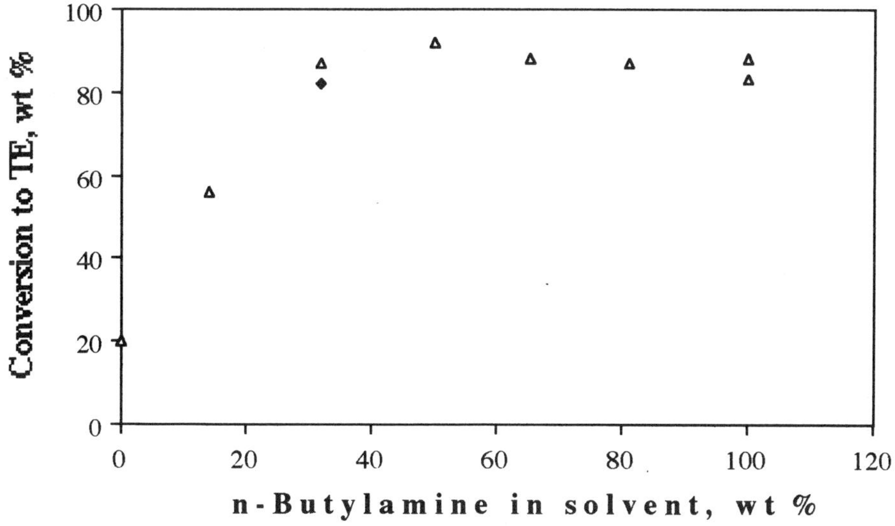

Figure 2. Liquefaction of Taiheiyo coal at 300°C for 10 hr using the mixture of n-butylamine-tetraline (Δ) and n-butylamine-9,10-dihydroanthracene (♦), [after Tagaya et al. (3)].

One method of improving solubilization of coal is to extract with supercritical fluids, which has been reported to give higher yields of tar than conventional carbonization or flash pyrolysis. Toluene and other organic solvents have been used and the higher yields have been attributed to solubilization of thermal fragments generated by the supercritical fluids used. Recently, Egiebor and Herath (5) subjected several coals to supercritical toluene extraction at 380°C and 20 MPa (197 atmospheres). Although they conducted a limited number of experiments, data were collected in a systematic way and multivariate analysis applied to the results. This allowed them to determine the factors which had the greatest effects on conversion, both positive and negative, using partial least square (PLS) regression analysis. They found that hydrogen and sulfur content of the coals had the greatest positive effects on liquefaction yields, whereas, increasing oxygen, ash and fixed carbon had negative effects on liquefaction.

While solubilization of coal is considered as a combination of thermal and solvent extraction reactions dispersed catalysts have been used for many studies and in liquefaction process development. Some catalysts used effectively were nickel acetate (6), iodine and triflic acid (7) and hydroxides and chlorides of Ni, Co, Zn, Cr, Sn, Fe and Mo (8).

Coal solubilization was enhanced by Miyake and Stock (9) through base-catalyzed C-alkylation. They found that base strength was significant in increasing the pyridine soluble yield of a low volatile bituminous coal but even more important was the chain length of the alkyl group introduced during the alkylation reaction. Table 3 shows these effects.

They concluded that the high solubilization was a result of the introduction of alkyl groups on the cross-linking and side-chain groups of the aromatic structural elements in the coal. Favorable polarization interactions between butyl and octyl groups apparently exceed those between the structures in the coal and solubilization is realized.

Many solubilization techniques have been used to essentialy put the entire organic matter in coal into solution. With some systems there is still a question whether there are true solutions or colloidal suspensions present. Most investigators recognize that conversion yields to soluble material decreases with increasing rank for high rank coals (medium volatile bituminous to anthracite). There is still controversy about whether highest solubilization yields are obtainable with lignites or subbituminous coals or from high volatile bituminous coals. Part of the reason for disagreement may be that coals from different provinces have properties which affect solubility as much as rank does. There are also questions about the influence of petrographic composition on solubilization, although most agree that vitrinites and exinites are easiest to put into solution. However, there is evidence that low-rank vitrinites do not react as readily as high-rank vitrinites do, or as completely (2). Inertinites are usually considered as unreactive components but some have reported substantial yields of even micrinites and semifusinites (10).

To obtain complete, or nearly complete, solubilization of coal either as a step in coal liquefaction or to study the structural and chemical aspects of coal, considerable hydrogen must be provided. Many studies show that coal is a strong hydrogen donor. In fact about 30 percent liquefaction has been achieved with little or no outside hydrogen (11). Eventually hydrogen must be supplied to assist in the solubilization. Hydrogen donors, hydrogen-transfer reactants and hydrogen shuttlers (or hydrogen shufflers) have all been. Deuterium labeled solvents have been used to obtain information on the reactions between them and coals during solubilization (12). The most dramatic effects on promoting complete solubilization have come from the application of catalysts.

TABLE 1. Effect of swelling and swelling agent on coal liquefaction yields for Illinois No.6 coal (bituminous), Wyodak (subbituminous) and Kinneman Creek (lignite), [after Joseph (4)].

Coal Swelling Agent	Swelling ratio[a]	Conversion, wt % Oil (Hexane sols)	Asphaltenes (Toluene solubles)	Preasphaltenes (THF solubles)	Total
Illinois No.6 coal (hvab)					
None	-	22	31	16	69
TBAH[(b)](15%)	2.3	37	33	13	83
THF[c]	2.0	28	37	24	89
Methanol	1.5	28	31	29	88
Wyodak coal (Subbituminous)					
None	-	32	22	13	67
TBAH[(b)] (15%)	3.1	50	22	6	78
THF [(c)]	1.5	52	13	9	74
Methanol	1.4	44	13	10	67
Kinneman Creek (Lignite)					
None	-	33	15	8	56
TBAH [(b)](15%)	3.5	51	12	5	68
THF[(c)]	1.3	33	16	7	56
Methanol	1.5	34	15	7	56

a Selling ratio-weight of packed bed of coal swollen in a tube / height of unswollen coal.

b Reagent grade tetrabutylammonium hydroxide.

c Unstabilized tetrahydrofuran.

TABLE 2. Effect of particle size on swelling and liquefaction of Wyodak (subbituminous) coal, [after Joseph (4)].

Particle Size, mesh	Swelling, (TBAH)	Particle Size, (after swelling) mesh	Liquefaction Field,wt%
-325	No	-325	69
-8+70	No	-8+70	57
-325	Yes	-325	78
-8+70	Yes	-8+70[a]	77
-8+70	Yes	-325[b]	77

a 97% of this sample was +70 mesh but crumbled easily.

b For this experiment the preswollen -8+70 mesh coal was ground to -325 mesh.

TABLE 3. Effects of electrophilic agent and alkylation species on solubilization of lower Kittanning seam (PSOC 1197).

mmol of $NaNH_2$	Reagent	Product, wt %	Alkylation group/100C	Solubilization in pyridine, wt %
0	None	-	-	5
100	NH_4Cl	102	-	9
45	methyl iodide	109	-	10
45	1-butyl iodine	109	2.4	50
45	1-octyl iodine	121	2.9	90

Recently, Derbyshire (13) has summarized the findings of catalytic reactions and catalysts, including dispersed hydrogenation catalysts. He indicates that there is a need for developing catalysts with higher hydrogenation activities if promotion of dissolution can be carried out as a catalytically driven thermal process. Catalysts for dissolution or solubilization required intimate coal-catalyst contact and must have quite different properties than those used for typical refining or upgrading reactions of liquids. Advances in analytical capabilities which can be applied to coal-derived materials and increased interest in coal solubilization should lead to significant further advances in this important area of coal science.

3. References

1. Oele, A. P., Waterman, H. I., Goedkoop, M. L., van Krevelen, D. W. (1951) 'Extractive Disintegration of Bituminous Coals', Fuel, 30, 169-178.
2. Berkowitz, N. (1985) The Action of Solvents on Coal, The Chemistry of Coal, Elsevier Science Publ., Chapter 8, 275-316.
3. Tagaya, H., Sugai, J., Onuki, M. and Chiba, K. (1987) 'Low Temperature Coal Liquefaction Using n-Butylamine as a Solvent', Energy and Fuels , 1, No. 5, 397-401.
4. Joseph, J. T. (1991) 'Liquefaction Behaviour of Solvent-Swollen Coals', Fuel, 70, 139-144.
5. Egiebor, N. O. and Herath, B. (1991) 'A Multivariate Analysis of Property-Reactivity Relationship of Coal during Supercritical Fluid Extraction', Fuel Science and Techn. Internat. 9, No. 6, 769-789.
6. Yasuhiro, T. and Okada, K. (1988) 'Coal Hydroliquefaction Activity of Nickel Catalyst Using a Nickel Acetate Precursor,' Fuel, 67, 1548-1553.
7. Fraenkel, D., Pradhan, V. R., Tierney, J. W. and Wender I. (1991), 'Liquefaction of Coal Under Mild Conditions; Catalysis by Strong Acid, Iodine and Their Combination', Fuel, 70, 64-73.
8. Kuznetsov, P. N. Beregovtsova, N. M. Ivanchenko, Korniyets, E. D., Tarabanko, V. E., Rubaylo, A. I. and Truckhacheva, V. A. (1990) 'Catalytic Liquefaction of Kansk-Achinsk Lignite in Methanol at 63 K' , Fuel, 69, 985-991.
9. Miyake, M. and Stock, L. M. (1988) 'Coal Solubilization. Factors Governing Successful Solubilization through C-Alkylation', Energy and Fuels, 2, 815-818.
10. Whitehurst, D. D., Mitchell, T. O. and Farcasiu, M. (1980) 'Coal Liquefaction', Academic Press. Chapter 5, Coal Rank and Liquefaction, 134-137.
11. Kang, D. (1979) 'Elucidation of Coal Structural Components by Short Residence-Time Extractive Liquefaction', Ph. D. Thesis, University of Utah, pp. 68, 71.
12. Ratto, J. J., Heredy, L. A. and Skowvonski, P. (1980) ACS Symp. Series 139, 347.
13. Derbyshire, F. J. (1988) 'Catalysis in Coal Liquefaction: New Directions for Research', IEA Coal Research, London, Chapter 3, 16-28.

COAL LIQUEFACTION KINETICS

L. L. ANDERSON
Department of Fuels Engineering
University of Utah
Salt Lake City, UT 84112
U.S.A.

ABSTRACT. Coal liquefaction reactions include several types which are related to the thermal or degradation reactions of the coal itself or to subsequent reactions of the coal fragments. When coal liquefaction is conducted in a liquid phase, i.e., in the presence of a solvent, dissolution is usually considered an essential first step. While kinetic evaluation for individual reactions is possible for chemical models, the same is not true for coal. Generally, kinetic models and data for coal liquefaction must be considered as composites of many reactions occurring. This is a consequence of the complex nature of coal and the many thermal and catalytic pathways involved in the liquefaction process. Kinetic data have been obtained for all of the main routes to liquid products from solid coal; namely, pyrolysis, hydrogenation and dissolution. Mechanisms for these processes have been formulated based on kinetic studies and the reaction orders and activation energies obtained. It is interesting to note that the activation energies have been found to be similar for all these processes and fall in the range of 30-50 Kcal/mole regardless of the reaction order assumed by the researchers.

1. Introduction

Coal liquefaction is a very general term which can involve the breakup of solid coal to soluble fragments. Subsequent reactions to even smaller species or in the case of pyrolysis it may involve only thermal degration of the solid to liquid, solid and gaseous products. Liquids may also be formed by synthesis of coal products (indirect liquefaction) but these reactions will not be considered here.

Kinetic studies on coal liquefactions have been made by many workers who have described the processes involved and attempted to use their results to predict reaction times, yields or even the expected products. Even for a complex process like coal liquefaction, if one can use kinetic information to improve the yield or quality of products or to decrease the times required to form products, then such studies are valuable. Whether kinetic studies can actually determine the mechanism of liquefaction reactions has been extensively debated without universal agreement. In order to summarize our knowledge and present thinking on coal liquefaction kinetics specific examples on solubilization, pyrolysis and hydroliquefaction will be given followed by discussion of the general topic.

2. Dissolution of Coal

Conversion of coal to material soluble in pyridine, benzene, tetrahydrofuran (THF) and hexane have been used to define liquefaction yields. Such solubility determination are operational and may even depend on the method of determination. The use of yield data based on solubility have largely been justified or interpreted

Y. Yürüm (ed.), Clean Utilization of Coal, 39–48.

as direct reactions of coal to fragments which are liquid-size molecules. Conversely, one can base the conversion rate on the amount of insoluble material remaining at any particular time. Examples of kinetic rates based on solubility in benzene, toluene, pyridine or THF are extensive and consider insoluble material in these solvents as "unreacted coal". Anderson and Hill (1) found the rate expression for hydrosolvation in the temperature range 400-500°C to be:

$$\text{rate of dissolution} = r_A = k\, C_{SO}\, (1 - x)\, (C / S)$$

where,

- x : conversion fraction (weight of dissolved products / weight of organic material in the coal)
- C_{SO} : weight fraction of organics in the untreated coal
- k : rate constant for dissolution
- C / S : weight of coal/weight of solvent.

The integrated equation is:

$$\ln (1 - x) = k' \theta$$

θ was defined as $(C_{SO} / C_{AO})\, (t)\, (C / S)$

where,

- t : reaction time
- C_{AO} : grams of ash-free coal per cc of reactor volume (numerically = 0.245).

Correlations were made from plots of $(1 - x)$ vs θ and evaluated by Wen and Han (2) with good agreement. The activation energy found from an Arrhenius plot was very low (11 Kcal/mole) indicating physical processes or diffusion as rate controlling steps in the dissolution reaction.

From dissolution studies one may follow the formation of a particular soluble product, but this is more diffucult than determination of insolubles, which are not easily lost by evaporation. Kang (3) followed the mineral matter in the coal and determined the conversion by:

$$\%\ \text{conversion} = [(M_t - M_o) \times 10^4] / [M_T (100 - M_o)]$$

where,

- M_T : weight % mineral matter in the residue at any reaction time, t.
- M_o : weight % mineral matter in the original unreacted coal.

Extraction yields are not simple primary products and probably involve thermal fragmentation, followed by capping of these fragments by hydrogen donor or hydrogen transfer solvents or molecular hydrogen. Conversion values usually are expressed as the sum of gas and liquid conversion. In the case of dissolution, the liquefaction yields are more diffucult to determine because the analysis of liquid product is complicated by the presence of solvent and solvent products. The same can be said of coal hydrogenation or hydroliquefaction when vehicle oils or solvents are used.

It appears that while kinetic data and correlations are useful in providing insight for the rate controlling steps in the dissolution or extraction process, the determination of exact mechanisms is not possible. The order of the reaction is similarly useful only in generally suggesting the overall process reactions and limitations. In many studies the order of reaction has not even been obtained. The data from several studies indicate that the dissolution process is first order with respect to unreacted coal and first order with respect to solvent. The latter determination requires careful evaluation at coal / solvent ratio over a range of values (4, 5). While there is considerable controversy over the rate controlling step in the coal dissolution process as well as the applicable mechanism there is some agreement that both physical and chemical processes are important in the dissolution process. The rate determining step probably changes during the course of the reaction and is also probably different for different conditions (turbulence, reaction temperature, etc.). The most probable rate controlling step at temperatures below the softening temperature of the coal (350°C) is diffusion. At higher temperatures the activation energy value indicates that chemical reactions are rate controlling. These controlling reactions could be the thermal cleavage of certain bonds in the coal (6), the transfer of hydrogen to coal radical (7) or even rehydrogenation of hydrogen donor solvent (8). It must also be recognized that coal dissolution or extraction is a complex process which includes not only dispersion, solution and depolymerization of coal components but also many secondary reactions of the solvent, such as, adduction repolymerization and isomerization which can distort the kinetic measurements obtained (9).

Since the kinetics of coal dissolution are so complex, one must ask whether there is actual value in pursuing such studies or whether one can learn anything significant from the kinetic results of such studies. In general terms there has certainly been advancement in our understanding as the result of kinetic studies. At the time Oele (10) studied the process it was very diffucult to measure the extent of reaction, especially during the first few minutes, when most of the dissolution took place (Figure 1). The reaction was assumed to be a zero order forward reaction and a first order backward reaction:

$$dx / dt = k_f - k_b x .$$

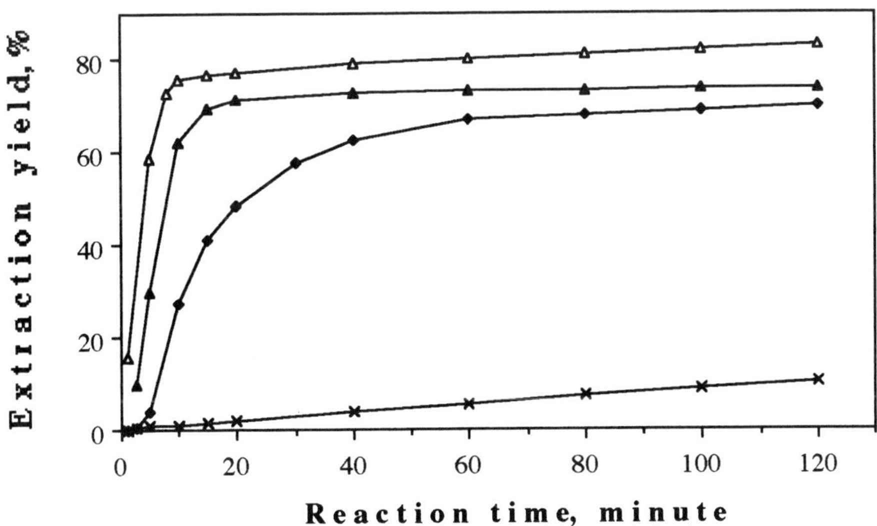

Figure 1. Extraction of Limburg coking coal (VM= 28.1%) with anthracene oil at 200-350°C, x : 200°C, ♦ : 250°C, ▲ : 300°C, Δ : 350°C, [after Oele et al. (10)].

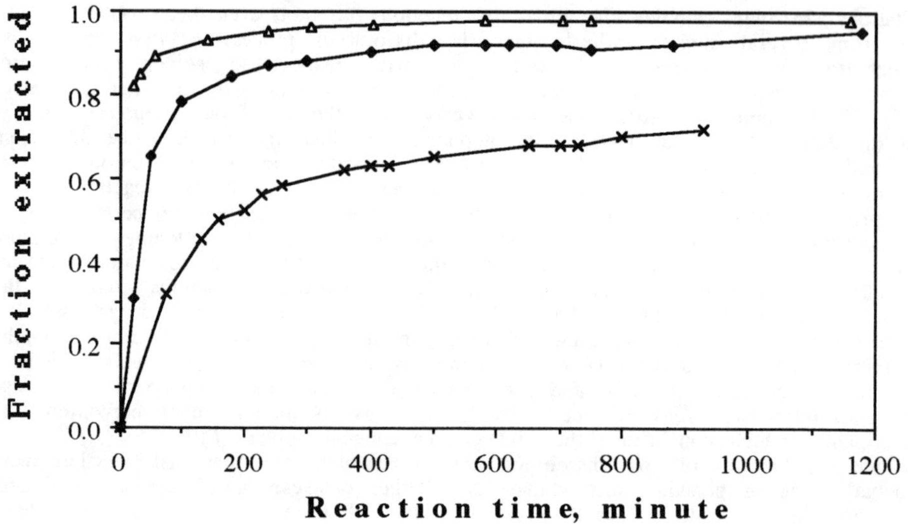

Figure 2. Time-yield curves for thermal dissolution of coal in tetralin, x : 350°C, ♦ : 400°C, Δ : 440°C, [after Hill et al. (4)].

When there was no change in the amount extracted, the rate was zero and the two terms were considered to be equal. Later Hill et al. (4) were able, by injecting coal into solvent already at the reaction temperature, to measure the fraction extracted, even at short residence times and thus, able to obtain kinetic data during the whole extraction process (Figure 2). The kinetic expression contained dependence on both the amount of coal available for reaction as well as the solvent:

$$dx / dt = k_1 (a - x) + k_2(a - x) (b - x)$$

where,
- a : the initial concentration of coal.
- b : the initial concentration of solvent (tetralin).

It was recognized that many reactions were taking place and that the most easily reacted and dissolvable fractions reacted first. As the reaction proceeded, the unreacted coal was visualized as more and more unreactive. The activation energy for reaction thus increased with reaction time and a series of reactions were proposed:

$$\text{Coal} + \text{Solvent} \xrightarrow{k_1} R_1 + L_1 + G_1$$

$$R_1 + \text{Solvent} \xrightarrow{k_2} R_2 + L_2 + G_2$$

$$R_2 + \text{Solvent} \xrightarrow{k_3} R_3 + L_3 + G_3$$

.

.

$$R_{n-1} + \text{Solvent} \xrightarrow{k_n} R_n + L_n + G_n$$

where, R_i is the solid residue or unreacted coal, L_i is the liquid product (extract) and G_i is the gaseous product. It was assumed that

$$k_1 > k_2 > k_3 \cdots\cdots > k_n$$

The main reaction is 1 at the beginning but shifts to 2, 3,...n as the reaction proceeds. Thus, a first-order rate expression can be defined with a variable rate constant:

$$dx / dt = k_v (1 - x)$$

It was found that k_v varied linearly with the extent of reaction (x), thus:

$$k_{vb} = C_1 - C_2 x = k_0 (1 - a x)$$

and "a" was found to be $1/x_m$ where x_m is the maximum possible conversion at the temperature being studied. Rearrangement of the rate expression gives:

$$dx / dt = k_m (x_m - x)(1 - x)$$

this is a pseudo second order rate expression. This expression fits the data over most of the reaction studied; at high conversions, a first-order expression was found to be better, probably because the reaction near completion was a more simple process.

One conclusion that can be drawn from these kinetic studies is that while precise mechanisms cannot be elucidated from them, general information on the process limiting the reaction (the slow step in the dissolution process) can be. Additionally, it may be valuable to know rate constant ranges for the formation or disappearance of certain species such as preasphaltenes and asphaltenes as was determined by Gangwer(11).

3. Pyrolysis

Many studies have been done to determine the kinetic parameters for coal pyrolysis not only because pyrolysis can produce liquids but also because pyrolysis reactions are involved in combustion and gasification. Kinetic data are also easier to obtain for pyrolysis since simple weight loss changes can be followed. Pyrolysis reactions have been studied isothermally at temperatures as low as 180°C (devolatilization) and as high as 1100°C (high temperature carbonization).

When pyrolysis studies are conducted using weight loss measurements the weight fraction at infinite time is critical in arriving at a correct rate equation. Since the sample weight of a coal sample undergoing pyrolysis asymptotically approaches its final value it is impractical to carry the experiment to actual completion. The final weight fraction pyrolyzed at a particular temperature has been determined in different ways by different groups. For example, Berkowitz and den Hertog (12) used a procedure which involved fitting a polynomial of varying degrees to each weight loss versus time curve and computing the coefficients of each polynomial. Rate equations are usually found to be of the form

$$dx / dt = k_n (a - x)^n$$

where,

a : weight loss fraction at $t = \infty$

x : weight loss fraction at time t and $n \neq 1$.

Wiser found that the apparent reaction order, n can vary considerably, depending on the value of "a" assumed or determined (13). He found pyrolysis kinetics best explained as second-order reactions in isothermal studies in the range 408-497°C.

Several other groups have agreed with second-order pyrolysis explanations but some have found that first-order or fractional order treatments have led to better fit of the data (14). One reason for the lack of agreement on pyrolysis mechanisms is that during pyrolysis many reactions are occurring simultaneously. There are also differences in kinetics when different heating rates are used. Like other coal conversion processes pyrolysis may also be complicated by secondary reactions of the initial pyrolysis products.

Some workers have interpreted pyrolysis kinetics as involving a series of slow overlapping reactions (15). The integrated equations obtained from the kinetic data and first-order assumption are expressed as:

$$k = (1 / t) \ln [a / (a - x)]$$

where,

k : pyrolysis decomposition constant at the temperature studied

a : the maximum value of x possible at the same temperature

x : the fraction of the coal pyrolyzed (volatilized) at any time, t,

as determined from weight loss.

However, such a treatment was not found to be satisfactory by Wiser (4) or Berkowitz (16). One problem is an implicit assumption that the volatile matter evolves as fast as it is formed by the decomposition of the coal (no secondary reactions). Berkowitz has postulated that isothermal pyrolysis is a diffusion-controlled process but proving diffusion control or reaction rate control has not been conclusive since similar equations result from both.

Recently, Johnson et al. studied pyrolysis of coal at different heating rates (19). They fitted data to single reaction model:

$$\text{Coal} \xrightarrow{k_i} x_i V + (1 - x_i) S$$

where,

i : 1 (or 1, 2 if a double reaction model is assumed)

V : volatile products

S : solid

$k_i = A_i \exp(-E / RT)$.

Their best fit to experimental data was obtained by assuming that the activation energy was linearly dependent on temperature ($E_i = E_{oj} + a_j T$). E_{oj} was found to be quite low (0.5 to 7.2 Kcal/mole) with a mean value of 3.1 Kcal/mole. They also discussed several other model reactions and concluded that the following was about the minimum to give a good representation of the observations made on their pyrolysis experiments:

1. Coal $\xrightarrow{k_i}$ x Radicals + (1-x) Solid

2. Coal $\xrightarrow{k_2}$ Radicals

3. Coal $\xrightarrow{k_3}$ Radicals

4. Radicals $\xrightarrow{k_4}$ Tars

5. Radicals $\xrightarrow{k_5}$ Gases (H_2O, CO, CH_4, H_2, etc.)

6. Tars $\xrightarrow{k_6}$ Solid

7. Tars $\xrightarrow{k_7}$ Gases

The first four reactions can be represented by the overall reaction:

$$\text{(Coal component 1)} \xrightarrow{R_{T1}} \text{Intermediate} \xrightarrow{R_{T2}} \text{Tar vapor}$$

where,

R_{T1} is the rate of a first-order reaction with a pre-exponential factor of $1x10^{13}$ s^{-1} and an activation energy of 35.85 Kcal/mol.

R_{T2} is the rate of vaporization of tars (assumed to be equal to the rate of heat flux into the coal divided by the latent heat of vaporization of the tars).

They also postulated the presence of an additional coal component and a first-order reaction to simulate the total of the measured hydrocarbon gas yield:

$$\text{(Coal component 2)} \longrightarrow \text{Hydrocarbon gases.}$$

The pre-exponential factor and activation energy were taken as $1x10^{13}$ s^{-1} and 50 Kcal/mole, respectively. However, these investigators recognized and illustrated the wide diversity of pyrolysis results in the literature from different groups.

4. Hydrogenation (Hydroliquefaction)

Most of the reactions applicable to hydrogenation (or hydroliquefaction) of coal are considered in the dissolution of coal. However, the temperatures may be higher for coal hydrogenation and in many research studies, as well as pilot plant and commercial operations, much higher pressures have been used in hydrogenation. The hydrogenation of coal by molecular hydrogen has not been considered as appreciable unless a catalyst is used, especially at temperatures below 500°C. The conversion under these conditions is essentially the result of pyrolysis reactions, although hydrogen increases the yield due to capping of radicals and other reactive species. However, the role of molecular hydrogen is more significant than formerly thought, as has been shown by Vernon(18). In his studies, molecular hydrogen participation in reactions of models dibenzyl and diphenyl were significant at hydrogen pressures commonly used for coal hydrogenation reactions (10 MPa).

The determination of rates of reaction for coal hydrogenation to soluble products was carried out by Wiser et al. (7). The data were correlated with a second-order rate expression over essentially all of the reaction time (8 hours), yielding an activation energy of 39 Kcal/mole. Several other investigators have assumed first-order kinetics for hydrogenation reactions, but activation energies are close to the value given by Wiser and indicate that the rate controlling step in the reaction is chemical in nature. For a hydrogenation reaction, one could presume that the mass transfer rate could be the controlling step. At least for some hydrogenation processes this appears not to be the case (19).

Some limited data have been obtained on the kinetics of hydrocracking of coal in an ebullated bed (20). The authors assumed that the catalytic action involved only the solvent and primary liquefaction products. Hydrocracking kinetics were represented by a "pseudo" first-order equation of the form:

$$F / V = kE (1 - a) / a$$

where,

F / V : space velocity , gram extract / hr.cm^3 (sump)

k : "pseudo" first-order constant (hr^{-1})

E : extract concentration, gram / cm^3

a : conversion of extract to distillate plus gas.

Using the experimental data, the following equations resulted:

$\ln (kE) = -16{,}700 / T + 23.0$ $\quad P_{total} = 239$ atm, $\quad P(H_2) = 197$ atm

$\ln (kE) = -20{,}000 / T + 28.4$ $\quad P_{total} = 409$ atm, $\quad P(H_2) = 340$ atm.

This study, like many others, involved reaction in the presence of a catalyst. However, it was concluded that the reactions were relatively insensitive to catalyst concentration but that thermal pyrolysis of hydrogenated extract played a major role in the hydrocracking mechanism.

When a high volatile bituminous or similar rank coal is heated in the presence of hydrogen, but in the absence of a catalyst, product yields are similar in quantity and composition to those obtained by pyrolysis. An appropriate catalyst can significantly increase the yield of soluble products and significantly changes the product composition. Hydrogenation of coals with catalysts present have supplied much of the data for kinetic studies of coal hydroliquefaction. This fact alone helps us to draw some conclusions about the rate determining reactions. As pointed out by Wiser et al. (7), hydrogen reacts with thermally produced reactive fragments, which tends to lead to the production of relatively smaller molecules than are found in pyrolysis liquids. The kinetics of such catalytic hydrogenation of a bituminous coal was correlated with a second-order rate expression and yielded an activation energy of 39 Kcal/mole, indicating a chemical controlling slow step for the reaction. This work is typical of kinetic studies in bituminous coals.

Petrakis and Grandy (21) have observed that most of the coal liquefaction literature contains information about yields or molecular structure of products but relatively little information on kinetics. Table 1 gives their summary of results on dissolution kinetics showing kinetic parameters determined by different groups.

TABLE 1. Summary of gross coal dissolution kinetics, [after Petrakis and Grandy (21)].

Coal	Solvent and Solvent/Coal Ratio	T, °C	Definition of conversion	Kinetics Parameters, Kcal/mole
Utah HV	Tetralin 10:1 (pyrolysis)	409-497 (pyrolysis) 350-450 (liquid)	Weight loss (pyrolysis) Benzene sol. (liquid)	35.6, (2)+ 4.1, (1)e2 28.8, (2)i 13.1, (1)e
Pitt Seam	Tetralin 4:1	324-387	Xylenol, Cyclohexane, Benzene, cresol	$E_a[K_1$(fast)]=30.0 $[K_2$(slow)]=38.0
Belle Ayr (sub)	HAO* HPH*	400-470	Pentane, Benzene, Pyridine	Coal—>Product E_a=16.7 (HAO) 20.5 (HPH)
Pitt Seam (Bruceton)	None	400-440	Benzene	Coal—>Asphaltene E_a=37.0
Miike	Process (330-380) MoO_3 cat	350-450	Hexane, Benzene	Coal—>Asphaltene E_a(calc)=16.4 Asphaltene Oil E_a(calc)=16.0
Utah Spring Canyon	Tetralin 10:1	350-450	Benzene	H = 37.2-85.5 from 0.90% reacted
Big Horn	Process 3:2	413-440	Boiling ranges heavy oil>343°C	Coal—>091 E_a=55.0 Coal—>Furnace Oil E_a=40.5
Bruceton	Synthoil	400	Benzene	
Makum	Tetralin	380-410	Benzene	Initial R_x E_a=79.0 Step 2 E_a=47.0 Step 3 E_a=35.0

*HAO - hydrogenated anthracene oil
HPH - hydrogenated phenanthrene

5. Coal Depolymerization Kinetics and Reaction Pathways

These results are similar to those already discussed and show most activation energies for the dissolution process are in the range of (30-40 Kcal/mole). Since these values are, in general, greater than the bond dissociation energies for C-C bonds, it has been postulated that the bonds broken in the coal liquefaction process produce radical species that are stabilized by extensive delocalization by p electron systems of large aromatic species (24). The activation energies for such reactions would be an average of the bond dissociation energies in the chemical species involved and would vary with the rank and chemical nature (maceral composition) of the coal, if the activation energy in coal liquefaction depends on thermal bond rupture.

6. References

1. Anderson, L. L. and Hill, G. R. (1974) 'Hydrogenation of Western Coal in Dilute Phase under Medium Pressures for Production of Liquids and Gases' Qtrly. Prog. Rept. (March) for Office of Coal Research, Contract No.14-32-0001-1200.
2. Wen, C. Y. and Han, K. W. (1975) ACS Div. Fuel Chem., Preprints 20, No. 1, 216.
3. Kang, D. (1979) Ph. D. Thesis, University of Utah, Salt Lake City, UT, U.S.A.
4. Hill, G. R., Hariri, H., Reed, R. I. and Anderson, L.L. (1966) in 'Coal Science', R. F. Gould (ed.), Washington, D. C., ACS Adv. in Chem. Series, 55.
5. Anderson. L. L., Shifai, M. Y., and Hill, G. R. (1974) Fuel, 53, 33.
6. Curran, G. P., Struck, R. T. and Gorin, E. (1967) Ind. Eng. Chem. Proc. Des. and Dev., 6, 166.
7. Wiser, W. H., Anderson, L. L. Qader, S. A. and Hill, G. R. (1971) J. Appl. Chem. Biotechn. 21, 82.
8. Guin, J. A., Tarrer, A. R., Pitts, W. S. and Prather, J. W. (1977) Liquid Fuels from Coal, R. T. Ellington (ed.), London, U. K., Academic Press, p.133.
9. Pullen, J. R. (1981) Solvent Extraction of Coal. IEA Coal Research, London, U. K. , IEA Rept. No. ICTIS/ TR16.
10. Oele, A. P., Waterman, H. I., Goedkop, M. L. and van Krevelen, D. W. (1951) Fuel, 30, 169.
11. Gangwer, T. (1980) Brookhaven National Laboratory Rept., BNL 27279, 58 pp.
12. Berkowitz, N. and den Hertog, M. W. (1962) Fuel 41, 502.
13. Wiser, W. H. (1991) private communication.
14. Solomon, P. R. and Hamblen, D. G. (1983) Progr. Energy Combust. Sci., 9, 323.
15. Fitzgerald, D. (1956) Trans. Fara. Soc. 1, 52, 362-369.
16. Berkowitz, N. (1960) Fuel, 39, 47, and Proc. Symp. on the Nature of Coal, Central Fuel Res. Inst., Gealogra, INDIA, 284.
17. Johnson, G. R., Murdoch, P., and Williams, A. (1988) 'A Study of the Mechanism of the Rapid Pyrolysis of Single Particles of Coal' Fuel, 67, 834-842.
18. Vernon, L. W. (1980) Fuel, 59, 102.
19. Gorin, E. (1977) in 'Chemistry of Coal Utilization', Second Suppl. Vol., M. A. Elliot (ed.), Willey Interscience, New York, N. Y., Ch. 27, 1890 pp.
20. Anon, K. (1971) Pilot Scale Development of the CFS Process, U.S.D.I., OSRR and D Rept., No. 39, 4, Book 3, Consolidation Coal Co.
21. Petrakis, L. and Grandy, D. W. (1983) Free Radicals in Coals and Synthetic Fuels, Coal Science and Technology 5. Elsevier Sci. Publ., Amsterdam, 211 pp.

CATALYSIS IN DIRECT COAL LIQUEFACTION

L. L. ANDERSON
Department of Fuels Engineering
University of Utah
Salt Lake City, Utah 84112
USA

ABSTRACT. Coal liquefaction can be defined as the conversion of solid coal to products which are liquid at ambient temperatures. Usual methods for direct liquefaction processing are pyrolysis, solvent extraction or hydrogenation. These processes can all be enhanced by the use of catalysts (for example, catalytic pyrolysis or catalytic hydropyrolysis, Exxon Donor Solvent Process and catalytic two-stage liquefaction). Catalysis is necessary because there are many reactions possible at the conditions used for liquefaction and only certain ones are desirable, such as bond cleavage and chain termination. Since coal is such a chemically heterogeneous material, no single-function catalyst appears able to satisfy all of the requirements for successful implementation of catalysts to full-scale coal liquefaction. If suitable catalyst combinations can be identified and developed for coal liquefaction this fact alone could lead to greatly increased possibilities for a coal liquefaction industry to compete with decreasing and eventually more expensive liquids from petroleum. Since coal liquefaction was first commercialized in Germany, many improvements have been made in the use of catalysts. Although not complete, our understanding of the chemistry of coal and the reactions whicw must be catalyzed has advanced considerably. The application of both dispersed and supported catalysts is now possible and the feasibility of direct coal liquefaction more viable.

1. Introduction

Since its beginning, direct coal liquefaction has been hampered by the difficulties inherent in the chemical reaction limitations of treating a material that is both heterogeneous and relatively stable (thermodynamically). In addition coal liquefaction must be carried out at relatively low cost if it is to be commercially feasible. In Germany in World War II times, and in South Africa in more recent times, coal liquefaction was carried out at large scale without regard to its economic feasibility. It was recognized that such processing would require the application of catalysts just to be technically feasible. It is accepted that for coal liquefaction to be competitive in producing distillates (transportation) fuels successful application of suitable catalysts will be necessary. This is true whether the liquefaction route is dissolution, hydrogenation, pyrolysis, or indirect liquids production. New catalysts, more mild reaction conditions or other changes which can lead to more favorable process economics will depend on their ability to 1) increase reaction rates and improve product selectivity, 2) lower the temperatures and pressures of liquefaction, or 3) to accomplish both of the above.

Since Fischer-Tropsch and other indirect methods of producing liquids from coal involve specific chemical processes where the coal is no longer present these processes will not be considered here.

Y. Yürüm (ed.), Clean Utilization of Coal, 49–64.

Catalysis for direct coal liquefaction has such an important significance because rates of the uncatalyzed liquefaction reactions are too low for consideration in commercially feasible processes. Although there are many reactions that can be considered "liquefaction reactions" the major reactions to be catalyzed involve bond cleavage, the addition of hydrogen to the liquids produced or the removal of heteroatoms from the liquid products. The high cost of producing coal-derived liquids (CDL) is mostly attributed to the large amounts of hydrogen necessary to convert a solid material that is about 5% by weight hydrogen to a liquid with about 11-14% by weight hydrogen. Conventionally, coal liquefaction reactions have been carried out at severe conditions of temperature and pressure. The presence of inorganic material in the coal also poses material handling problems that have not as yet been completely solved. As shown in Table 1, there has been significant progress in reducing the severity, particularly the pressure, of the reaction conditions and in achieving better selectivity (i.e., higher liquids/gas ratios).

TABLE 1. Progress in catalysis and process configuration on operating conditions and selectivity for bituminous coal liquefaction.

Process	Temperature, °C	Pressure, MPa	Liquid/gas ratio
Single-stage			
I.G. Farben	480	30-70	2.4
Rurhkohle	475	30	2.3
H-coal	450	12	4.0
Two-stage (noncatalytic/catalytic)			
British coal	400-425	20	4.8
Lummus	410-460	18	10.8
Two-stage (catalytic/catalytic)			
HRI	400-440	17	12.0
SEI	432-404	18	6.0

Even in the early development of direct coal liquefaction in Germany it was recognized that a single-stage reaction to accomplish production of distillate fuels compatible with petroleum liquids was impractical. At least two stages have been found to be required to give the yields and quality of product to make coal liquefaction of interest for commercial development. In recent developments catalysts have been used in both stages of the reaction with favorable results (Table 2).

TABLE 2. Process development and performance for bituminous coal liquefaction.

Single-stage noncatalytic (1982)	Single-stage catalytic (1982)	2-stage noncatalytic/ catalytic (1985)	2-stage catalytic/ catalytic (1986)	2-stage catalytic/ catalytic (1990)
Distillate yield (wt. % coal maf)				
41	52	62	70	74
Distillate gravity (°API)				
12.3	20.2	20.2	26.8	27.5
Nonhydrocarbons (wt %)				
S 0.33	0.20	0.23	0.11	0.03
O 2.33	1.00	1.90	<1	0.44
N 1.00	0.50	0.25	0.16	0.09

Both the elements in the catalytic material and the way the catalyst is applied for coal liquefaction are as diverse as the many processes that have been studied. Soluble catalysts such as zinc or tin chloride have been used as well as solid, insoluble, supported catalysts such as Co, Ni, Mo and W. Because of its relatively low cost iron in many forms has been extensively studied, tested and used. One of the few generalizations that can be made regarding catalyst application to coal liquefaction is the difference between dispersed and supported catalysts. According to Derbyshire (1) dispersed catalysts have been applied primarily to promote dissolution of coal and supported catalysts have been mostly used in upgrading and refining of solubilized coal liquids. Exceptions to this observation are rare since dispersed catalysts are not suited to realize the selectivity possible with supported catalysts which were developed specifically for refining processes and the difficulties of putting a supported catalyst in the vicinity of a reaction site on a solid (coal) preclude its use to influence the initial reactions of coal. In some reactions where coal is present as a solid, use of supported catalysts actually gave results no better than uncatalyzed reactions (Tables 3 and 4).

TABLE 3. Comparison of catalysts for LEFCO Process using 1 / 8 in. (ID) x 30 in. Long Last Chance Kaiparowitz Coal*, [after Derbyshire (1)].

Catalyst	Conversion,%	Catalyst	Conversion,%
$ZnBr_2$	58.5	Sn (powder)	7.9
ZnI_2	46.3	$SnCl_2.H_2O$	7.8
$ZnCl_2$	41.1	$FeCl_3.6H_2O$	7.2
$SnCl_2.2H_2O$	40.5	Zn (powder)	7.0
$SnCl_4.5II_2O$	25.8	$ZnSO_4.7H_2O$	5.4
LiI	16.6	$(NH_4)_5Mo_7O_{24}.4H_2O$	5.4
$CrCl_2$	12.3	$FeCl_2$	3.3
$Pb (C_2H_3O_2)_2.3H_2O$	11.7	$CaCl_2.H_2O$	no reaction
NH_4Cl	11.0	$Na_2CO_3.H_2O$	no reaction

* P = 2000 psi, T = 650°C. Where no catalyst was present, essentially no reaction took place. Catalyst concentration used was 10% of the coal.

TABLE 4. Comparison of catalysts for LEFCO Process using 5/16 in. (ID) x 54 ft Long Kaiparowitz Coal*, [after Derbyshire (1)].

Catalyst	Conversion,%
$ZnCl_2$	65.9
Molecular sieves	21.5
Cobalt molybdate on silica-alumina	17.5
No catalyst added	19.1
WS_2	21.5
Fe_2O_3	17.7

* P = 1800 psi, T = 482°C. Catalyst concentration used was 5.8% of the coal weight.

2. Historical Perspective

In order to understand why certain materials have been used as catalysts for hydrogenation or other direct liquefaction processing, it is helpful to know something about the history of catalytic liquefaction, especially in Germany during the period 1913-1945. A personal summary of this period has been given from first-hand knowledge by Donath and Hoering (3) and some facts from their account will be used here.

The conditions which led investigators to look for hydrogenation catalysts for coal processing grew out of the work of Matthias Pier who discovered selective, oxidic catalysts that were not poisoned by sulfur in the methanol synthesis (1923). Before this, even though Bergius had demonstrated that coal could be liquefied with hydrogen at high pressures, the situation seemed hopeless since essentially all coals contain sulfur, oxygen and nitrogen. At the time sulfur alone was considered to be a prohibitive poison for all hydrogenation catalysts.

The first sulfur resistant coal hydrogenation catalysts prepared by Pier in 1924 were sulfides and oxides of molybdenum, tungsten and the iron group metals. The use of these catalysts made it possible to add hydrogen, split carbon-carbon bonds and eliminate S, N and O from coals and oils.

The work and time spent to develop processes and suitable catalysts were major investments by Badische Anilin und Soda Fabrik (BASF) in Germany. The initial catalysts used for coal conversion were those found to be resistant to sulfur poisoning in the methanol synthesis. Later as a result of thousands of experiments, both small and large scale, and in-depth discussions and collaboration with others, Pier and associates developed other catalysts and processes. From the discovery of molybdenum-zinc oxide catalysts in 1925 to the discovery of dual function catalysts (1930 and later) the work was greatly assisted by the operation of large-scale plants. At the same time more than 2500 catalysts were tested in high pressure flow reactors and about five times as many catalysts were actually prepared though many were not tested at high pressure due either to "unpromising composition" or unsatisfactory physical properties. This extensive testing was only a part of the catalyst development since the catalysts mentioned here are vapor phase catalysts.

After the discovery of Mo-ZnO catalyst, which gave 100% yield of gasoline from coal tar, it was found that the throughputs were too small and the hydrogen recycle rates too high for commercial feasibility. At higher throughputs the catalyst deactivated severely and Pier and coworkers concluded that the hydrogenation process needed to be conducted in two stages:

1. A "liquid phase" where coal and liquid residues were converted to a middle oil

(boiling below 325°C) using a suspended catalyst, and

2. A "vapor phase" where the middle oil was converted into motor fuels using a fixed-bed catalyst.

Liquid phase catalyst at first were molybdenum or ammonium molydate. Because of cost and supply problems with molydenum, other catalysts were developed, including low-cost iron ores and especially Bayermasse. With such low cost catalysts or catalysts used in small concentrations, these could be discarded after passing through the reactor, greatly simplifying this part of the process.

Vapor phase catalysts at first were hydrorefining types and the requirements for a fixed-bed catalyst were obviously quite different from the dispersed liquid phase catalysts, even in physical properties such as physical strength, shape, size and melting point. In 1927 a catalyst was adopted which consisted of equimolar amounts of MoO_3, ZnO and MgO (catalyst 3510). Later (1930) a superior catalyst (catalyst 5058) composed of tungsten disulfide was adopted. This catalyst, a dual function catalyst, performed well at full scale and gave conversion rates of middle oil to gasoline three times as great as did catalyst 3510. Results are shown for catalyst 5058 in Table 5.

One problem with 5058 was that in the production of gasolines the aromatic content and thus, the octane number, were low. In 1934 montmorillonite clay (Terrana) was found which, when treated with hydrochloric acid and used with 10% catalyst 5058, gave gasolines with higher octane number but the same yield at the same temperature as catalyst 5058.

Further developments on vapor phase catalysts were necessary because of the presence of nitrogen compounds and their effect on some catalysts. The vapor phase process was eventually divided into two steps, "prehydrogenation" and "splitting hydrogenation or benzination" (hydrocracking). Catalysts such as metal sulfides were developed and much work was done on the effects of acidic and basic support materials.

Much of the work on the catalysis of coal hydrogenation have found more application in the petroleum refining industry. The work by coal scientists laid a foundation for many of the catalysts and processes presently used in hydrorefining and hydrocracking. It still remains to be seen if these catalysts and processes will be used for direct liquefaction of coal on a commercial scale.

3. Catalyst Reactions of Importance in Coal Hydrogenation

From the early German work and subsequent research and development on direct liquefaction of coal, it becomes obvious that heterogenous catalysis, where the reaction takes place on a solid catalyst surface, involved liquid products from coal decomposition or thermal reactions. The reactions taking place after the initial coal decomposition are numerous. This variety has been summarized by Cusumano et al. (4) and is shown in Table 6.

Some of the general reactions and products listed in Table 6 are only "expected" since coal liquefaction, except utilizing the Fischer-Tropsch (Sasol) process is not presently commercial.

In addition to the general reactions and products from coal conversion to liquids there are a multitude of effects and interactions which can affect the catalytic process involved. Some of these are catalyst particle size, effects of catalyst-support interactions, surface effects of catalysts, poisoning, regeneration, and recovery of catalyst materials. Cusumano et al. (4) have also summarized (Table 6) some of the significant advances of importance to coal conversion catalysis.

TABLE 5. Hydrogenation with Tungsten Disulfide Catalyst (Catalyst 5058), [after Hessley et al. (3)].

		Reaction Conditions			
Feed	Product	Temperature ^oC	Pressure atm	Rate kg/l.h	Conversion %
Diisobutane	Isooctane	216	250	2.0	98.0
Naphthalene	Decalin	336	200	0.9	90.0
Brown coal	Prehydrogenated M.O.	380	200	1.0	
Middle oil					
Phenols 18.0%					99.5
N-Bases 4.0%					99.5
Sulfur 2.5%					99.5
n-Butane	Isobutane	400	200	0.5	35.5
Paraffins 260-320°C	Gasoline 180°C EP	408	220	1.0	90.0

4. Catalysts Used in Direct Liquefaction

The least expensive catalyst available for coal conversion is the mineral matter in the coal. The catalytic effects of mineral matter on the SRC processes and the Exxon Donor Solvent process have been investigated by several researchers, including Gangwer et al. (5), Mukherjee and Chowdhury (6) and Gray (7).

Both homogeneous and heterogeneous catalysts have been used for coal liquefaction. Homogeneous catalysts for coal and coal-derived liquids have included hydrogenation transition metal complexes of cobalt, ruthenium, rhodium, palladium and platinum (2, 8). Examples of such complexes are dicobalt octacarbonyl (9) and the rhodium complex of N-phenylanthranilic acid (10). A major problem with these homogeneous catalysts is recovery from the product mixture. The catalysts mentioned above are expensive and consequently, anything less than close to complete recovery results in a process too expensive to be feasible. Additionally, some of these catalysts are quickly poisoned by heteroatoms, especially sulfur.

Heterogeneous catalysts which have been tested for coal hydrogenation, hydrocracking or hydrogenolysis include almost all solid catalytic materials. Over the years essentially all solid metals and metal compounds have been at least considered. The most often tested and utilized in process developments have been sulfided nickel/molybdenum catalyst supported on alumina, molybdenum catalysts, or iron sulfides. Many other materials have been tested as catalysts which are directly related to the transition metal catalysts listed above. In a recent survey of the literature 1977-1986 on catalysts and catalytic liquefaction of coal the catalysts listed in Table 6 were cited.

As indicated earlier the reactions desirable in coal liquefaction are primarily hydrogen addition, molecular weight reduction (depolymerization) and heteroatom removal. Catalysts applied to the liquefaction process are usually directed mainly to depolymerization since upgrading (heteroatom removal) can be carried out in a subsequent process without mineral matter. Molecular weight reduction or depolymerization by hydrocracking can be effectively catalyzed by such catalysts as $ZnCl_2$ for either coal or coal-derived liquids (11, 12). The hydrocracking action of metal halides, where smaller organic molecules are produced by the thermal cleavage (pyrolysis) of larger entities in the coal structure needs to be assisted by a catalyst and using hydrogen, takes place at lower temperatures (400-430°C) with $ZnCl_2$. Some other metal halides are effective catalyst for hydrocracking as shown by Wood et al. (13) and probably act as homogeneous catalysts at the temperature applied, as shown by their vapor pressures. The use of $ZnCl_2$ and other halides results in high conversions but some, such as $AlCl_3$, are too active, producing smaller molecules which are gases instead of liquids. The metal halides, particularly Zn and Sn, have been extensively tested. Sn is far too expensive for application unless 100 percent can be removed. $ZnCl_2$, which is approximately one order of magnitude less expensive than $SnCl_2$, has received more attention but can also be too expensive for application without high recovery rates as indicated by Wood et al. (15). These authors carried out liquefaction runs utilizing small concentrations of catalyst (5%); much of the work reported using metal halides used massive amounts of catalyst (1:1 catalyst:coal ratios). This leads to major problems of recovery especially since during the hydrocracking of coal some $ZnCl_2$ is converted to ZnS and other species. These compounds must then be converted back to $ZnCl_2$ in order to recycle the catalyst. There are also significant engineering problems associated with handling the $ZnCl_2$ which is corrosive (2).

TABLE 6. Catalytic reactions in coal conversion, [after Donath and Hoering (4)].

Process	General Reactions	Specific Reactions/Products
Direct liquefaction	Hydrogenation Cracking Hydrofining	Aromatic liquids Hydrodesulfurization (HDS) Hydrogenitrogenation (HDN)
CO/H_2 synthesis	Fischer-Tropsch	Methane Hydrocarbon liquids Alcohols Chemicals
Water-gas shift		Hydrogen
Direct gasification	Hydrogasification Oxidative gasification	Methane Synthesis gas
Liquids refining and upgrading	Cracking Reforming Hydroforming	Hydrogenation Dehydrogenation Dehydrocyclization Isomerization Hydrogenolysis Hydrodesulfurization (HDS) Hydrodenitrogenation

The most commonly utilized heterogeneous catalyst for direct coal liquefaction is sulfided cobalt molybdate on alumina and the current literature still includes more papers on this catalyst than any other. Co/Mo on alumina is an active hydrogenation catalyst that is not easily deactivated by the heteroatoms in coal-derived liquids. Deactivation is usually due to clogging of the catalyst pores by mineral in the coal. The best results (yields) are obtained when the catalyst is used in combination with hydrogen and a donor solvent (16).

According to Gun et al. (16), the donor solvent gives up hydrogen to free radicals formed from the coal thermolysis reactions to stabilize molecules. Then the dehydrogenated solvent goes to the catalyst surface to react with dissociated hydrogen and regain its hydrogen donating ability.

5. Mild Temperature Catalytic Reactions

Especially since about 1962, major efforts of many investigators who have been interested in coal liquefaction, have been directed at "depolymerization of coal". Coal does not have a simple repeating structure so the term depolymerization does not imply the production of monomers from a polymeric substance. The main objective of many coal liquefaction reactions is to cleave the minimum numbers of chemical and physical bonds which will result in the maximum number of liquid range products. One of the obstacles to systematic studies to accomplish this objective has been the conditions, especially temperature, at which liquefaction reactions have been studied. This dilemma has prompted many studies during the last two decades where catalysts were designed to attack certain linking groups or bonds thought to be present in coal. In order to avoid the indiscriminate thermal bond ruptures which occur above the softening temperature of bituminous and subbituminous coals (350°) many experiments have now been conducted on catalytic liquefaction of coal at temperatures below $300^{o}C$. These conditions have opened whole families of new catalyst possibilities since many materials stable below $300^{o}C$ decomposed, became unstable or vaporized at ordinary coal liquefaction conditions (>$400^{o}C$). Representative studies and their catalysts will now be discussed which show the types of catalysts, reactions and products from these lower temperature (often called mild conditions) studies. As far back as 1962 Heredy and Neuworth (17), showed that coal could be "depolymerized" and rended soluble by reaction with phenol and boron trifloride at $100^{o}C$. The reactions of this application have been studied using model compounds. The following reactions are visualized for coal depolymerization (18-22):

$$(AR)_1\text{-}CH_2\text{-}(AR)_2\text{-}CH_2\text{-}CH_2\text{-}(AR)_3 + 2C_6H_5\text{-}OH \rightarrow C_6H_5\text{-}CH_2\text{-}(AR)_1 + C_6H_5\text{-}CH_2\text{-}CH_2\text{-}(AR)_2H + H(AR)_3$$

$$C_6H_5\text{-}CH_2\text{-}(AR)_1 + C_6H_5\text{-}OH \rightarrow C_6H_5\text{-}CH_2\text{-}C_6H_5\text{-}OH + H(AR)_1$$

These depolymerization reactions dramatically affect the solubility of the components of coal as shown in Tables 8 and 9 the original coals (being less than 5 percent soluble in single solvents such as benzene and even tetralin at its normal boiling point). The depolymerization reactions listed are clearly not only cleavage reactions taking place when coals are treated with phenol/boron trifloride since it has been shown that at least some C-O, C-N and C-S bonds are also affected (23).

Berkowitz (24), Ouchi et al. (25) and Yürüm et al. (26-28) have applied similar reactions to catalyze the depolymerization of coals, usually with the objective of rendering a large fraction of the coal soluble so that more information may be obtained on the chemical structures present. Ouchi (24), for example, was able to solubilize all but a few coals having greater than 85 percent carbon to more than 90 percent using p-toluene sulfonic acid/phenol at $185^{o}C$ (Table 10).

TABLE 7. Advances of importance to the catalytic conversion of coal, [after Donath and Hoering (4)].

Subject	Representative Areas of Impact
Multimetallic catalysis	Upgrading of coal liquids Methanation Fischer-Tropsch synthesis
Catalyst-support interactions	Thermal and chemical stabilization of catalyst and supports
Catalyst characterization	Improvement of coal liquefaction catalysts Upgrading of coal liquids Determination of catalyst intrinsic activity
Catalyst preparation	New catalyst formulations Controlled variations in catalyst properties Higher surface area catalysts
Poisoning and regeneration	Effect of sulfur in coal conversion catalysis Prevention of carbon deposition, removal of carbon
Mechanism and surface chemistry	Better understanding of the important steps in coal conversion Identification of rate limiting process, directions for improvement
Reaction engineering catalyst testing	Development of effective catalyst testing procedures Data interpretation
Inorganic chemistry	New catalytic materials and compositions with improved poison tolerance and activity
Materials science	Novel support materials and structures Novel refractory compounds Understanding of sintering phenomena (catalyst deactivation)
Surface science	New characterization techniques

Although the depolymerization reactions cited here are referred to as "acid catalyzed depolymerization" (3), they must be considered as chemical reactions involving the acid components as well as phenol. This could also be said of some other solubilization reactions recently applied to coal, including reductive alkylation. However, for catalytic considerations, acid materials must be recognized as facilitating many reactions at mild conditions which may be applied for solubilization but might also serve as liquefaction reactions. These reactions would include nonreductive alkylation (29), Friedel-Crafts alkylation and acylation. Similarly, based-catalyzed depolymerization reactions might serve as liquefaction steps; hydrolytic depolymerization (30), and hydrogenolysis (31-34).

In looking at the total range of acid- and base-catalyzed reactions of coal at mild conditions, Anderson and Miin looked at over 100 materials (36). Tables 10 through 16 give results for Bronsted acids and a whole range of compounds which may act as acidic or basic catalysts. Their conclusions included recognition of the different classes of acids and their properties which could be related to the extent of coal conversion. Only one property was shown to be directly related to the catalyst's ability to catalyze the production of liquids from a high volatile bituminous coal. That property was the HSAB hardness of the acid material. Only borderline catalysts gave more than 34 percent conversion of coal to liquids and gases (mostly liquids).

TABLE 8. Coal liquefaction catalysts (1977-1986).

Catalyst	Support	Form of Preparation
MoO_3	α-alumina, γ-alumina, zirconia, silicon dioxide refractory oxides, clays, silicate sludge	Impregnated in coal, sulfided
$(NH_4)_6Mo_7O_{24}$ Mo naphthenate		Micro encapsulated
MoS_{23}		Impregnated in coal
Co-Mo,Co-W	alumina, silicon dioxide	
Sn-Mo Mo-Ti, Ni-Mo-Ti, Sn-Fe-Mo	alumina	
Sn-Co-Mo	alumina	Stabilized with La-Si, La-Zr
Sn-Co-Ti-Mo Sn $SnCl_2$		Molten Molten
$SnCl_2$ SnO_2 ZnO $ZnCl_2$ ZnO-TiO_2	alumina alumina alumina	Mixed with coal, molten
Bi_2O_3 Ga_2O_3	alumina alumina	
Se Pd hydrous tetanide TiO_2, MnO_2, MnO Mo,Co,Ni,Mn,Sa (oil sol. catalyst)		Mixed with coal
Co boride Ni boride Ni		From $CoBr_2$+NaB_5H_8 From $NiBr_2$+NaB_5H_8 Electroplated on coal
SRC mineral residue SRC mineral residue ash		
Fe		Impregnated on coal Mol. dispersed
Fe dust		From coal gasification and electric furnace
Fe pres magnetite,geothite, limonite,hematite, bauxite,laterite,red mud		Oxided,sulfided, reduced with Co
Fe sulfides FeS Pyrite(FeS_2), Pyrite + H_2S Triolite(FeS), Pyrrhotite($Fe_{1-x}S$) Iron III sulfide(Fe_2S_3), Fe_7S_8		Absorbed(collodial from $FeSO_4$+Na_2S(ag)
Fe oxides FeO Fe_2O_3, Fe_2O_3+S powder Fe_2O_3-TiO_2, Fe_2O_3-MoO_3		Treated with CO, H
$FeSO_4$ H_2S		Impregnated in coal

Although these mild temperature catalytic reactions of coal are being investigated vigorously by some, the practical application of the catalysts and conditions have not been demonstrated. This area of coal processing has not only opened up new possibilities for catalytic coal liquefaction but also made possible more exact identification of the bonds and structures involved in the liquefaction reactions. The application of organic chemistry principles to coal depolymerization reactions in a major liquefaction scheme has been demonstrated by Shabtai et al. (36) and consists of the following steps (Figure 1):

1. Intercalation of coal with a Lewis acid catalyst material. Zinc chloride, ferric chloride or other metal halide, followed by mild hydrotreatment (HT),
2. Base-catalyzed depolymerization (BCD) of the product from step 1, at supercritical conditions followed by,
3. Hydroprocessing (HPR) of the depolymerized product with a sulfided Co-Mo catalyst.

The authors of this scheme visualize the process in the following way:

Step 1 consists of partial depolymerization of the coal by preferential hydrogenolytic cleavage of methylene, benzyletheric and some activated aryletheric linkages in the coal framework.

Step 2 is designed to complete the depolymerization of the product of Step 1 by based-catalyzed hydrolysis or alcoholysis of the diaryletheric, dibenzofuranic and the other bridging groups.

Step 3 is hydroprocessing step consisting of heteroatom removal, partial hydrogenation and C-C hydrogenolysis reactions.

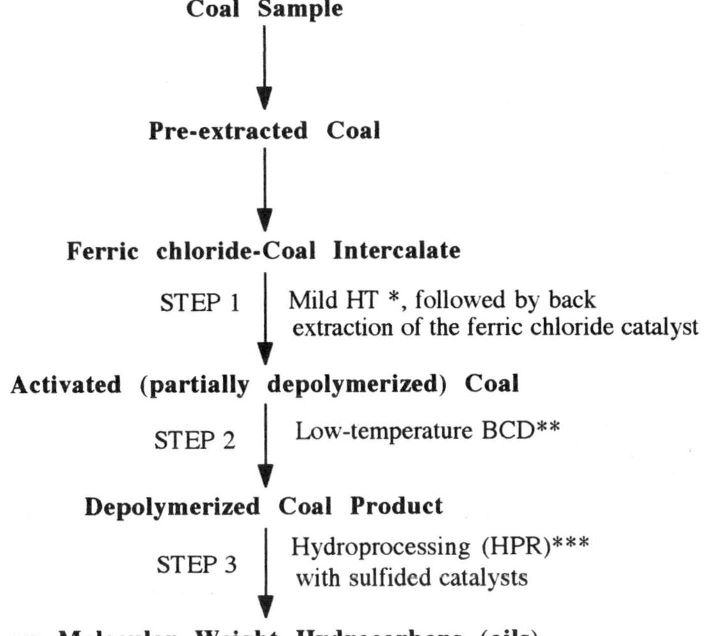

Figure 1. Procedure for production of low molecular weight hydrocarbons by base-catalyzed depolymerization [after Shabtai (37)].

* HT: hydrotreatment (240-275°C; H_2 pressure, 70-100 atm)

** BCD: base-catalyzed depolymerization (250-275°C, initial N_2 pressure, 70 atm)

*** HPR: hydroprocessing (350-370°C, H_2 pressure, 180 atm)

TABLE 9. Phenolation of (Japanese) coals with phenol/ p-toluene sulphonic acid at 185°C, [after Ouchi et al., (25)].

% C, dmmf	Composition/100 C-atoms	%Soluble in pyridine
69.7	$C_{100}H_{88.9}O_{26.2}N_{0.6}S_{0.2}$	90+
70.9	$C_{100}H_{79.0}O_{24.5}N_{1.0}S_{0.3}$	90+
75.8	$C_{100}H_{73.9}O_{18.0}N_{0.9}S_{0.2}$	90+
78.0	$C_{100}H_{99.3}O_{13.9}N_{0.9}S_{trace}$	90+
81.9	$C_{100}H_{85.9}O_{9.8}N_{2.2}S_{0.2}$	90+
83.1	$C_{100}H_{87.5}O_{6.2}N_{1.9}S_{1.0}$	90+
84.6	$C_{100}H_{80.2}O_{6.7}N_{1.7}S_{0.04}$	90+
86.2	$C_{100}H_{81.6}O_xN_yS_z$	90
89.5	$C_{100}H_{65.2}O_{2.5}N_{1.8}S_{0.3}$	30
92.9	$C_{100}H_{43.6}O_xN_yS_z$	10

TABLE 10. Yields of reaction products (THF soluble = gases), wt %, from hydrotreatment of Hiawatha (Utah) hvbb coal at P=13.5 MPa-H_2 and T=300°C for 3 hours. Catalysts: Transition metal compounds.

	Yields, (wt %)		
Catalyst	Gas	Liquid	Total
$RuCl_3.nH_2O$	13	19	32
$CoCl_2.6H_2O$	1	23	24
CoS	1	23	24
NiS	1	16	17
$NiCl_2.6H_2O$	3	23	26
Cu_2Cl_2	2	20	22
$CuCl_2.2H_2O$	3	17	20

TABLE 11. Yields of reaction products (THF soluble + gases), wt %, from hydrotreatment of Hiawatha (Utah) hvbb coal at P=13.5 MPa-H_2 and T=300°C for 3 hours. Catalysts: Group VA compounds.

	Yields, (wt %)		
Catalyst	Gas	Liquid	Total
$SbCl_3$	2	32	34
BiF_3	3	19	22
$BiCl_3$	4	55	59
$BiBr_3$	8	64	72
BiI_3	9	34	43

TABLE 12. Yields of reaction products (THF soluble + gases), wt %, from hydrotreatment of Hiawatha (Utah) hvbb coal at P=13.5 MPa-H_2 and T=300°C for 3 hours. Catalysts: Group IVA compounds.

Catalyst	Yields, (wt %)		
	Gas	Liquid	Total
$GaCl_4$	4	82	86
SnF_2	3	17	20
$SnCl_2.2H_2O$	9	78	87
$SnBr_2$	13	81	94
SnI_2	13	80	93
SnO	3	13	16
SnF_4	12	31	43
$SnCl_4.5H_2O$	2	73	75
$SnBr_4$	4	77	81
SnI_4	4	93	97
PbF_2	2	19	21
$PbCl_2.5H_2O$	1	21	22
$PbBr_2$	2	46	48
PbI_2	3	38	41

TABLE 13. Yields of reaction products (THF soluble + gases), wt %, from hydrotreatment of Hiawatha (Utah) hvbb coal at P=13.5 MPa-H_2 and T=300°C for 3 hours. Catalysts: Iron compounds and hard acids.

Catalyst	Yields, (wt %)		
	Gas	Liquid	Total
FeS	1	17	18
FeF_2	2	19	21
$FeCl_2.4H_2O$	2	27	29
$FeBr_2$	2	45	47
FeI_2	3	41	44
$FeCl_3$	14	19	33
$AlCl_3$	13	17	30
$CrCl_2.6H_2O$	14	26	40
$CeCl_3.7H_2O$	1	4	5

TABLE 14. Yields of reaction products (THF soluble + gases), wt %, from hydrotreatment of Hiawatha (Utah) hvbb coal at P=13.5 MPa-H_2 and T=300°C for 3 hours. Catalyst: Group IIB compounds.

Catalyst	Yields, (wt %)		
	Gas	Liquid	Total
ZnF_2	1	15	16
$ZnCl_2$	8	71	79
$ZnBr_2$	12	82	94
ZnI_2	12	73	85
$ZnSO_3$	2	32	34
$ZnSO_4.H_2O$	2	18	20
Zn_3P_2	1	23	24
$Zn_3(PO_4)_2$	1	19	20
$Zn(OAc)_2$	1	25	26
ZnO	1	15	16
CdF_2	2	26	28
$CdCl_2.2.5H_2O$	1	15	16
$CdBr_2$	1	29	30
CdI_2	3	18	21
$HgCl_2$	1	23	24

TABLE 15. Yields of reaction products (THF soluble + gases), wt%, from hydrotreatment of Hiawatha (Utah) hvbb coal at P=13.5 MPa-H_2 and T=300°C for 3 hours. Catalysts : Bronsted Acids.

Catalyst	Yields, (wt%)		
	Gas	Liquid	Total
Benzoic Acid (anhydrous)	3	11	14
Benzoic Acid (2.N)	7	19	26
Phenol	2	19	21
HOAc (glacial)	7	13	20
HOAc (15.N)	6	20	26
H_3PO_4 (85%)	2	18	20
HCl (4.9 N)	12	38	50
HBr (4.9N)	18	42	60
HI (4.9N)	45	43	88

Although there have been some significant improvements in catalysis of coal liquefaction, there are still many problems and opportunities for changes which could make production of liquids from coal a worldwide industry. One might ask why there has not been more progress in this important area. Part of the answer is that the reactants and products of coal liquefaction are highly complex materials and are not easily amenable to analysis and characterization. Another aspect is that while there has been much effort devoted to the problem, it has been cyclic in nature with long periods of relative inactivity. Much of the time when there has been adequate interest in the problem, research and development teams have had to start from scratch in the education process of the new members of such teams. The basis for this cyclic effort is the question of the economic feasibility and the considerable uncertainty regarding supplies of liquids from petroleum, their availability, and their price.

While coal liquefaction is an extremely complex process and there are many problems to be solved, if it is to be significantly improved the application of appropriate catalysts will be one of the keys. Further advancements in our analytical and characterization capabilities and a better understanding of the dissolution process and how we can treat at the high molecular weight fragments of that process will be necessary.

6. References

1. Derbyshire, F. (1990) 'Catalysis in Direct Coal Liquefaction', Chemtech, 439-443.
2. Anderson, L. L., Wood, R. E. and Wiser, W. H. (1976), Soc. Min. Eng.Trans., 260, 321.
3. Hessley, R. K., Reasoner, J. W. and Riley, J. T. (1986), Coal Science, Wiley Interscience, NY, p. 143.
4. Donath, E. E. and Hoering, M. (1977), Fuel Proc. Technol., 1, 3.
5. Cusumano, Dalla Beta, R. A. and Levy, R. B. (1978) Catalysis in Coal Conversion, Academic Press, NY.
6. Gangwer, T. E. and Prasad, H. (1979), Fuel, 58, 1577.
7. Mukherjee, D. K. and Chowdhury, P. B. (1976), Fuel, 55, 4.
8. Gray, D. (1978), Fuel 57, 213.
9. Cox, J. L., Wilcox, S. A. and Robert, G. L. (1978) in J. Larsen (ed.), 'Organic Chemisty of Coal', ACS Symp. Series 71, ACS, Washington, D. C., p. 186.
10. Friedman, S., Metlin, S., Swedi, A. and Wender, I. (1959), J. Org. Chem., 24, 1287.
11. Holy, N. T., Nalesnik, N. and Mc Clanahan, S. (1977), Fuel, 56, 47.
12. Zielke, C. W., Struck, R. T., Evans, J. M., Costanza, C. P. and Gorin, E. (1966), Ind. Eng. Chem., Process Des. Devel., 5, 158.
13. Zielke, C. W., Struck, R. T., Evans, J. M., Costanza, C. P. and Gorin, E. (1966), Ind. Eng. Chem., Process Des. Devel., 5, 151.
14. Wood, R. E., Wiser, W. H., Anderson, L. L. and Oblad, A. G.(1977), US Pat.4, 134, 822.
15. Hill, A. (1976) Zinc Chloride Catalyst Recovery from Char, M.Sc. Thesis, University of Utah, Salt Lake City, Utah.
16. Wood, R. E., Anderson, L. L. and Hill, G. R. (1970) Quarterly of the Colorado School of Mines, Golden, CO., 65, 201.
17. Gun, S. R., Sama, S. K., Chowdhury, P. B., Mukherjee, S. K. and Mukherjee, D. K. (1979), Fuel, 58, 171.
18. Heredy, L. A. and Neuworth, M. B. (1962), Fuel, 41, 221.
19. Heredy, L. A., Kostyo, A. E. and Neuworth, M. B. (1963), Fuel, 42, 182.
20. Heredy, L. A., Kostyo, A. E. and Neuworth, M. B. (1964), Fuel, 43, 414.
21. Heredy, L. A., Kostyo, A. E. and Neuworth, M. B. (1965), Fuel, 44, 125.
22. Heredy, L. A., Kostyo, A. E. and Neuworth, M. B. (1966), Adv. in Chem. Series No. 55, 493.

23. Larsen, J. W. and Kuemmerle, E. (1976) Fuel, 55, 162.
24. Berkowitz, N. (1965) The Chemistry of Coal, Elsevier Sci. Publ., Amsterdam, The Netherlands, p. 182.
25. Ouchi, K., Imuta, K. and Yamashita, Y. (1965), Fuel, 44, 205.
26. Yürüm, Y., Yiğinsu, I. and Zümreoğlu, B. (1979) Fuel, 58, 75.
27. Yürüm, Y. and Yiğinsu, I. (1981) Fuel, 60, 1027.
28. Yürüm, Y., (1981) Fuel, 60, 1031.
29. Yürüm, Y., Yiğinsu, I. (1982) Fuel, 61, 1138.
30. Ignasiak, B. S., Carson, D., Szladow, A. J. and Berkowitz, N. (1980) ACS Symp. Series, No. 139, 97.
31. Dryden, I. G. C. (1963) in H. H. Lowry (Ed.) 'Chemistry of Coal Utilization' Suppl. Vol., , J. Wiley, NY, p. 292.
32. Reggel, L. Zahn, C., Wender, I. and Raymond, R. (1965) U.S. Bureau of Mines Bull., No. 615.
33. Makabe, M. and Ouchi, K (1979) Fuel, 58, 43.
34. Ouchi, K., Fuwata, K., Makabe, M. and Itoh, H. (1979) ACS, Fuel Chem. Preprints 24, 185.
35. Anderson, L. L., Chung, K. E., Pugmire, R. G. and Shabtai, J. S. (1981) in 'ACS Symp. Series', No. 169, 223.
36. Anderson, L. L. and Miin, T. C. (1986) Fuel Proc. Technol. 12, 165.
37. Shabtai, J. S., Skulthai, T. and Saito, I. (1986) ACS Fuel Chem. Div. Preprints, 31, 15.

CATALYTIC AND CHEMICAL BEHAVIOR OF COAL MINERAL MATTER IN THE COAL CONVERSION PROCESS

HAROLD H. SCHOBERT
Fuel Science Program
Department of Materials Science and Engineering
The Pennsylvania State University
University Park, PA 16802 USA

ABSTRACT. Most synthetic fuel production processes that have been envisioned rely in some way on catalysts to facilitate one or more steps of the overall process. Catalysts added externally (that is, catalytic materials not inherently present in one or more of the reactants) represent an increased operating cost for the process and, usually, the necessity to incorporate operations for catalyst recovery and recycle into the process flowsheet. Coals contain many kinds of inorganic species that potentially can serve as catalysts - sulfides (notably pyrite), clays, and alkali or alkaline earth cations. If these catalysts inherent in the coal could be taken advantage of in conversion processing, the need for external catalysts could be lessened or eliminated. Examples of the catalytic effects of these kinds of species are discussed.

1. Introduction

Coals contain a diverse array of inorganic species. The prospect of taking advantage of some of these inorganic components as catalysts for coal conversion reactions offers several potential advantages. Since the inorganic components are already present in the coal, the need for catalysts added to the reactor as separate materials is reduced or eliminated. A related point is that the inorganic species in the coal are free, and do not add an economic burden to the process, as would result from catalysts especially purchased for addition to the reactor. If the inorganic components of the coal are to be used in the conversion process, then less attention must be directed to removal of inorganics as a pre-treatment step prior to the conversion reaction. In some conversion processes, most notably direct liquefaction, the importance of catalyst dispersion is well recognized. It is desirable that the catalyst should be available at "the smallest and most remote particle of coal" [1] as the conversion reactions begin. Coals having catalytic inorganic components well dispersed through the carbonaceous matrix very likely offer superior dispersion to that attainable with externally added catalysts, regardless of the steps taken to insure good coal-catalyst mixing. A complicating factor in considering the behavior of the inorganic components of coals as prospective catalysts is their possible interaction with other catalysts that have been added deliberately to the reaction [2]. Such possible interactions have not been extensively studied.

This chapter discusses aspects of the catalysis of conversion reactions by some of the inorganic components of coal. Liquefaction, gasification, and pyrolysis processes are of interest. There is of course a fascinating and complex chemistry of the inorganic components of coal in combustion processes, which is not germane to the present discussion. That topic is well served by the books by Raask [3] and Vorres [4].

2. Inorganic Components of Coals

The total number of minerals that have been identified in coals exceeds 100. Many of these occur in very small amounts in most coals, and their possible role as catalysts is not well understood. The major interest in mineral components of coals as catalysts has focused on two families of minerals. The sulfides, most notably pyrite and its dimorph marcasite, are potential hydrogenation catalysts. The clays may function as cracking catalysts.

In addition to inorganic components occurring as discrete mineral grains, the low-rank coals (that is, coals of subbituminous and lower rank) also contain some inorganic species as cations. These cations are primarily those of the alkali and alkaline earth elements, with some coals

Y. Yürüm (ed.), Clean Utilization of Coal, 65–73.

also having small amounts of iron and aluminum cations [5] It is well established that these ions are associated with the carboxylate functional groups in the coal structure [6]. The carboxylate groups are thermally labile; their destruction on heating releases the cations, which may then form well-dispersed particles of oxides on the coal or char surface. All will be discussed below, the dispersion of inorganic particles in this fashion is of particular interest for the possible catalysis of gasification reactions. Over the years bits and pieces of evidence have accumulated suggesting that some of the transition elements may occur in coals as coordination compounds [7]. However, the nature of coordination compounds in coals seems to be the least understood aspect of the inorganic chemistry of coals, and the contribution of such species to catalysis is unknown.

3. Catalysis by Sulfides

The most important by far of the sulfide minerals in coal is pyrite, FeS_2. Marcasite is dimorphous with pyrite [8]. In many instances in the coal literature no effort is made to distinguish between the two, a convention that will also be adopted for this chapter, so the term "pyrite" as used here refers to the specific mineral pyrite, its dimorph marcasite, or both. Pyrite seems to be the coal mineral that has the most significant effect on the catalysis of direct liquefaction reactions [9]. Pyrite itself, however, may have no catalytic activity, and it is generally thought that in the strongly reducing conditions typical of direct liquefaction processes the pyrite may be converted to pyrrhotite [9]. Thus it is important to recognize the distinction between the mineral matter in coal itself being catalytic and the mineral matter being converted, during processing, into a catalytic species [2]. Pyrrhotite is a non-stoichiometric sulfide commonly represented by the molecular formula $Fe_{1-x}S$, where x varies in the range 0-0.2 [8]. The sulfur content of naturally occurring pyrrhotite varies from 50 to 56 atomic percent; however, it is shown on structural grounds that the structure must in fact be iron-deficient and not sulfur-rich [8]. A good review of the factors involved in transformation of pyrite to pyrrhotite is available [10].

The addition of pyrite to coal is known to increase the formation of benzene-soluble products, conversion increasing with increased levels of pyrite addition [11-13] Production of distillable liquids rather than solids in the early versions of the Solvent Refined Coal (SRC) processes is attributed to the catalytic behavior of pyrite and marcasite in the coal [14]. Recycling the insoluble portion of the product from the SRC-II process caused an increase in distillate yield as the concentration of recycled solids in the feed stream was increased [15]. Since it is generally thought that the insoluble organic fraction of the coal has no catalytic behavior, the increase in liquids yields can be attributed to the increased concentration of catalytically active minerals in the recycled solids. These results were obtained for a Pittsburgh seam (U.S.) coal which typically contains 1.5-2.5% pyritic sulfur. The same enhancement of liquids yields was not observed for a Wyodak (U.S.) subbituminous coal, which had a low concentration of pyritic sulfur [15]. This is circumstantial evidence suggesting the importance of pyrite (or its reduction products) in enhancing liquids yields during direct liquefaction.

Pyrite is an effective catalyst for the hydrogenation of creosote oil at 425°C, as is "reduced pyrites" (presumably pyrrhotite) [16]. Both the physical and chemical state of the pyrite appeared to be important in this reaction. Pyrite added to the liquefaction Kentucky (U.S.) bituminous coals improved the conversion to benzene-solubles, but pyrrhotite added to the reactor did not have the same effect [2]. This result was surprising in light of the generally presumed conversion of pyrite to pyrrhotite at liquefaction conditions. However, naturally occurring pyrites have different physical structures, and it may well be that pyrite purchased as a mineral specimen for addition to a reactor will will differ from the pyrite found naturally in the coal [2]. For example, pyrite in some coals occurs in a framboidal structure different from the massive crystals typical of minerals [5]. These different forms of pyrite may be converted to pyrrhotites of different composition (i.e., different values of x in the formula $Fe_{1-x}S$), which in turn may have different activities as catalysts [2]. In this connection, it is known that molybdenum sulfide, a useful hydrogenation catalyst, can exist in either type of defect structure: $Mo_{1+x}S_2$ or $Mo_{1-x}S_2$. The composition of the solid is determined by the relative amounts of H_2 and H_2S [17]. It is possible that in coal conversion systems the composition and catalytic activity of pyrrhotite produced from pyrite could be influenced by the partial pressures of H_2 and H_2S; controlling these values may facilitate formation of a catalytically active pyrrhotite. Indeed the partial pressure of H_2S, or the fraction of coal sulfur converted to H_2S during processing, does affect the composition of pyrrhotite formed in the reaction [18-20]

Furthermore, the behavior of iron sulfide surfaces is has been shown to be dependent on the exact stoichiometry of the iron sulfide, with iron-deficient surfaces being more reactive than surfaces of less deficiency [21]. The iron vacancies provide the centers for dissociation of H_2 or H_2S for subsequent reaction.

One way in which the pyrite may facilitate liquefaction is by enhancing the hydrogenation of aromatics to hydroaromatics [22]. For example, pyrene hydrogenation is enhanced by pyrite [23]. Dihydropyrene is certainly an effective hydrogen donor. Indeed, many other hydroaromatics, such as tetralin and dihydrophenanthrene, are efficient hydrogen donors in coal liquefaction, and are certainly more beneficial than either the parent aromatic compound or the fully hydrogenated cycloalkane. The most common example is the naphthalene-tetralin-decalin system, in which tetralin is a hydrogen donor whereas neither naphthalene nor decalin donates hydrogen. Hence the sulfides may help maintain a desirable level of hydroaromatic hydrogen donors in the liquid vehicle [22]. There is little evidence for pyrite catalysis of coal decomposition to preasphaltenes, but pyrite (or its reaction product pyrrhotite) does appear to catalyze the further decomposition (or hydrogenation) of preasphaltenes to asphaltenes [24]. Other other observations show an increase in asphaltenes and oils with added pyrite [25]. If a liquefaction reaction proceeds via an initial thermal decomposition to heavy products (i.e., preasphaltenes) which then must be upgraded by hydrogenation to asphaltenes and oils, this finding [24] may be consistent with the concept that that the role of pyrite is to maintain a supply of hydrogen donors in the system [22]. Furthermore, pyrite may help to catalyze the hydrogen transfer process itself, that is from the hydrogen donor to the heavy, condensed aromatics in the preasphaltenes or asphaltenes [26]. Pyrite also appears to facilitate the removal of some functional groups from the coal or liquids (as, for example, the dehydroxylation of phenols) [26].

The decomposition of pyrite in a reducing atmosphere leads to the formation of hydrogen sulfide as well as pyrrhotite [27,28]. Since hydrogen sulfide itself is known to enhance liquefaction, some investigators suggest that the catalytic benefit of pyrite or pyrrhotite is in fact actually due to the hydrogen sulfide produced as the by-product of conversion of pyrite to pyrrhotite [20]. Indeed the hydrogen sulfide by-product has been shown to lead initiate radical chain decomposition processes, as in the decomposition of 1,3-diphenylpropane to toluene and C_2-benzenes [29]. Nevertheless, liquefaction of a relatively unreactive Kentucky bituminous coal shows that only pyrite (and not H_2S provides the catalytic ability to facilitate the hydrogenolysis needed to increase conversion [30].

Pyrite has been noted to affect the pyrolysis behavior of a Pittsburgh seam bituminous coal in helium by reducing the methane yield [31]. Iron-sulfur minerals seemed to have little effect on the hydropyrolysis of this coal [31].

4. Clay Minerals

For most coals, the clay minerals are the dominant mineral family. Clays may account for up to 70% of the total mineral matter [9]. A remarkable variety of clay minerals is known; however, the two most commonly found in coals are kaolinite and illite. Kaolinite is generally represented by the molecular formula $Al_4(Si_4O_{10})(OH)_8$. It has several polymorphs which are occasionally reported in coals (e.g., halloysite) but they are of much less importance than kaolinite. The composition of kaolinite shows little variation, partly because it has less ion exchange capacity than other clays. Kaolinite undergoes thermal dehydration in the range 400-525°C, and at higher temperatures experiences structural rearrangements first to meta-kaolin and then to mullite. The latter transformation occurs above 800°C. Illite is much more variable in composition. Its composition is sometimes represented as $K_{1-1.5}Al_4(Si_{7-6.5}Al_{1-1.5}O_{20})(OH)_4$ [32]. Dehydration of this clay occurs over a very wide temperature range, evidently in a number of separate stages. At high temperatures illite transforms to a spinel structure. The detailed structural chemistry of the clays is outside the scope of this chapter, but ample information is available in the geological literature [8,34] However, it is interesting to note that in illite, the potassium ions, which are located between aluminosilicate layers, appear to block the entry of organic liquids into the structure [32].

Naturally occurring clays were probably the first solid acid catalysts to be used [33]. The Houdry process for improving gasoline quality by catalytic cracking involves an acid-washed montmorillonite clay [34].However, natural clays have largely been superseded by synthetic silica-alumina catalysts and zeolites. Nickel-substituted montmorillonites have been shown to be useful in the hydroisomerization of alkanes [35].

Although clays are common constituents of coals, and in many cases are in intimate mixture with the carbonaceous portion of the coal [9], their role in coal conversion chemistry is much less studied than that of pyrite. Indeed, the role of clays in the organic reaction chemistry of coals is virtually unknown [2]. Kaolinite appears to have a catalytic effect in liquefaction [12], although at very severe reaction conditions (9.8 MPa H_2 at 400°C for 3 h). Clays appear to have no significant effect on pyrolysis yield structure from a Pittsburgh seam bituminous coal [31]. Some reduction in tar and oil yields from pyrolysis was attributed to secondary cracking reactions enhanced by the clays. Char-forming reactions catalyzed by clays (or acidic sites on coal mineral matter) may limit conversion in some liquefaction processes, such as conversion in carbon monoxide / water systems [36].

5. Alkali and Alkaline Earth Cations

In many low-rank coals, an appreciable portion of the total amount of sodium, calcium, and magnesium occur as cations associated with the carboxylate functional groups. Because of this association, these ions are easily exchanged; for example, they can be removed by treating the coal with solutions of ammonium acetate. In addition, some fraction of the potassium occurs in exchangeable form, as well as, in some cases, small amounts of some other elements, such as aluminum and iron. Carboxyl groups are destroyed by heating coals into the range 300-400°C. Loss of the carboxyls leaves the associated cations behind in the resulting char, often in very highly dispersed form. The alkali and alkaline earth cations can be effective gasification catalysts [37]. The *in situ* formation of highly dispersed catalysts is one of the factors contributing to the increased gasification reactivity of low-rank coal chars relative to bituminous coal chars [38] For example, lignites and subbituminous coals were found to be more reactive by factors of 8 to 10, relative to an Illinois (U.S.) bituminous coal, for steam gasification at 700-800°C [40]. (However, it is not absolutely necessary that the carboxylates decompose to observe a catalytic effect of calcium, at least for low-temperature gasification in air [41]. The exchange ability of the coal also makes it possible to treat the coal with aqueous solutions of desired ions (for example, Fe^{+2}), in order to effect dispersion of these elements during carbonization of the coal. Differences in reaction kinetics for low-rank coal char gasification in carbon dioxide may reflect differences in the content of inorganic species, that could act as catalysts, in the parent coals [42].

The presence of metal cations in a coal significantly affects its weight loss behavior during entrained flow pyrolysis [43,44]. The weight loss on pyrolysis is decreased relative to that observed for an acid-washed coal under the same reaction conditions. (Acid washing, usually with dilute HCl, is an effective way of removing the exchangeable cations.) That is, when the metal cations are removed from a coal, an increased pyrolysis weight loss is observed. Generally the increased weight loss is accompanied by a significant increase in tar yield [43], suggesting that in some fashion the cations affect tar yield. In this context, calcium and magnesium appear to behave similarly, provided that their effects are compared on a similar molar concentration basis [43].

At very short pyrolysis times (<0.1 s) in an entrained flow reactor, calcium cations appear to enhance the yield of volatiles [45]. However, at residence times >0.1 s there is a definite reduction in release of volatiles caused by the presence of exchangeable calcium [45]. The exchangeable cations could have two possible roles in reducing the volatiles yield or weight loss. One effect is physical, in that the cations may block egress of volatiles simply by blocking off pores. A chemical role of cations is their catalysis of secondary char-forming reactions by enhancing repolymerization of the organic species liberated from the coal. Rapid pyrolysis of a Wyodak subbituminous coal indicated a decreased tar yield, as well as decreased yields of methane and ethane, when cations were present [46].

A variety of reactions are involved in steam gasification, such as

$$C + H_2O \rightarrow CO + H_2$$

$$CO + H_2O \rightarrow CO_2 + H_2$$

$$C + CO_2 \rightarrow 2\,CO$$

$$C + 2\,H_2 \rightarrow CH_4$$

Since the desired products of gasification are usually synthesis gas or methane, the carbon-steam and carbon-hydrogen reactions are among the most important of the gasification reactions. However, these reactions also happen to be among the slowest [46]. If the rates of these two reactions can be enhanced by catalysis, a potential process improvement is available through the reduced size of the gasifier vessel needed to accommodate a given gas production (due to the higher throughput rate) and through possible improvements in thermal efficiency [46]. Alkali and alkaline earth cations are good catalysts for char gasification in steam, as well as in air and carbon dioxide [47]. For coals of less than about 80% carbon [48] and greater than 10% oxygen [49], catalysis of the carbon-steam reaction by sodium, potassium, or calcium cations appears to be the dominant factor determining reaction rate. Char reactivity in steam tends to increase linearly with calcium content [50-52]. Reactivity in air increases with calcium and magnesium contents [53]. However, magnesium is a poor catalyst for char gasification in air, relative to calcium [54]. For steam gasification the general order of effectiveness as catalysts is K > Ca > Na > Fe > Mg [52].

Pyrolysis of lignite chars results in a decrease of oxygen anions in the coordination sphere around a calcium ion [55]. The reduced oxygen anion coordination could provide sites for incorporation of oxygen or carbon dioxide, initiating a subsequent reaction with char. In rapid pyrolysis, even to 1000°C, there is no transformation of the original calcium carboxylates to a well-defined calcium-oxygen structure [56]. However, slow pyrolysis forms bulk CaO [56]

Several concepts have been suggested to explain the catalysis of gasification reactions by the alkali and alkaline earth cations. One example is the mechanism for potassium catalysis [57]. Potassium is superior to sodium in this regard [58]. The potassium carboxylates originally in the coal would transform to potassium carbonate on pyrolysis. Then

$$K_2CO_3 + 2C \rightarrow 2\,K + 3\,CO$$

$$2\,K + 2\,H_2O \rightarrow 2\,KOH + H_2$$

$$CO + H_2O \rightarrow CO_2 + H_2$$

$$2\,KOH + CO_2 \rightarrow K_2CO_3 + H_2O$$

The net reaction in this system is then

$$2C + 2\,H_2O \rightarrow 2\,CO + 2\,H_2$$

An alternative mechanism involves the formation of metal-carbon complexes [59]. Again the first step is the reduction of potassium carbonate,

$$K_2CO_3 + C \rightarrow 2\,K + CO + CO_2$$

then followed by formation of the potassium-carbon complex

$$K + n\,C \rightarrow KC_n$$

Reaction in steam generates surface oxides

$$KC_n + H_2O \rightarrow (KC_n)O + H_2$$

which decompose to regenerate the metallic potassium

$$(KC_n)O \rightarrow K + (n-1)\,C + CO$$

At high temperatures (~1000°C) during both pyrolysis and gasification, calcium oxide particles formed from thermal decomposition of calcium carboxylates appear to agglomerate [60] . A decrease in calcium oxide dispersion can reduce the catalytic effect obtained from the cations. For example, the reactivity of lignite chars decreases as the final heat treatment temperature of char formation is increased, or as the heating time at the heat treatment temperature is increased [47]. The decrease in reactivity correlates well with a decrease in calcium oxide dispersion, as followed by x-ray diffraction. In comparison, potassium carboxylates transform to potassium carbonate, which does not then agglomerate [60].

To sum up this section, calcium, potassium and sodium are excellent catalysts for the reaction of carbon with oxygen, steam, or carbon dioxide [51]. Magnesium is generally a poor catalyst for these reactions [51]. Calcium also catalyzes the carbon-hydrogen reaction [51]. Oxides of iron are poor catalysts for carbon-oxygen and carbon-carbon dioxide reactions. However, metallic iron is a good catalyst for these reactions, so if a reduction step could convert the iron compound formed from decomposition of iron carboxylates to metallic iron, a useful catalyst could be formed.

6. Summary

Sulfide catalysts - pyrite or substances derived *in situ* from pyrite - offer the potential for improving liquefaction processes. Alkali and alkaline earth cations, which decompose to oxides, hydroxides, or carbonates on pyrolysis of the parent coal, can provide catalysis of char gasification reactions. In fact, the presence of these ions on ion-exchange sites in low-rank coals makes it relatively easy alter the population of exchangeable cations in ways that enhance the possible catalysis. For some applications, particularly combustion processes, high-sulfur, high-mineral-matter coals are considered "dirty" and undesirable without extensive cleaning. However, for coal conversion processing the coals that may be undesirable from the standpoint of combustion might prove to be loaded with catalysts that make such coals highly desirable conversion feedstocks.

7. Literature Cited

1 Hawk, C.O., and Hiteshue, R.W. (1965) "Hydrogenation of coal in the batch autoclave", Bulletin 622, U.S. Bureau of Mines, Washington.

2 Mraw, S.C., DeNeufville, J.P., Freund, H., Baset, Z., Gorbaty, M.L., and Wright, F.J. (1983) "The science of mineral matter in coal", in M.L. Gorbaty, J.W. Larsen, and I. Wender (eds.), Coal Science. Volume 2., Academic Press, New York, pp. 2-64.

3 Raask, E. (1985) Mineral Impurities in Coal Combustion, Hemisphere Publishing Corporation, Washington.

4 Vorres, K.S. (1986) Mineral Matter and Ash in Coal, American Chemical Society, Washington.

5 Schobert, H.H., Karner, F.R., Olson, E.S., Kleesattel, D.R., and Zygarlicke, C.J. (1987) "New approaches to the characterization of lignites: a combined geological and chemical study", in A. Volborth (ed.), Coal Science and Chemistry, Elsevier Science Publishers, Amsterdam, Chapter 14.

6 Huffman, G.P., and Huggins, F.E. (1984) "Analysis of the inorganic constituents of low-rank coals", in H.H. Schobert (ed.), The Chemistry of Low-Rank Coals, American Chemical Society, Washington, Chapter 10.

7 Swaine, D.J. (1990), Trace Elements in Coal, Butterworths, London.

8 Berry, L.G., and Mason, B. (1959) Mineralogy, W.H. Freeman and Company, San Francisco.

9 Renton, J.J. (1982) "Mineral matter in coal", in R.A. Meyers (ed.), Coal Structure, Academic Press, New York, Chapter 7.

10 Narain, N.K., Cilli, D.L., Stiegel, G.F., and Tischer, R.E. (1987) "Disposable catalysts in coal liquefaction", in A. Volborth (ed.), Coal Science and Chemistry, Elsevier Science Publishers, Amsterdam, Chapter 4.

11 Moroni, E.C., and Fischer, R.H. (1980) "Disposable catalysts for coal liquefaction", Amer. Chem. Soc. Division of Fuel Chemistry Preprints 25(1), 16-17.

12 Mukherjee, D.K., and Chowdhury, P.B. (1976) "Catalytic effects of mineral matter constituents in a north Assam coal on hydrogenation", Fuel 55, 4-13.

13 Given, P.H., Cronauer, D.C., Spackman, W., Lovell, H.L., Davis, A., and Biswas, G. (1975) "Dependence of coal liquefaction behavior on coal characteristics. 1. Vitrinite-rich samples", Fuel 54, 34-39.

14 Wright, C.H., and Severson, D.E. (1972), "Experimental evidence for catalyst activity of coal minerals", Amer. Chem. Soc. Division of Fuel Chemistry Preprints 16(2), 68-92.

15 Tsai, S.C. (1982) Fundamentals of Coal Beneficiation and Utilization, Elsevier Scientific Publishing Company, Amsterdam.

16 Pullen, J.R. (1983) "Solvent extraction of coal", in M.L. Gorbaty, J.W. Larsen, and I. Wender (eds.), Coal Science. Volume 2., Academic Press, New York, pp. 174-288.

17 Swaddle, T.W. (1990) Applied Inorganic Chemistry, University of Calgary Press, Calgary.

18 Lambert, J.M. (1982) "Transformation of pyrite to pyrrhotite and its implications in coal conversion processes", Ph.D. Dissertation, The Pennsylvania State University, University Park, PA.

19 Montano, P.A., and Granoff, B. (1980) "Stoichiometry of iron sulfides in liquefaction residues and correlation with conversion", Fuel 59, 214-223.

20 Lambert, J.M., Simkovich, G., and Walker, P.L. (1980) "Production of pyrrhotites by pyrite reduction", Fuel 59, 687-691.

21 Montano, P.A., Lee, Y.C., Yeye-Odu, A., and Chien, C.H. (1986) "Iron sulfide catalysis in coal liquefaction", in K.S. Vorres (ed.), Mineral Matter and Ash in Coal, American Chemical Society, Washington, Chapter 29.

22 Guin, J.A., Tarrer, A.R., Prather, J.W., Johnson, D.R., and Lee, L.M. (1978) "Effects of coal minerals on hydrogenation, desulfurization, and solvent extraction of coal", Industrial and Engineering Chemistry Process Design and Development 17, 118-127.

23 Derbyshire, F.J., Varghese, P., and Whitehurst, D.D. (1981) "Interactions between solvent components, molecular hydrogen, and mineral matter during coal liquefaction", Amer. Chem. Soc. Division of Fuel Chemistry Preprints 26(1), 84-93.

24 Gray, D., and Neuworth, M.B. (1981) "Disposable catalysts in coal liquefaction: the effect of iron sulfides", Report No. WP-81W00427, U.S. Department of Energy, McClean, Virginia.

25 Garg, D., Givens, E.N., Schweighardt, F.K., Clinton, J.H., Tarrer, A.R., Guin, J.A., Curtis, C.W., Huang, S.M., and Shridharani, K. (1980) "The role of non-ferrous coal minerals and by-product metallic wastes in coal liquefaction", Report No. DOE/ET-146806-T1, U.S. Department of Energy, Springfield, Virginia.

26 Derbyshire, F.J., Varghese, P., Whitehurst, D.D. (1981) "The promotion of coal liquefaction by mineral matter catalysis", Proceedings, International Conference on Coal Science, pp. 356-361.

27 Kawa, W., Hiteshue, R.W., Anderson, R.B., and Greenfield, H. (1960) "Reactions of iron and iron compounds with hydrogen and hydrogen sulfide", Report No. 5690, U.S. Bureau of Mines, Washington.

28 Thomas, M.G., Padrick, T.D., Stohl, F.V., and Stephens, H.P. (1982) "Decomposition of pyrite under coal liquefaction conditions: a kinetic study", Fuel 61, 761-766.

29 Bockrath, B.C., and Schroeder, K.Y. (1983) "Iron sulfide promoted organic reactions: models of pyrite-assisted coal liquefaction", Proceedings, International Conference on Coal Science, pp. 94-97.

30 Baldwin, R.M. (1983) "Coal liquefaction catalysis: dissimilar behavior with iron pyrite and hydrogen sulfide", Proceedings, International Conference on Coal Science, pp. 98-101.

31 Franklin, H.D., Peters, W.A., and Howard, J.B. (1981) "Mineral matter effects on the rapid pyrolysis and hydropyrolysis of a bituminous coal", Amer. Chem. Soc. Division of Fuel Chemistry Preprints 26(3), 35-43.

32 Deer, W.A., Howie, R.A., and Zussman, J. (1966) An Introduction to the Rock Forming Minerals, Longman, Harlow, Essex, UK.

33 Haag, W.O., and Chen, N.Y. (1987) " Catalyst design with zeolites", in L.L. Hegedus (ed.), Catalyst Design - Progress and Perspectives, John Wiley and Sons, New York, Chapter 6.

34 Satterfield, C.N. (1991) Heterogeneous Catalysis in Industrial Practice, McGraw-Hill, New York.

35 van Santen, R.A., Röbschläger, K.H.W., and Emeis, C.A. (1985) "The hydroisomerization activity of nickel-substituted mica montmorillonite clay", in R.K. Grasselli and J.F. Brazdil (eds.), Solid State Chemistry in Catalysis, American Chemical Society, Washington, Chapter 17.

36 Ross, D.S., Green, T.K., Mansani, R., and Hum, G.P. (1987) "Coal conversion in water. 1. Conversion mechanism", Energy and Fuels 1, 287-291.

37 Miura, K., Nakamura, H., and Hashimoto, K. (1991) "Analysis of a two-step reaction observed in air gasification of coal through a temperature-programmed reaction technique", Energy and Fuels 5, 47-51.

38 Walker, P.L. (1978) "Gasification rates as related to coal properties", in B.R. Cooper (ed.), Scientific Problems of Coal Utilization, U.S. Department of Energy, Morgantown, West Virginia, pp. 237-247.

39 Sears, R.E., Timpe, R.C., Galegher, S.J., and Willson, W.G. (1986) "Catalyzed steam gasification of low-rank coals to produce hydrogen", Amer. Chem. Soc. Division of Fuel Chemistry Preprints 31(3), 166-175.

40 Hegermann, R., and Hüttinger, K.J. (1989) "Kinetics of brown coal gasification", Amer. Chem. Soc. Division of Fuel Chemistry Preprints 34(1), 22-29.

41 Kwon, T.W., Kim, S.D., and Fung, D.P.C. (1988) "Reaction kinetics of char-CO_2 gasification", Fuel 67, 530-535.

42 Morgan, M.E., and Jenkins, R.G. (1984) "Role of exchangeable cations in the rapid pyrolysis of lignites", in H.H. Schobert (ed.) The Chemistry of Low-Rank Coals, American Chemical Society, Washington, Chapter 13.

43 Morgan, M.E., and Jenkins, R.G. (1986) "Pyrolysis of a lignite in an entrained flow reactor. 1. Effect of cations on total weight loss", Fuel 65, 757-763.

44 Morgan, B.A., and Scaroni, A.W. (1984) "Cationic effects during lignite pyrolysis and combustion", in H.H. Schobert (ed.) The Chemistry of Low-Rank Coals, American Chemical Society, Washington, Chapter 16.

45 Franklin, H.D., Cosway, R.G., Peters, W.A., and Howard, J.B. (1981) "Effects of cations on the rapid pyrolysis of a Wyodak subbituminous coal", Proceedings, International Conference on Coal Science, pp. 725-729.

46 Nayak, R.V., and Jenkins, R.G. (1983) "Coal gasification - fundamentals", in S.K. Majumdar and E.W. Miller (eds.), Pennsylvania Coal: Resources, Technology, and Utilization, Pennsylvania Academy of Science, Easton, Pennsylvania, Chapter 12.

47 Hengel, T.D., and Walker, P.L. (1984) "Catalysis of lignite char gasification by exchangeable calcium and magnesium", in H.H. Schobert (ed.) The Chemistry of Low-Rank Coals, American Chemical Society, Washington, Chapter 17.

48 Hashimoto, K., Miura, K., Xu, J.J., Tezen, Y., and Nagai, H. (1987) "Reactivities of various coals and demineralized coals under steam gasification", Proceedings, International Conference on Coal Science, pp. 507-510.

49 Serio, M.A., Solomon, P.R., and Bassilakis, R. (1989) "The effects of minerals and pyrolysis conditions on char gasification rates", Amer. Chem. Soc. Division of Fuel Chemistry Preprints 34(1), 9-21.

50 Linares-Solano, A., Hippo, E.J., and Walker, P.L. (1986) "Catalytic activity of calcium for lignite char gasification", Fuel 65, 776-779.

51 Walker, P.L., Matsumoto, S., Hanzawa, T., Muira, T., and Ismail, I.M.K. (1983) "Catalysis of gasification of coal-derived cokes and chars", Fuel 62, 140-149.

52 Hippo, E.J., Jenkins, R.G., and Walker, P.L. (1979) "Enhancement of lignite char reactivity to steam by cation addition", Fuel 58, 338-344.

53 Fung, D.P.C., and Kim, S.D. (1984) "Chemical reactivity of Canadian coal-derived chars", Fuel 63, 1197-1201.

54 Hengel, T.D., and Walker, P.L. (1984) "Catalysis of lignite char gasification by exchangeable calcium and magnesium", Fuel 63, 1214-1220.

55 Huggins, F.E., Shah, N., Huffman, G.P., Lytle, F.W., Greegor, R.B., and Jenkins, R.G. (1988) "In situ XAFS investigation of Ca and K catalytic species during pyrolysis and gasification of lignite chars", Fuel 67, 1662-1667.

56 Huggins, F.E., Huffman, G.P., Shah, N., Jenkins, R.G., Lytle, F.W., and Greegor, R.B. (1988) "Further EXAFS examination of the state of calcium in pyrolyzed char", Fuel 67, 938-941.

57 Hessley, R.K., Reasoner, J.W., and Riley, J.T. (1986) Coal Science, John Wiley and Sons, New York.

58 Spiro, C.L., McKee, D.W., Kosky, P.G., and Lamby, E.J. (1983) "Catalytic CO_2 gasification of graphite vs. coal char", Fuel 62, 180-184.

59 Jüntgen, H., and van Heek, K.H. (1984) "Kinetics and mechanism of catalytic gasification of coal", Amer. Chem. Soc. Division of Fuel Chemistry Preprints 29(2), 195-205.

60 Shah, N., Huggins, F.E., Shah, A., Huffman, G.P., Jenkins, R.G., and Piotrowski, A. (1989) "Studies of the agglomeration and reactions of Ca and K during pyrolysis and gasification", Amer. Chem. Soc. Division of Fuel Chemistry Preprints 34(1), 30-35.

DETERMINATION OF COAL BEHAVIOR FOR PRACTICAL COAL CONVERSION PROCESSES

P.J.J. TROMP, F. KAPTEIJN and J.A. MOULIJN*
Department of Chemical Engineering, University of Amsterdam,
Nieuwe Achtergracht 166, 1018 WV Amsterdam, The Netherlands
* *Department of Chemical Engineering, Delft University of Technology,*
Julianalaan 136, 2628 BL Delft, The Netherlands.

ABSTRACT. Rapid pyrolysis of four coals (particle size: 38-53 micron), ranging from lignite to semi-anthracite, was performed in an entrained flow reactor (temperature range: 800-1300 K, residence time: 0.3 and 0.6 s) to simulate the initial stage in realistic coal conversion processes at high heating rates. Weight losses and residual volatile matter contents of the collected coal chars were determined. Carbon dioxide gasification reactivities of coal chars (pyrolysis conditions: 1200 K, 0.6 s) were measured in a thermobalance at atmospheric pressure. When compared with data obtained under slow heating conditions, the results show that the heating rate often significantly influences coal behavior with respect to coal weight loss and char gasification reactivity. The effect of heating rate on the thermoplastic behavior of a coal appears to be of major importance for the char reactivity.

1. Introduction

Coal pyrolysis is an important aspect of coal behavior as it is the initial step in most coal conversion processes. Depending on the kind of process and the type of reactor used distinction should be made between slow (heating rate $< 1\ K\ s^{-1}$) and rapid (heating rate $> 1000\ K\ s^{-1}$) pyrolysis. Whereas in slow heating dense coal samples, like packed beds, are used, rapid pyrolysis is achieved with small coal particles ($< 100\ \mu m$) in a dilute gas phase. Rapid pyrolysis of coals is of interest in connection with pulverized coal combustion, fluidized bed combustion or gasification, and entrained flow gasification.

One of the instruments used in research on rapid pyrolysis of coal is the entrained flow reactor system [1-5]; the combination of injector and reactor in such a system constitutes a realistic simulation of practical coal conversion processes. However, the disadvantage to perform experiments in an entrained flow reactor is that the instrumentation is expensive, and more difficult to operate than, for instance, a thermobalance.

With the above in mind, the pyrolysis behavior of four coals of different rank during rapid heating was investigated in an entrained flow reactor to simulate the initial stage in practical, high heating rate coal conversion processes. Important aspects in these processes are coal weight loss upon pyrolysis and reactivity of the char formed after the pyrolysis stage. The results are compared with data obtained under slow heating conditions in a thermobalance to investigate the necessity of performing experiments in an entrained flow reactor.

Y. Yürüm (ed.), Clean Utilization of Coal, 75–84.

2. Experimental

2.1. THE ENTRAINED FLOW REACTOR SYSTEM

The general experimental set-up is shown schematically in figure 1. In brief: At the top of the reactor coal particles are carried by a primary helium gas flow through a water-cooled injector and are introduced into the hot reactor. The coal particles are entrained in, and rapidly heated by, a much larger preheated secondary N_2 gas flow. The secondary N_2 gas with the coal particles flows downwards along the reactor under laminar flow conditions. The pyrolysis process is stopped abruptly by collecting the char particles and volatiles in a cooled probe. The residence time of the particles in the hot reactor is controlled by the position of the collector probe, which can be moved along the reactor axis. The particles are separated from the gas stream by a cyclone.

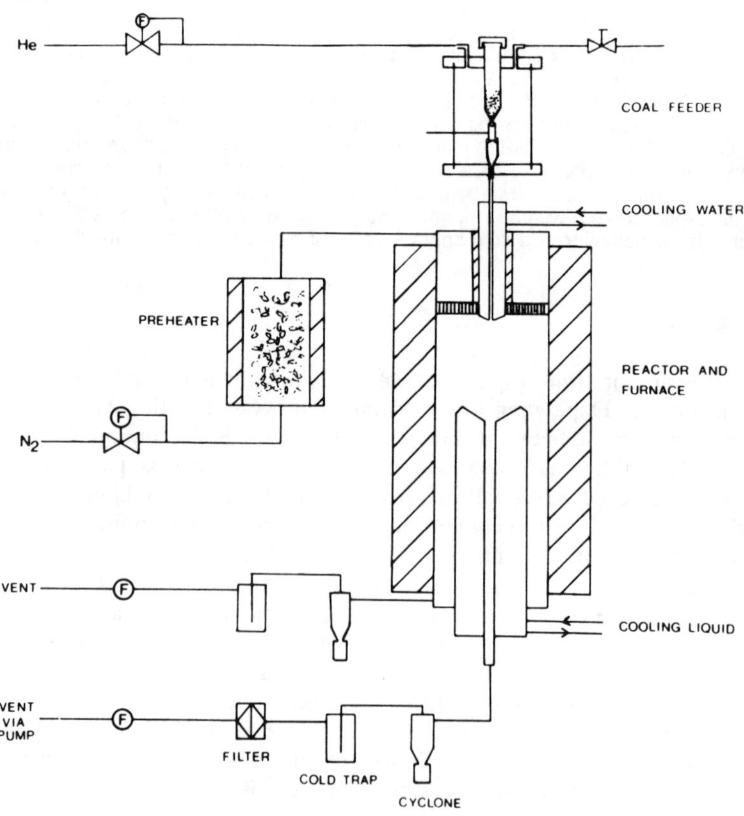

FIGURE 1. Schematic diagram of the entrained flow reactor system.

For every experiment 2 g of coal (particle size: 38-53 μm; for analyses of the coals, before they were grinded and sieved, see table 1) was used with a constant coal feed rate of 0.5 mg s^{-1}. The reactor temperature varied between 780 K and 1310 K. Two positions of the collector probe were employed: 0.175 m and 0.35 m distant from the injector probe.

Model calculations showed that for the small particles used convective heat transfer is the dominant mode of heating the coal particles [6]. The particles reach the surrounding gas temperature within 10 K in a time never longer than 0.05 s. This result indicates initial heating rates much larger than 10 K s^{-1}. Inside the coal particles heat is transferred by conduction. For the particle size range used the temperature gradients in the particles are negligibly small.

The hot gas with the coal particles flows under laminar flow conditions (Reynolds number: $\approx$ 100) through the pyrolysis zone. Laminar flow conditions are needed to avoid radial dispersion of the coal particles and to enable the estimation of a more reliable and uniform coal pyrolysis time. From gas and particle transport models, developed by Morgan [7], it emerged that the pyrolysis zones of 0.175 and 0.35 m corresponded with residence times of 0.34 and 0.64 s, respectively.

A more detailed description of the entrained flow reactor system has been given elsewhere [6].

TABLE 1. Analyses of the coals.

SBN code	DE10	BE30	DE38	DE53
country	FRG	Belgium	FRG	FRG
mine	Anna	Kempen	Ensdorf Saar	RWE-Köln
ASTM rank	semi-anthr.	hvA bit.	hvB bit.	lignite
Proximate analysis (%)				
moisture(as recv'd)	1.7	2.5	4.0	14.0
ash (dry)	8.0	3.2	7.3	3.6
VM (daf)	11.6	31.9	41.1	53.6
fixed carbon (daf)	88.4	68.1	58.9	46.4
Ultimate analysis (%)				
C (daf)	89.8	86.6	82.1	65.9
H (daf)	3.6	5.1	5.3	4.5
N (daf)	1.2	1.4	1.6	0.6
S (daf)	0.5	0.9	1.1	0.3
O (by diff.)	4.8	6.0	9.5	28.6

2.2. WEIGHT LOSS MEASUREMENTS

As the char collecting efficiencies in the entrained flow reactor were less than 100% the weight loss had to be measured indirectly by the ash tracer method [1, 3, 4]. This method relates the relative amount of ash present in the coal char with that present in the original coal.

The samples, pyrolysed in the entrained flow reactor, contained residual volatile matter. In our study both the residual volatile matter content and the ash content were determined in one experiment. 10-40 mg of char was heated in a thermobalance at a heating rate of 0.5 K s^{-1} to 1223 K under a purified argon flow. The weight loss up to 1223 K was taken to be the residual volatile matter content of the sample. The ash content of the sample was determined by slowly adding oxygen to the argon flow at 1223 K in order to burn-off the organic material. The same procedure in the thermobalance was applied to the original coal (38-53 μm size fraction). In this way both the weight loss as a function of the temperature under slow heating conditions and the ash content of the original coal was obtained. The weight loss of the original coal sample as a function of the temperature in the thermobalance was compared with the weight loss data of the coal in the entrained flow reactor.

2.3. MEASUREMENT OF THE CHAR REACTIVITY

Gasification reactivities of the chars, which were pyrolysed in the entrained flow reactor for 0.6 s (reactor length: 0.35 m) at 1200 K, were measured in a thermobalance. Gasification was performed isothermally in a pure CO_2 gas flow (110 μmol s^{-1}, 1 bar). The char samples (sample size: $\approx$ 4 mg) were contained in a macroporous sample pan manufactured from a sintered quartz frit (o.d. 5 mm; height 2 mm) to enhance mass transfer between gas phase and sample. The samples were heated at 1.6 K s^{-1} to the gasification temperature. Gasification temperatures were in the range 1073-1373 K. Normalised gasification rates, based on the initial weight of char at the gasification temperature (rn: g s^{-1} g_i^{-1}), were measured as a function of the char conversion (% burn-off).

The reactivities were compared with those of chars produced under slow heating conditions. In the latter case samples of the original coals were heated at 1.6 K s^{-1} to the gasification temperature in pure CO_2 (110 μmol s^{-1}, 1 bar), and subsequently gasified.

3. Results

3.1. WEIGHT LOSS MEASUREMENTS

The weight loss versus temperature data for the four different coal samples, when pyrolysed in the entrained flow reactor (EFR), are shown in figure 2A-D. For the high-volatile A bituminous coal (BE30, figure 2C) no data were obtained in the temperature range 873-973 K as in these experiments the coal particles were in a liquid state when they reached the collector probe. As a consequence, a major part of the softened coal particles sticked to the collector and was not collected in the cyclone. At 1310 K the char of this coal exhibited unrealistic high ash contents. The reason of this phenomenon is not known yet. Therefore, for this coal results of lower temperature conditions (1210 K) are presented.

For comparison the weight loss versus temperature curves of the original coals, obtained under slow heating conditions in the thermobalance, are included in the figures.

The weight losses of the coal samples increased progressively as a function of temperature. In comparison with pyrolysis under slow heating conditions, decomposition of the coals in the entrained flow reactor shifted to higher temperatures. With increasing temperatures the weight loss versus temperature curves for the two different residence times in the entrained flow reactor converged (e.g. see figure 2B). In all cases higher weight losses were obtained under rapid heating conditions. However, for the lignite a higher temperature than the final temperature in the thermobalance was needed to achieve this.

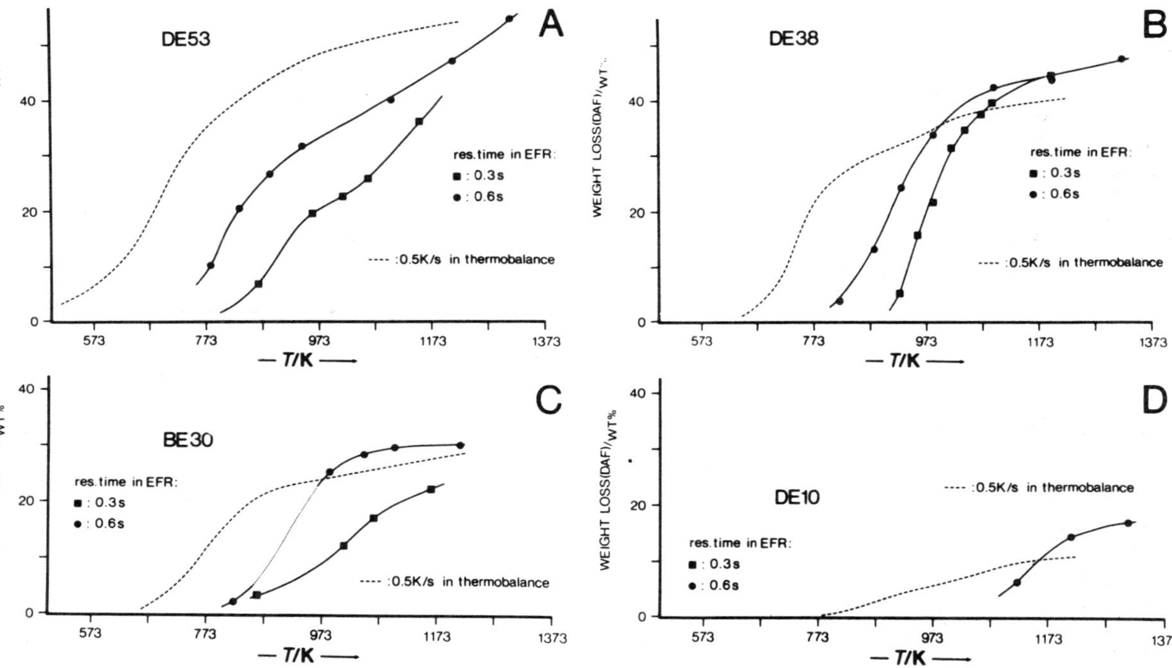

FIGURE 2. Coal weight loss as a function of the temperature in the entrained flow reactor (EFR) at two different residence times, ■ : 0.3 s, ● : 0.6 s, for **A**: the lignite (DE53), **B**: the hvB bituminous coal (DE38), **C**: the hvA bituminous coal (BE30), **D**: the semi-anthracite (DE10). Included is the coal weight loss versus temperature curve, as obtained under slow heating ($0.5 \ K \ s^{-1}$) in a thermobalance. Coal particle size range: 38-53 μm.

TABLE 2. Weight losses of the coal samples under slow (ΔW_1) and rapid (ΔW_2) heating conditions. The samples pyrolysed under rapid heating conditions contained residual volatile matter (VM_{res}). For further explanation see text.

Coal	ΔW_1	ΔW_2*	VM_{res}
	wt%, DAF basis		
DE53 (lignite)	54.1	54.6	12.0
DE38 (hvB bit.)	40.1	47.4	6.0
BE30 (hvA bit.)	29.3	30.6*	6.3
DE10 (semi-anthr.)	10.9	17.0	3.5

* weight loss at 1310 K and 0.6 s, for the BE30 coal 1210 K and 0.6 s.

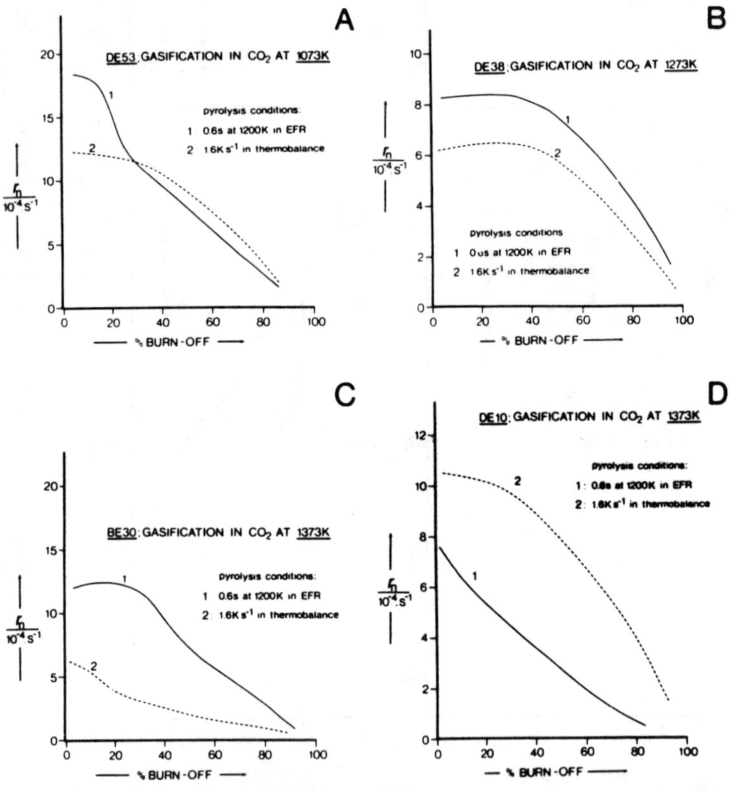

FIGURE 3. Normalised CO_2 gasification reactivities as a function of the char conversion for **A**: the lignite char, **B**: the hvB bituminous coal char, **C**: the hvA bituminous coal char, and **D**: the semi-anthracite coal char. Curve 1: after pyrolysis in N_2 for 0.6 s at 1200 K in the entrained flow reactor, Curve 2: after pyrolysis in CO_2 in the thermobalance at a heating rate of 1.6 K s^{-1} to the gasification temperature.

After pyrolysis for 0.6 s in the entrained flow reactor at the highest temperature the chars still contained residual volatile matter ("VM_{res}"). The values of VM_{res}, together with the weight losses of these chars in the entrained flow reactor (ΔW_2), are given in table 2. For comparison the total weight loss of the original coal samples under slow heating to 1223 K in the thermobalance (ΔW_1) are also shown in the table.

3.2. CHAR REACTIVITY

The lignite chars were by far the most reactive. Whereas isothermal gasification of these chars was performed at 1073 K, much higher temperatures were needed for the higher rank coal chars. The hvB bituminous coal chars were gasified at 1273 K, the hvA bituminous and semi-anthracite coal chars at 1373 K. The normalised gasification rates as a function of the char conversion are shown in figures 3A-D for the four different types of coal. In each figure the reactivity of the char formed under rapid heating conditions in the entrained flow reactor at 1200 K and a residence time of 0.6 s (curve 1) is compared with that of a char, which was pyrolysed under slow heating conditions in the thermobalance (curve 2).

The influence of heating rate during the pyrolysis stage on the gasification reactivity of the char differed for the various types of coals used. Figure 3A shows that for the lignite the char pyrolysed under rapid heating conditions was slightly more reactive up to about 30% burn-off. For the hvB bituminous coal (figure 3B) a slightly higher gasification reactivity of the char, collected from the entrained flow reactor, was observed over the entire burn-off range. More significant effects of heating rate on char gasification reactivity were observed for the hvA bituminous coal (figure 3C) and for the semi-anthracite (figure 3D). However, whereas for the hvA bituminous coal high-heating rate conditions affected the char gasification reactivity positively, the opposite effect was measured for the semi-anthracite.

4. Discussion

4.1. DEVOLATILIZATION BEHAVIOUR

Due to the much shorter heating time scale, decomposition of the coals in the entrained flow reactor shifts to higher temperatures, compared with pyrolysis in the thermobalance. More interesting, however, is that larger weight losses can be attained under these rapid heating conditions. This is explained by the explosive release of volatiles, thereby reducing the residence time of the volatiles within the coal particles. These shorter residence times minimize secondary char forming reactions. Moreover, the sample geometry in the thermobalance (bed of loosely packed coal particles) opposes the release of volatiles to the bulk gas phase and, thus, char forming reactions within the coal bed are also possible. Extra volatile yields are highly profitable in coal gasification and combustion as, in general, gas phase reactions are orders of magnitude faster than gas/solid reactions.

Although the weight loss versus temperature curves at the two different residence times converge with increasing temperatures (e.g. see figure 2B) no complete devolatilization of the coals was achieved in the entrained flow reactor. Similar results were reported by Scaroni et al. [3] for the pyrolysis of a Texas lignite in an entrained flow reactor. They concluded that coal pyrolysis is a multistage process, consisting of rapid initial devolatilization followed by slower stage devolatilization.

From high temperature studies, performed up to 2200 K, it was proposed that devolatilization of coals under high heating rate conditions consists of a low and a high temperature stage [2, 8-10]. According to this two-stage model temperatures much higher than the maximum temperature used in our study are needed to achieve the rapid release of the residual volatile matter of the coals. In these rapid pyrolysis studies it has generally been observed that the weight losses of the coal particles exceed the standard ASTM volatile matter contents of the coals [1-4, 8-10]. Hence, the Q-factor has been introduced to quantify the fractional enhancement in weight loss of a coal over its ASTM volatile matter content [1, 10]. Under the assumption that the weight losses of the original coal samples (table 2, ΔW_1) and their chars (table 2, VM_{res}) at 1223 K in the thermobalance equal their ASTM volatile matter contents, the following Q-factors are obtained by us:

coal	DE53	DE38	BE30	DE10
Q-factor	1.3	1.4	1.3	2.3

Similar enhancement factors have been reported by others [1, 3, 7-9]. These Q-factors are used in the modeling of coal devolatilization to determine the maximum potential devolatilization weight loss ΔW_{max} [1]. When the temperature history of a coal particle in a conversion process is known, ΔW_{max} is a better parameter to calculate coal weight loss with time due to pyrolysis.

4.2. CHAR REACTIVITY

The most important intrinsic factors that determine the reactivity of coal chars are:

- concentration of active gasification sites
- porosity of the char
- catalytic effects of inorganic constituents

Usually, the last factor is only of importance for lignite chars. Lignites generally contain high concentrations of calcium associated with the carboxyl groups. Due to this catalytic effect of calcium it was possible to gasify the lignite chars at 200-300 K lower temperatures, compared with the higher rank coals.

Under identical pyrolysis conditions char reactivity generally decreases with increasing rank of the precursor coal. This is primarily due to the higher degree of condensation of the char structure with coal rank, resulting in a lower concentration of active gasification sites. On the other hand, it was found by us that thermoplastic behavior of a coal affects the reactivity of the char formed; employing a low heating rate in the pyrolysis stage (0.08 K s^{-1}), highly plastic coal chars showed a lower gasification reactivity than non-plastic coal chars of higher rank [11]. This phenomenon is also observed for the hvA bituminous coal char and the semi-anthracite if the samples were pyrolysed at 1.6 K s^{-1} in the thermobalance (curves 2 in figures 3C and 3D). The lower reactivity of the hvA bituminous coal is explained by a better alignment of the aromatic planar units in the coal (cristallization) during the softening of the coal. The alignment reduces both the porosity and the concentration of active sites.

Alignment is opposed by rapid heating conditions [12]. This explains the higher reactivity of the hvA bituminous coal char, which was pyrolysed in the entrained flow reactor. On the other hand, softening of coal can be induced by rapid heating. This occurs in part for the semi-anthracite. Whereas this coal is non-plastic under slow heating conditions, Scanning Electron Microscopy showed that some softening of coal particles occurred after pyrolysis in the entrained flow reactor and might explain the decreasing char reactivity. Likewise, the small effect of heating rate on the reactivity of the lignite char is reasoned by the absence of thermoplastic behavior.

In order to compare the reactivities of the chars of different rank, the time to gasify 90 % of the char has been calculated from the normalised gasification reactivities versus %burn-off curves (figures 3A-D). These reaction times are based on a gasification temperature of 1273 K, using an activation energy for the C-CO_2 reaction of 250 kJ mol^{-1}, assuming chemically controlled reaction. The results are shown in table 3 and clearly illustrate that the the lignite chars are orders of magnitude more reactive, compared with the higher rank coal chars.

TABLE 3. Reaction times [s] for the different coal chars to obtain 90% b.o. in CO_2 at 1273 K. For pyrolysis conditions see figure 3.

	DE53	DE38	BE30	DE10
EFR	$1.6x10^1$	$1.4x10^3$	$8.8x10^3$	$2.4x10^4$
Thermobalance	$1.7x10^1$	$2.0x10^3$	$2.9x10^4$	$7.3x10^3$

Comparison of the reaction times in Table 3 of the hvA bituminous coal char (BE30) with those of the semi-anthracite (DE10) shows the former coal char to be the most reactive of the two chars, when the coals have been pyrolysed in the entrained flow reactor (EFR). However, when the coals are pyrolysed in the thermobalance, the semi-anthracite char is the most active of the two chars. This emphasizes the need to perform experiments under realistic process conditions, like in an entrained flow reactor, to determine coal behavior for practical, high heating rate coal conversion processes. No predictions should be based on data obtained under often applied, low heating rate conditions.

5. Conclusions

- Under rapid heating conditions in a dilute phase in an entrained flow reactor larger weight losses of coal particles are attained, compared with slow heating of a loosely packed bed of coal particles, like in a thermobalance. However, even up to 1300 K no complete devolatilization of the coal particles takes place in the entrained flow reactor.
- For thermoplastic coals no weight loss data are obtained in an entrained flow reactor at conditions, at which the particles are in a liquid state when they reach the collector.
- Lignite chars are by far the most reactive coal chars due to the catalytic action of calcium.
- The influence of heating rate during the pyrolysis stage on the gasification reactivity of the char formed depends on the type of coal used. In this respect, the effect of heating rate on thermoplastic behavior of coal plays a major role.
- To determine and predict coal behavior (coal weight loss, char reactivity) for practical, high heating rate, coal conversion processes it is necessary to perform experiments under realistic process conditions, like in an entrained flow reactor.

6. References

1 Badzioch, S. and Hawksley, P.G.W. (1970) 'Kinetics of thermal decomposition of pulverized coal particles', Ind. Eng. Chem. Process Des. Dev. 9, 521-530.

2 Kobayashi, H., Howard, J.B. and Sarofim, A.F. (1977) 16th Symp. (Int.) Combust., [Proc.], The Combustion Institute, Pittsburgh, pp. 411-425.

3 Scaroni, A.W., Walker Jr., P.L. and Essenhigh, R.H. (1981) 'Kinetics of lignite pyrolysis in an entrained flow, isothermal furnace', Fuel 60, 71-76.

4 Morgan, M.E. and Jenkins, R.G. (1986) 'Pyrolysis of a lignite in an entrained flow reactor', Fuel 65, 757-763.

5 Solomon, P.R. and Hamblen, D.G. (1985), 'Pyrolysis' in R.H. Schlosberg (ed.), Chemistry of Coal Conversion, Plenum Press, New York, pp. 121-251.

6 Tromp, P.J.J. (1987) Coal Pyrolysis; Basic Phenomena Relevant to Conversion Processes, Ph.D. Thesis, University of Amsterdam, Amsterdam, The Netherlands.

7 Morgan, M.E. (1983) Role of Exchangeable cations in the Pyrolysis of Lignites, Ph.D. Thesis, The Pennsylvania State University, University Park, PA, U.S.A.

8 Jamaluddin, A.S., Truelove, J.S. and Wall, T.F. (1986) 'Devolatilization of small coal particles', Combust. Flame 63, 329-337.

9 Morgan, M.E. and Roberts, P.A. (1987) 'Coal combustion characterisation studies at the International Flame Research Foundation', Fuel Proc. Technology 15, 173-187.

10 Kimber, G.M. and Gray, M.D. (1967) 'Rapid devolatilization of small coal particles', Combust. Flame 11, 360-362.

11 Tromp, P.J.J., Karsten, P.J.A., Jenkins, R.G. and Moulijn, J.A. (1986) 'The thermoplasticity of coals and the effect of K_2CO_3 addition in relation to the reactivity of the char in gasification', Fuel 65, 1450-1456.

12 Ashu, J.T., Nsakala, N.Y., Mahajan, O.P. and Walker Jr., P.L. (1978) 'Enhancement of char reactivity by rapid heating of precursor coal', Fuel 57, 250-251.

ORGANIC REACTIONS AT HIGH TEMPERATURES IN COAL TECHNOLOGY

HAROLD H. SCHOBERT
Fuel Science Program
Department of Materials Science and Engineering
The Pennsylvania State University
University Park, PA 16802 USA

ABSTRACT. This chapter discusses some of the aspects of the organic chemistry of coals at temperatures above 300°C. The intention is to provide an introduction to some of the concepts of organic reaction chemistry that may be important in the behavior of coals during high temperature processing in reductive (e.g. hydrogenation) or neutral (pyrolytic) environments. To maintain a focus on general aspects of the *chemistry* of coal, the discussion centers mainly on reaction mechanisms, and not on details of specific coal conversion processes.

1. Introduction

Being asked to prepare a brief book chapter on the high-temperature organic chemistry of coal seems akin to, say, being asked to prepare a brief review of the significant events of the twentieth century. To address this challenging task I have made a somewhat arbitrary selection of some boundary conditions. First, I take the concept of high-temperature organic reactions as being ones that are important above 300°C. For most coals the onset of significant breakdown of the macromolecular structure seems to begin somewhere in the range of 300-400°C; hence 300° might be taken as a lower limit of a high-temperature reaction regime. Furthermore, most of the common organic chemistry that is well-served by many excellent introductory textbooks seems to be concerned with reactions well below this temperature. Thus 300° might be considered a dividing point between the chemistry practiced by, and taught to, most organic chemists and the chemistry encountered in the reactions of coals. This temperature limit is not intended to imply that reactions of coals at low temperatures are unimportant or unworthy of study; indeed, I would argue the contrary. Second, I have largely ignored the high-temperature reactions of coals in oxidizing environments, i.e., combustion reactions. Again, I do not intend to slight the importance of what has been left out; indeed combustion processes consume probably 90% of all the coal used in the world. Rather, I have chosen to stay within the focus of this Advanced Study Institute, being concerned with reactions that might lead to production of synthetic fuels instead of the combustion of the fuels.

This chapter does not pretend to be a comprehensive review of the literature of the high-temperature organic chemistry of coals, nor even of the *recent* literature of the field. The chapter begins by considering some aspects of free-radical chemistry, and then moves into other mechanisms that are less commonly encountered in the coal literature, but which may nevertheless be applicable to high-temperature processing of coals. The intent is to stimulate thinking about applying the concepts of organic chemistry to developing reaction schemes for conversion of coals to synthetic fuels, not to provide an encyclopedic review of coal chemistry at high temperatures. Occasionally one finds in the literature comments to the effect that coal has an infinite number of possible structures and consequently its chemistry will be horribly complex, the implication of these statements being that since we can never hope to figure out the structure of coal or details of its chemical transformations, we might as well get on with empirical, "shake and bake" testing. While there is considerable complexity to questions of structure and mechanisms of coal chemistry, I disagree with the assertions just stated, and believe that much useful knowledge can come from the continued probing of coal chemistry and the applications of the concepts of organic reaction mechanisms to coals. Finally, I have focused mainly on the fundamentals of chemistry and reaction mechanisms, rather than on specific details of process configurations or unit operations, in the hope that the material presented in this chapter will be broadly applicable.

Y. Yürüm (ed.), Clean Utilization of Coal, 85–96.

2. Free Radical Chemistry

2.1 THE CONVENTIONAL VIEW

Thermally induced homolysis of covalent bonds leads to the production of free radicals. In the simplest sense, this reaction can be represented by

$$R{:}R' \rightarrow R\bullet + \bullet R'$$

where R and R' are used as symbols for otherwise undefined organic moieties and the "dot" (•) is used to indicate a free radical. The spontaneous bond breakage of coal macromolecules by this process is the classical mechanism of coal depolymerization [1,2]. The reaction of ether bonds, for example, proceeds via thermal cleavage to radicals [3]. These reactions could involve diaryl, aryl alkyl, or dialkyl ethers (and their sulfur-containing thioether analogs) [4]. Benzyl ethers may be particularly important in this regard [5]. The chemistry of the ether bond appears to be significant contribution to the high-temperature chemistry of coal [6]. In alkylaromatic systems which can be considered to represent the aromatic units in coal with attached aliphatic bridges or crosslinks, the initiation of pyrolysis normally occurs through scission of the bond between the α and β carbons. Scission at this point generates the stable benzyl radicals. For example, this reaction has been demonstrated in the thermal decomposition of 2-(3-phenylpropyl)naphthalene [7]. The size of the aromatic portion of the molecule has an effect on the course of the reaction. In part this relates to greater resonance stabilization of radicals afforded by larger ring systems. Thus 2-(3-phenylpropyl)naphthalene is more reactive than the related compound 1,3-diphenylpropane [7]. Further, the decomposition of 1-dodecylpyrene and dodecylbenzene follows different paths [7]. Dodecylpyrene undergoes cleavage of the very strong aryl-alkyl C-C bond, via addition of a hydrogen to the ipso position of the pyrene ring [8], whereas dodecylbenzene [7] and hexadecylbenzene [8] cleave to preserve the benzylic carbon on the ring.

In the initial generation of free radicals, cleavage of the carbon-carbon bond is preferred to cleavage of a carbon-hydrogen bond, because of the lower bond strength of the former. The C-H bond is stronger by roughly 18 kcal/mol [9]. As a rule, the aromatic rings do not undergo C-C bond cleavage within the ring, and tend to be stable toward thermally induced bond cleavage [9]. Thus C-C bond cleavage occurs either in alkyl side chains (including alkyl chains that form linkages to other aromatic ring systems) or in hydroaromatic rings. The aliphatic carbon - aromatic carbon bond that attaches the alkyl group to the ring is usually quite stable.

A review of free radical chemistry is provided in a modest introductory textbook on fuel chemistry [10]. The principal reactions are discussed below.

Hydrogen abstraction occurs when a free radical removes a hydrogen from another molecule or radical.

$$R\bullet + CH_3CH_3 \rightarrow RH + CH_3CH_2\bullet$$

Radicals can be "capped" with hydrogen that does not necessarily arise from another organic molecule. Thus

$$R\bullet + H_2 \rightarrow RH + H\bullet$$

Disproportionation is a reaction of two radicals in which one, in effect, abstracts a hydrogen from another.

$$CH_3CH_2\bullet + CH_3CH_2\bullet \rightarrow CH_3CH_3 + CH_2{=}CH_2$$

β-Bond scission is the second-most important reaction for coal depolymerization or bond breaking, second only to the initial thermal bond homolysis itself. β-Bond scission involves cleavage of a C-C (or other) bond at a position β to the radical site. For example

$$CH_3CH_2CH_2CH_2CH_2\bullet \rightarrow CH_3CH_2CH_2\bullet + CH_2{=}CH_2$$

Recombination is the direct reaction of two radicals to form a new C-C bond by pairing the radical electrons.

$CH_3\bullet + CH_3\bullet \rightarrow CH_3CH_3$

Of the many possible reactions of free radicals, one of the most important is stabilization by hydrogen. This process serves several beneficial functions. It reduces molecular size, thus providing a step along the way from the macromolecular, solid coal structure to desired low-boiling liquids. It adds hydrogen to the system, also facilitating the transition from the low H/C atomic ratios typical of coals (~0.8) to the much higher values of petroleum-like materials (~2). The "capping" of the radical with hydrogen also stabilizes it and prevents undesirable reactions, particularly retrogressive recombination that could lead to an insoluble, unreactive char. Hydrogen capping of the radicals may proceed via the reaction of the thermally generated radicals with hydroaromatic structures [3]. The hydroaromatic structures involved in the reaction could be hydrogen donors present in the solvent, or could be hydroaromatic systems elsewhere in the coal. (Coal can be shown to behave like a hydroaromatic compound [11].) In the latter case no net hydrogenation occurs.

Hydrogen abstraction reactions are very important in the conventional view of the mechanism of coal liquefaction. In this case, radicals formed by thermally induced bond homolysis abstract a hydrogen from donor molecules (e.g., hydroaromatics) in the vehicle, stabilizing the reactive radical [12,13]

Hydrogen transfer processes also provide a mechanism for the rapid "relaying" of radical centers across a structure where mass transport is difficult or prohibited [14]. This kind of process might allow reactions to occur at locations in the coal structure that would seem to be inaccessible for reactant diffusion, or in environments of severely limited diffusion.

Disproportionation reactions are important in the depolymerization of the coal structure during primary carbonization [15]. For example, molecular disproportionation may be important in the alkyl-aryl C-C bond cleavage in the pyrolysis of 1,3-bis(1-pyrenyl)propane [7]. (An important alternative explanation is radical hydrogen transfer, discussed below.) In this instance the aliphatic bridges between aromatic units are cleaved. The portions of the structure that experience a net increase in hydrogen from the disproportionation are volatile and evaporate as the pyrolysis tar. The structural fragments suffering a net loss of hydrogen eventually undergo recombination reactions leading to the formation of char or semi-coke.

Recombination of the radicals formed in the bond homolysis is undesirable. No hydrogen is added to the coal in this process and its molecular weight is not reduced. The char or semi-coke resulting from the recombined radicals is usually unreactive and represents a "dead end" in the liquefaction process.

2.2. RADICAL HYDROGEN TRANSFER

Hydrogenolysis of strong bonds can be accomplished by radical hydrogen transfer. The hydrogen donating solvent can transfer hydrogen atoms to the ipso-positions on aromatic ring systems, at sites on the aromatic ring that already bear a bridge or crosslink to another aromatic system [8]. Cleavage results from bimolecular transfer of hydrogen from cyclohexadienyl radicals to the ipso-positions on aromatic systems [16]. The cyclohexadienyl radicals can arise either from the solvent or from another portion of the coal structure.

The conventional free radical pathways for coal liquefaction, outlined in the previous section, have one notable failing. In principle, there should be some sort of relationship between the effectiveness of a compound as a donor (i.e., its effectiveness in liquefaction) and its ability to act as a radical scavenger. In fact, this relationship is not observed in coal liquefaction systems. For example, the best radical scavenger of compounds likely to occur in typical liquefaction solvents is 9,10-dihydroanthracene [16]. However, this compound is not a particularly good solvent for liquefaction, being inferior to its isomer 9,10-dihydrophenanthrene. The dihydroanthracene is a superior scavenger of benzyl radicals (by a factor of ten), yet the dihydrophenanthrene is the better liquefaction solvent [16].

The mechanism proposed [16] for radical hydrogen transfer in liquefaction is

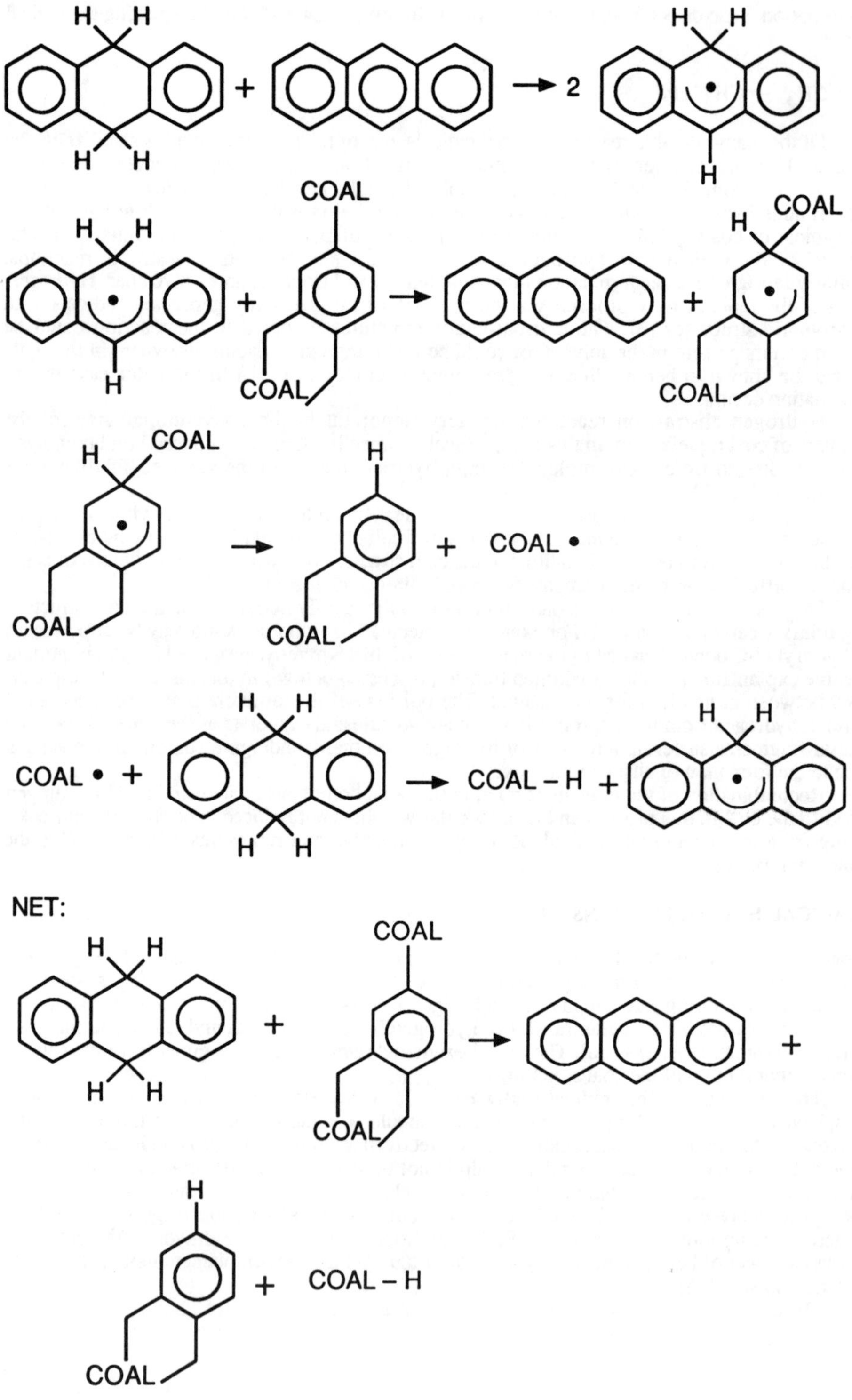

H H + → 2 H H

H H H

H H COAL + COAL → + H COAL

H COAL COAL

H COAL H

→ + COAL •

COAL COAL

COAL • + H H → COAL – H + H H

H H

NET:

H H COAL + → +

H H COAL

H

+ COAL – H

COAL

Bond scission by radical hydrogen transfer requires that some amount of radical formation has already occurred. The necessary radicals can be formed by reverse disproportionation or by thermal scission of weak bonds. However, once a population of radicals has been formed, as for example by thermal scission of weak bonds, bonds that would be too strong to break by thermal cleavage can be cleaved by radical hydrogen transfer [16].

Radical hydrogen transfer offers a mechanism to account for the cleavage of strong C-C bonds [7]. The addition of hydrogen at the ipso-position is followed by a rapid β-bond cleavage that essentially eliminates the aliphatic portion of the molecule. Radical hydrogen transfer, by attack at the ipso-position, represents a more selective hydrogenolysis mechanism than hydrogen atom addition to radicals formed by thermal homolytic bond cleavage [7]. Radical hydrogen transfer should preferentially dealkylate ring positions affording higher resonance stabilization energies [7]. As applied to coals, it is important to recognize that the stabilization of various positions on the ring (as would also be indicated by the free valence index [e.g., 17]) is different for different ring carbons for all structures of two or more rings [15]. In benzene all the ring bonds are equivalent and of the same bond order, and all carbon atoms have the same free valence index. Some differentiation in bond order (and free valence index) is apparent even in naphthalene, and the distinctions become more evident as ring size increases [15]. Chemical reactivity will be determined by the local concentration of π-electrons, not by average values of bond order or free valence [15]. Substitution will occur at bonds of maximum free valence, but addition will occur at bonds of high bond order between two carbon atoms of high free valence [15]. Hydrogen transfer to three- or four-ring aromatic systems is more favorable than transfer to one- or two-ring systems [8].

The selectivity with which the available donatable hydrogen can be used to cleave aryl-alkyl bonds depends on the extent of hydrogenation of the solvent [8]. Solvents containing abundant polycyclic aromatics use hydrogen very efficiently for hydrogenolysis of linkages. In some cases, the addition of polycyclic aromatics to solvents can improve conversion, as in the case of addition to perhydrophenanthrene or tetralin/decalin mixtures [8]. Solvents that have low concentrations of polycyclic aromatics still transfer hydrogen rapidly, but the transfer is less selective, and a greater proportion of hydrogen is used either for hydrogenation of rings or for formation of gaseous hydrogen. Hydrogen atom addition to the ipso-position generates a cyclohexadienyl radical that readily eliminates almost any linkage [8]. Some hydrogen may be transferred to positions that do not bear alkyl substituents. The efficiency of use of hydrogen is determined by what happens next.

The principal kind of linkage not cleaved by this process is the biaryl linkage, as for example in biphenyl [8,16]. The C-C bond linking the two rings in biphenyl is exceptionally strong (~115 kcal/mol). It is easier to break an aromatic carbon-hydrogen bond than it is to break the aromatic C-C bond in biphenyl-type structures. Consequently, rather than cleave the C-C bond, the intermediate cyclohexadienyl radical will lose the H•.

The radical formed by addition of hydrogen at positions other than those bearing an alkyl linkage (that is, a so-called "non-ipso" radical) can either abstract another hydrogen from the solvent or donate the added hydrogen back to molecules in the solvent. If the concentration of species able to abstract the hydrogen (e.g., polycyclic aromatic hydrocarbons) from the non-ipso radical is low, the radical will likely abstract a second hydrogen from solvent species. The initial product would be a dihydro- compound (e.g., formation of dihydronaphthalene from naphthalene). These compounds may be quite reactive and are readily hydrogenated further to the tetrahydro-derivative. Further hydrogenation of the tetrahydro- compound can lead to ring opening, cracking of the alkyl chain, and gas formation [8]. The alternative would be the donation of hydrogen from the non-ipso radical to the solvent, particularly to polycyclic aromatics. In that case the hydrogen transfer activity of the solvent is maintained. As long as there are species in the solvent capable of transferring hydrogen to the coal, there will still be opportunities for hydrogen addition at the key ipso-positions.

Diarylmethanes react by addition at the ipso-position of a hydrogen atom [18]. The source of the hydrogen may be a hydrogen donor molecule in the solvent or from hydrogen gas. For example, tetrahydrofluoranthene is a hydrogen donor that provides facile hydrogenation of the ipso-position [1,8]. The factors that affect attack at the ipso position include the size of the aromatic ring system, the nature of the hydrogen donor, the partial pressure of hydrogen, and the type of catalyst used [18]. The formation of partially reduced aromatic systems (i.e., hydroaromatics) is important for the subsequent thermal decomposition. The hydroaromatics appear to be more reactive than either the parent aromatic ring system or the fully hydrogenated cycloalkane system. For example, the hydrogenation of di(1-naphthyl)methane produces both mono- and di-tetralylmethanes. These compounds are more likely to undergo subsequent thermal decomposition than the parent naphthylmethanes [18]. The decomposition of the tetralyl methanes may be facilitated by intramolecular hydrogen transfer [18].

3. Survey of Other Reaction Processes in High-Temperature Coal Chemistry

3.1. KETO-ENOL TAUTOMERISM

The presence of phenolic groups on aromatic ring systems, and their specific substitution pattern on the ring, can have significant effects on thermal decomposition reactions [19]. The thermal decomposition of o- and p-cresol is four times faster than that of m-cresol at 543°C. The o- or p-hydroxyl group has an activating affect which is attributed to keto-enol tautomerism.

OH CH2 → O H CH2

O H CH2 → CH2• + O•

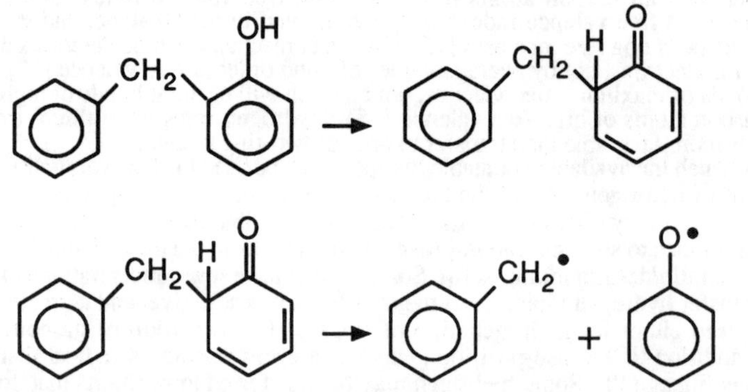

Keto-enol tautomerism may also be important in the conversion of coal in carbon monoxide / water systems [20]

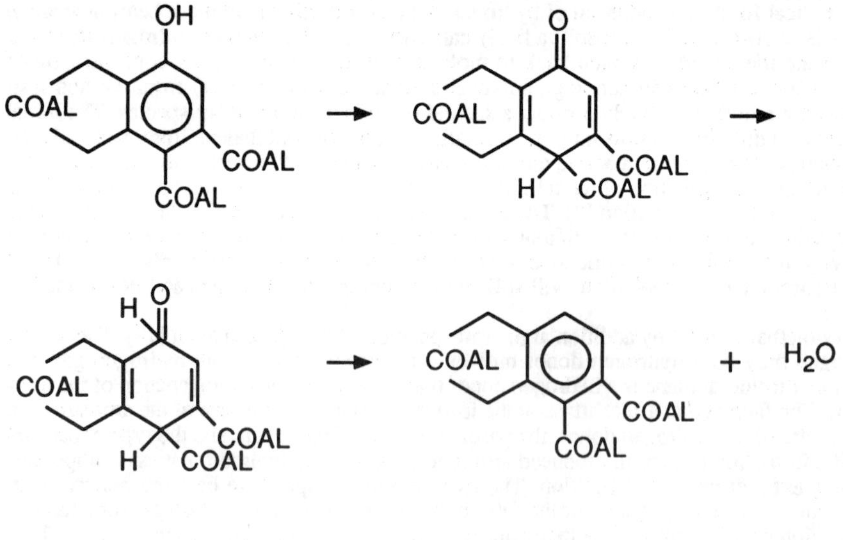

OH COAL COAL COAL → O COAL COAL H COAL →

H O COAL COAL H COAL → COAL COAL COAL + H_2O

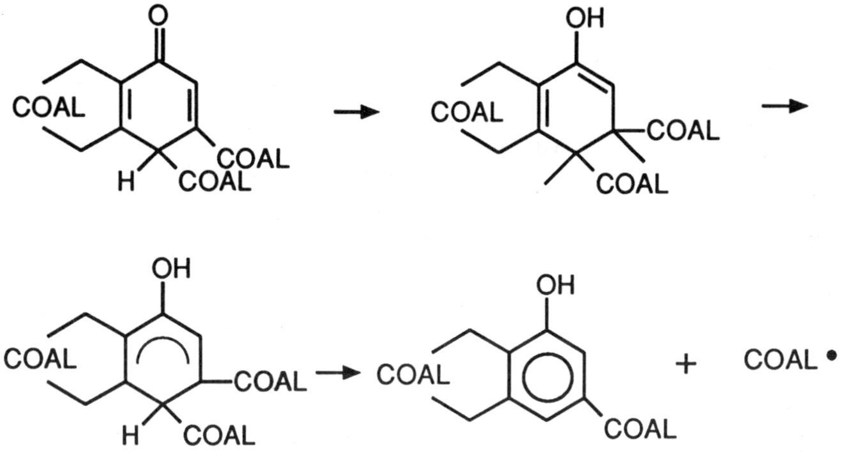

In this instance the keto-enol equilibrium is able to maintain a concentration of carbonyl groups at low, but steady, levels. In this reaction system the conversion is determined by the availability of the aromatic phenols in the coal structure. The keto-enol equilibrium produces some population of carbonyl groups in the coal. The carbonyl groups are then reduced by reaction with hydroaromatic groups or with formate ions. Two reactive sites are available for reduction, the carbonyl carbon or an alkyl-substituted carbon. When reaction occurs at the carbonyl carbon, an intermediate cyclohexadienol is formed, this intermediate then rapidly losing the elements of water. The decomposition of the cyclohexadienol results in a net loss of oxygen from the coal. If instead there is an attack at some other location on the ring system, the -OH functional group in the cyclohexadienol moiety becomes a vinylic -OH. The resulting structures are highly reactive, and can split off a radical, coal-derived fragment by one of two processes: bond homolysis (which is facilitated by the tendency of the cyclohexadienol to "re-aromatize", or β-bond scission. In either case, the coal-derived radical can be stabilized by H• sources. The overall result of this process is the breaking of strong aryl-aryl bonds in the coal [20].

Diphenylmethane is essentially inert in tetralin at 400°C, but a hydroxydiphenylmethane will react about as rapidly as bibenzyl, which has a weak benzyl carbon - benzyl carbon bond [21]. A possible mechanism to explain the difference in behavior between diphenylmethane and its hydroxy- derivative involves keto-enol tautomerism follows up the previously shown mechanism:

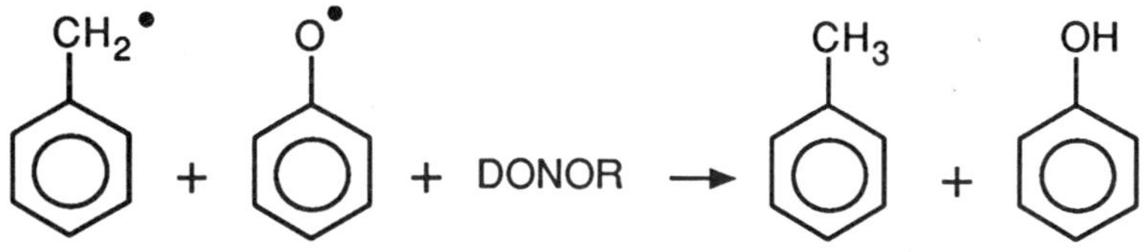

3.2. CONCERTED REACTION PATHWAYS

The interaction of hydroaromatic compounds with aromatic compounds, resulting in reduction of the latter, may occur via concerted reaction pathways [22,23]. (Useful introductory discussions of concerted reactions are available in recent monographs on organic chemistry [24,25] and an advanced text [26].) Transfer between certain reactants is allowed for thermal reactions by symmetry rules with suprafacial-suprafacial approach of reactants [22]. Suprafacial reactions are ones in which the atom or group transferring between reactants always remains on the same face of the conjugated π system throughout the reaction. Two examples of model compound reactions that involve a transition state that is allowed to occur by symmetry considerations are the reaction of anthracene with 1,4-dihydronaphthalene and the reaction of phenanthrene with 1,2-dihydronaphthalene [22]. On the other hand, some thermally driven suprafacial reactions are not able to occur thermally as a result of symmetry considerations. Examples of such symmetry-forbidden reactions include those of anthracene with 1,2-dihydronaphthalene and phenanthrene with 1,4-dihydronaphthalene [22]. (Phenanthrene and anthracene have opposite orbital symmetries [27].)

In essence, two classes of hydrogen donors and hydrogen acceptors can be recognized [23]. Hydrogen donors with 4n+2 electrons will transfer hydrogen to acceptors with 4n+2 electrons. This is perhaps the more common case involved in coal chemistry, since it involves familiar aromatic systems (so-called Hückel-type aromaticity, as in benzene, where n = 1 and the number of 4n+2 π electrons is 6) based on conversion of, say, benzene to cyclohexadiene intermediates. Transfer of hydrogen between systems with 4n electron components to 4n electron acceptors (a Möbius system, [26]) is also possible. Since a mixture of 1,2-dihydro- and 1,4-dihydronaphthalenes appears to be a more effective hydrogen donor to coal than either isomer is alone, it may be that coal contains both types of hydrogen acceptors [23].

It must be recognized that reactions that are forbidden, by symmetry considerations, to occur as suprafacial-suprafacial approach are not necessarily unable to occur at all, since other reaction pathways may be available. However, in the examples given above, the rate constants are higher for the allowed suprafacial reactions than for those forbidden [22]. This work suggests that a more detailed consideration of the nature of the specific ring structures in a coal feedstock and the structures of the likely hydrogen donor compounds might allow a "matching" of coal and donor structures to facilitate hydrogenation of the coal.

The pericyclic transition state for hydrogen transfer may be similar to the pericyclic transition state for cycloaddition [27]. (The Diels-Alder pericyclic pathway is discussed in detail in the literature [25].) The reaction of alcohol donors with aromatic compounds is illustrated by

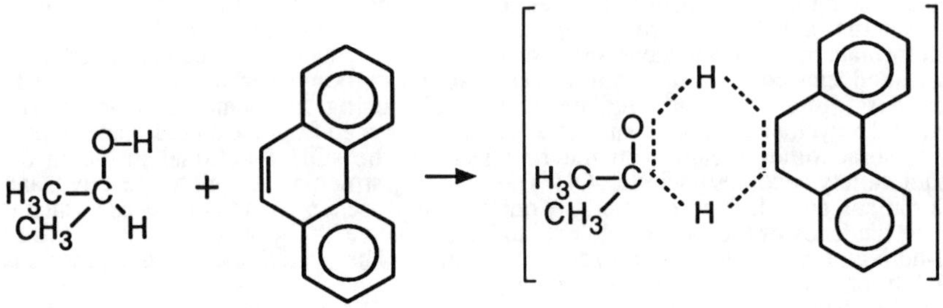

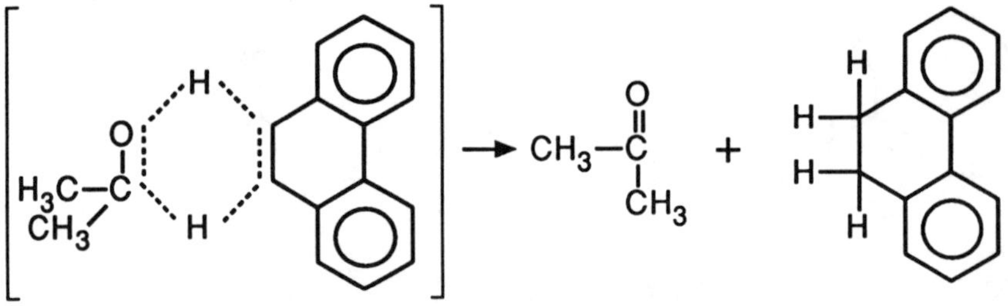

O–H
H3C–C
CH3 H
+
→
H
O
H3C–C
CH3 H

H
O
H3C–C
CH3 H
→ CH3–C
O
CH3
+
H
H
H
H

The transition state may resemble that for Diels-Alder cycloaddition [27]. Hydrogen elimination from dihydroanthracene may also proceed via a pericyclic transfer reaction [27].

Concerted pericyclic pathways may also be explain the formation of gaseous products in pyrolysis of low-rank coals [28]. For example, a guaiacyl group could undergo a six-electron elimination to produce methane

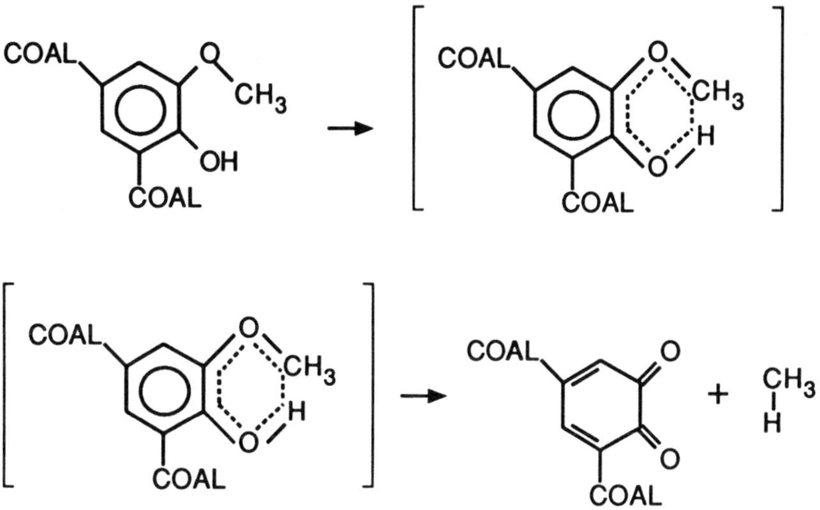

Cheletropic reactions represent a special class of reactions in which two σ bonds that terminate at the same atom are broken in a concerted reaction [25]. While in principle addition or elimination reactions can be cheletropic, the elimination reactions are more common, usually because the species being eliminated is of especially high thermodynamic stability. An example is the loss of a carbonyl unit from coniferaldehyde-like structures during pyrolysis, producing carbon monoxide [28]. Similarly, benzaldehyde may decompose via a cheletropic mechanism:

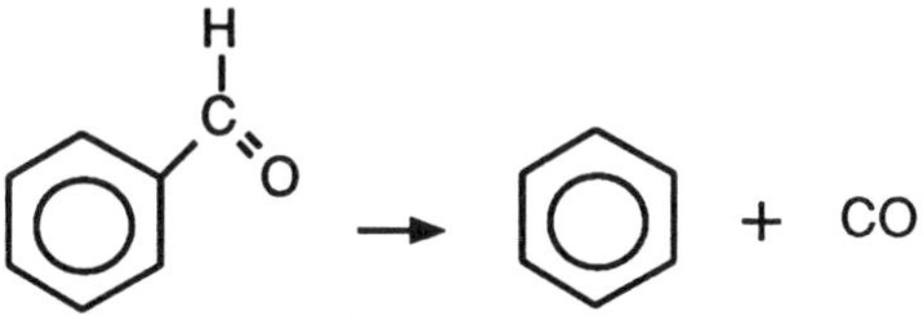

Retro-ene reactions are thermally induced elimination reactions resulting in an alkene [25]. Compounds containing heteroatoms are especially likely to undergo retro-ene reactions. Perhaps the best known example of a retro-ene reaction for organic synthetic work is the Chugaev reaction [29] in which a xanthate is pyrolyzed to form an alkene. Because of the importance of retro-ene reactions with heteroatom species, the liberation of so-called "chemical water" (that is, water formed by pyrolytic cleavage of oxygen functional groups, not water initially present in the coal as liquid H_2O) may proceed via retro-ene reactions of oxygen-rich assemblages such as guaiacyl-like structures [28].

In fact, the pyrolysis of guaiacol itself forms methane, carbon monoxide, catechol, and phenol, suggesting the following types of reactions

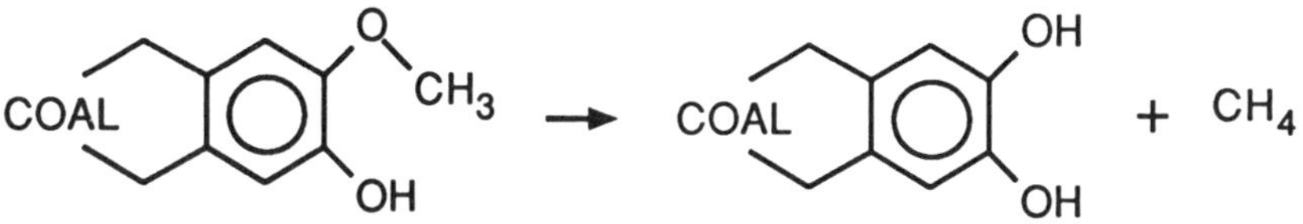

COAL [structure with OCH_3 and OH] → COAL [structure with OH] + CO

Compounds such as 2,6-dimethoxyphenol and iso-eugenol decompose in analogous fashion to methane and carbon monoxide [28]. Anisole also decomposes to methane, but at a rate two orders of magnitude lower than that of guaiacol. This distinction may be due to the fact that guaiacol can decompose via a pericyclic reaction pathway that would be unavailable to anisole [28].

3.3. IONIC REACTIONS

Thermal processing of coal (i.e., pyrolysis) and some coal reactions, such as hydrogenation, are generally considered to proceed via free radical pathways, albeit with some debate as to the exact radical mechanisms involved. However, ionic reaction mechanisms can certainly occur in coal processing, if initiated by the appropriate reagents. An example is the ionic mechanism postulated for the conversion of coal in carbon monoxide / water systems [30]:

$$CO + OH^- \rightarrow HCO_2^-$$

$$HCO_2^- + H_2O \rightarrow OH^- + H_2CO_2$$

$$H_2CO_2 \rightarrow H_2 + CO_2$$

$$\text{coal} + HCO_2^- \rightarrow CO_2 + \text{coal-H}^-$$

$$\text{coal-H}^- + H_2O \rightarrow OH^- + \text{coal-H}_2$$

$$CO_2 + OH^- \rightarrow HCO_3^- \rightarrow CO_3^{2-}$$

In these equations coal-H- represents the coal anion intermediate, and coal-H_2 represents the coal in which a portion of the structure has been reduced. The carbon monoxide / water system can be superior to hydrogen for conversion of coals [30]. In contrast, reactions of coals in carbon monoxide / hydrogen sulfide systems appear to proceed via radical mechanisms [13].

4. Summary

Most reactions of interest in high temperature coal liquefaction or pyrolysis technology proceed via free radical mechanisms. The customary view of this process involves generation of free radicals by thermally induced homolytic bond cleavage. The "traditional" free radical reactions - hydrogen abstraction or hydrogen capping, disproportionation, β-bond scission, and recombination - then suffice to account for much, but not all, of the conversion chemistry. The lack of a good correlation between the donor effectiveness and radical scavenging ability of model compounds, and the difficulty of explaining the scission of very strong bonds (as in diphenylmethane) by the traditional free radical reactions leads to the concept of the radical hydrogen transfer mechanism.

There exists a variety of mechanistic schemes that are worthy of consideration as possible reaction routes in high-temperature coal chemistry, and may be well worth exploiting as avenues to efficient coal conversion. Examples include the keto-enol tautomerism and its role in facilitating coal depolymerization, and a variety of pericyclic reaction mechanisms. Further, not all coal reactions are necessarily free radical reactions. Ionic mechanisms may also offer routes to facile coal conversion.

5. Literature Cited

1 Mochida, I., Takayama, A., Sakata, R., and Sakanishi, K. (1990) "Liquefaction of Australian brown coal with mixed solvents of different qualities and reactivities of transferrable hydrogens", Energy and Fuels 4, 398-401.

2 Sakata, R., Takayama, A., Sakanishi, K., and Mochida, I. (1990) "Roles of nondonor solvent in hydrogen transferring liquefaction of Australian brown coal", Energy and Fuels 4, 585-588.

3 Marzec, A., Czajkowska, S., Simmleit, N., and Schulten, H.R. (1990) "Coal reactive sites involved in hydrogen transfer from an H-donor studied by field ionization mass spectrometry", Fuel Processing Technology 26, 53-66.

4 Mayo, F.R., Huntington, J.G., and Kirshen, N.A. (1978) "Chemistry of coal liquefaction", in J.W. Larsen (ed.), Organic Chemistry of Coal, American Chemical Society, Washington, Chapter 8.

5 Deno, N.C., Curry, K.W., Cwynar, J.E., Jones, A.D., Minard, R.T., Potter, T., Rakitsky, W.G., and Wagner, K. (1980) "Chemical changes in coal liquefaction", Amer. Chem. Soc. Division of Fuel Chem. Preprints 25(4), 103-110.

6 Chakrabartty, S.K., and Berkowitz, N. (1978) "Early stages of coal carbonization: evidence for isomerization reactions", in J.W. Larsen (ed.), Organic Chemistry of Coal, American Chemical Society, Washington, Chapter 9.

7 Smith, C.M., and Savage, P.E. (1991) "Reactions of polycyclic aromatics. 2. Pyrolysis of 1,3-diarylpropanes", Energy and Fuels 5, 146-155.

8 McMillen, D.F., Malhotra, R., and Tse, D.S. (1991) "Interactive effects between solvent components: possible chemical origin of synergy in liquefaction and coprocessing", Energy and Fuels 5, 179-187.

9 Petrakis, L. (1983) Free Radicals in Coals and Synthetic Fuels, Elsevier Science Publishers, Amsterdam.

10 Schobert, H.H. (1990) The Chemistry of Hydrocarbon Fuels, Butterworths, London.

11 Sternberg, H.W. (1978) "Mechanism of coal liquefaction: a challenge to the organic chemist", in B.R. Cooper (ed.), Scientific Problems of Coal Utilization, U.S. Department of Energy, Morgantown, West Virginia.

12 Neavel, R.C. (1982) "Coal plasticity mechanism: inferences from liquefaction studies", in M.L. Gorbaty, J.W. Larsen, and I. Wender (eds.), Coal Science. Volume 1. Academic Press, New York.

13 Abdel-Baset, M.B., and Ratcliffe, C.T. (1980) "Novel approach to coal liquefaction utilizing hydrogen sulfide and carbon monoxide", Amer. Chem. Soc. Division of Fuel Chem. Preprints 25(1), 1-7.

14 Buchanan, A.C., Britt, P.F., and Biggs, C.A. (1990) "Thermolysis mechanisms. Evidence for an alternative pathway for radical migration in diffusionally constrained environments", Energy and Fuels 4, 415-417.

15 van Krevelen, D.W. (1981) Coal: Typology - Chemistry - Physics - Constitution, Elsevier Science Publishers, Amsterdam.

16 McMillen, D.F., Malhotra, R., Hum, G.P., and Chang, S.J. "Hydrogen-transfer promoted bond scission initiated by coal fragments", Energy and Fuels 1, 193-198.

17 Coulson, C.A. (1963) Valence, Oxford University Press, London.

18 Wei, X.Y., Ogata, E., Futamura, S., and Kamiya, Y. (1990) "Thermal decomposition and hydrocracking of hydrogenated di(1-naphthyl)methanes", Fuel Processing Technology 26, 135-148.

19 Gavalas, G.R. (1982) Coal Pyrolysis, Elsevier Science Publishers, Amsterdam.

20 Ross, D.S. (1984) "Coal conversion in carbon monoxide - water systems", in M.L. Gorbaty, J.W. Larsen, and I. Wender (eds.), Coal Science. Volume 3. Academic Press, Orlando, Florida.

21 McMillen, D.F., Ogier, W.C., and Ross, D.S. (1981) "Coal structure cleavage mechanisms: scission of diphenylmethane and diphenyl ether linkages to hydroxylated rings", Amer. Chem. Soc. Division of Fuel Chem. Preprints 26(2), 181-190.

22 Bockrath, B.C. (1983) "Chemistry of hydrogen donor solvents", in M.L. Gorbaty, J.W. Larsen, and I. Wender (eds.), Coal Science. Volume 2. Academic Press, New York.

23 Pullen, J.R. (1983) "Solvent extraction of coal", in M.L. Gorbaty, J.W. Larsen, and I. Wender (eds.), Coal Science. Volume 2. Academic Press, New York.

24 Sykes, P. (1986) A Guidebook to Mechanism in Organic Chemistry, Longman Scientific and Technical, Harlow, Essex, UK.

25 Gill, G.B., and Willis, M.R. (1974) Pericyclic Reactions, Chapman and Hall, London.

26 Carey, F.A., and Sundberg, R.J. (1990) Advanced Organic Chemistry. Part A. Structure and Mechanisms, Plenum Press, New York.

27 Garry, M.J., and Virk, P.S. (1980) "Hydrogen transfer from alcohol donors to aromatic substrates", Amer. Chem. Soc. Division of Fuel Chem. Preprints 25(4), 132-147.

28 Klein, M.T., and Virk, P.S. (1980) "Model pathways for gas release from lignites", Amer. Chem. Soc. Division of Fuel Chem. Preprints 25(4), 180-190.

29 Mundy, B.P., and Ellerd, M.G. (1988) Name Reactions and Reagents in Organic Synthesis, John Wiley and Sons, New York.

30 Ross, D.S., Green, T.K., Mansani, R., and Hum, G.P. (1987) "Coal conversion in water. 1. Conversion mechanism", Energy and Fuels 1, 287-291.

BIOCONVERSION OF COAL

A.TANYOLAÇ, T.DURUSOY, T.ÖZBAŞ, Y.YÜRÜM*
Hacettepe University
Department of Chemistry and*
Department of Chemical Engineering
Beytepe 06532 Ankara, Türkiye

ABSTRACT. An outline of lignite degradation, methane generation from lignite, bioprocessing of coal breakdown products and possible bioreactor configurations were presented under the topic of bioconversion of coal, a recently new development. Preliminary work has been done on the fungal and bacterial degradation of lignite and bacteria have been identified which decompose hard coals. The work is not likely to result in a useful economic process in less than a decade and, more likely, it will be 15 to 20 years before such a process is feasible, even given favorable progress with development work. Research has been continued on seeking new and mutant microorganisms for different biotransformations and on biodegradation of organic compounds as models for microbial studies such as destructions of lignin, organic sulfur compounds, carboxyl groups and organic nitrogen bonds.

1. Introduction

The traditional routes for coal conversion to high grade products usually involve some coal cleaning followed by treatment at high temperatures and pressures through a series of chemical processes and separation techniques. These require close control of reaction conditions and necessitate high capital and operating costs which make many processes uneconomic under current circumstances. The use of biotechnology could provide alternative processes for coal conversion and also could provide different techniques for coal cleaning prior to use.

In principle, biotechnology can be used to bring about a general breakdown of the coal molecule or to achieve the selective removal of particular components. Coal is basically carbonaceous material with mineral inclusions and it is such a heterogeneous substance that special microorganisms may be able to attack, degrade or use a number of different constituents. Bioprocess might have significant advantages over conventional methods of breakdown currently employed and might create the possibility of the economical production of new chemicals. This is because the reaction conditions involved would be much milder than those required for the equivalent chemical transformations i.e., lower energy input to the system per unit of derived energy and because the involvement of less expensive construction materials and finally because the production of specific products or the removal of certain impurities could be maximized.

Microorganisms may attack either the carbonaceous coal or the dispersed inorganic materials within the structure. One route is the depolymerization of the coal and the breakage of various key links which could provide the basis for liquefaction. A second route would be a reduction in oxygen content either through reduction of C=O to CH_2 or maybe by decarboxylation to CO_2

Y. Yürüm (ed.), Clean Utilization of Coal, 97–107.

which could improve the calorific value [1]. A third would be the removal of sulfur, nitrogen or metals from the coal prior to combustion which might break up the infrastructure as well as would reduce undesired emissions [2].

The literature is scarce and the work done is limited in this area. Because of the complex structure of coal and its heterogeneous nature, there is general agreement that the development of microorganisms to degrade coal directly will not be easy. Most of the work so far identified has been based on the possible degradation of lignite because it is in the younger coals that the original biological structures are better preserved [3]. Also the materials present have been subject to lower temperature and pressure during their history. Some microorganisms were identified which degrade hard coal [4] and there is evidence that subbituminous coals may also be amenable to microbial degradation [5].

The best approach in terms of seeking a potentially useful process for making degradation products is to use a preliminary chemical breakdown stage which would provide a feedstock of smaller molecular size and would also facilitate the separation of possible bactericides (bacterial poisons/inhibitors) and of other unwanted chemicals. Such treatment may also break bonds that organisms can not attack and/or may result in more active sites for microbial activity. It is thought that this approach will produce practical and possibly economic processes some considerable time before that of seeking organisms to degrade coal directly. Similarities in the chemical structure of liquefaction products from high temperature and pressure processes [6] enhanced the possibility of application if current research is developed successfully and the economics are attractive. The work on special microorganisms capable of living at high temperatures and severe conditions has led to speculations about microbial mining and even gasification in lower coal seams where conventional extraction techniques turn out to be extremely difficult because of the strata temperature and other environmental difficulties.

There is preliminary evidence that coal can be microbially degraded and biologically converted to methane [7]. While it was suggested that the huge reactors needed would be prohibitively expensive, the injection of microbes into coal seams deliberately to enhance methane production has also been proposed [8]. However this possibility may remain speculative and certainly there would be formidable problems to be tackled with before putting it into realization.

2. Lignite Degradation

Although coal conversion technology has been studied since 1780 [9], only in the past few years several species of fungi and bacteria have been shown to convert low rank coals into water soluble liquid products through bioconversion [10-12].

In these studies, surface culture techniques have generally been used to screen for the most efficient organism. This is done by first establishing a fungal mat on a growth medium then lignite particles are put on top and the mat is placed in a humid environment. If liquefaction takes place, droplets will form on the surface of the coal and, in the best situations, the coal particles are completely liquefied [13].

Of this work, first Krucher *et al.* [14] reported the growth of *Candida tropicalis* on coal substrates in 1977. The early study of Cohen and Gabriele [12] identified two species of fungi *Polyporus (Trametes) versicolor* and *Poria monticola* grown in minimal liquid or solid medium when supplemented with crushed leonardite. Leonardite is a naturally oxidized or weathered lignite with a typical oxygen content of 28-29% compared with 19-20% in the equivalent US lignite. The microorganisms also grow directly on crushed lignite but less well, no fungal growth occurred in the minimal agar medium without leonardite. Liquefaction starts within 24 hours after adding the coal, with the content of fungi excretion possibly enzyme solution until lignite is completely liquidified or growth ceased due to inhibition by toxic materials and/or material depleted.

In the course of these studies the laccase enzyme thought to be primarily responsible for the degradation has been isolated [5], and attempts were made to see whether the enzyme would increase the solubility of low rank coals alone. The results showed little or no increase, which would indicate that the action of *Polyporus versicolor* is not solely dependent on the laccase enzyme.

Coal is generally considered to be derived from wood. Many of the interunit structures in lignin are thought to resemble those occurring in coal [15]. Therefore wood decay microorganisms are prime candidates for coal liquefaction, in fact many of the cultures which have been shown to solubilize coal also degrade wood. In this regard, *Phanerochaete chrysosporium* has been extensively studied for its ligninolytic activities. Kirk *et al.* [16] described a ligninolytic oxygenase from *P. chrysosporium* which oxidatively cleaves $C\alpha$-β model compounds resembling interunit linkages thought to be similar to those in low rank coals.

Following preliminary analyses, it has been concluded that the products from the microbial degradation of the leonardite are water and methanol soluble, highly oxygenated, structurally heterogeneous compounds which appear to have a wide range of molecular weights. The absence of low molecular weight compounds appears to be a consequence of selective and limited breakage of specific bond types in the lignite structure [5]. It appears that the highly cyclic and insoluble structure of lignite is broken to form aliphatic, polar, water soluble compounds. Ultrafiltration and gel permeation chromatography show that the molecular weights of the extract falls primarily in the range from 30,000 to 300,000 [13].

Other possible microorganisms which might be studied have been identified from naturally occurring fungi growing on a lignite outcrops [17]. The fungi involved included one species each of *Aspergillus, Candida* and *Paecilomyces*, four species of *Penicillium* and two of *Mucor* , and two other fungi had been reported in an earlier work as *Penicillium waksmanii* and a *Candida sp.* The soil fungi grew on solid lignite, aqueous lignite extracts, ethanolic lignite extracts and on a number of alkanes which occur in lignite. However, growth of some isolates on coal was enhanced about 30% by supplementing the water with mineral salts.

A considerable amount of work has been carried out by Oak Ridge National Laboratory. Various species have been tested with a Mississippi lignite from an exposed, highly weathered location of which initial characterization of the liquid products from various tests is given in Table 1. An analysis of two of the lignite samples used is given in Table 2.

Nevertheless, microbial coal liquefaction is a slow process depending on the coal oxidation step and other chemicals present in the coal. In the case of *Candida* fungi (which is amongst the most productive) liquefaction starts quite aggressively after about 10 days and is essentially complete after two weeks when residual solids are primarily undesirable ash. Bacterial liquefaction tends to be quicker and dissolution starts within about two days of inoculation [13].

There has been work done with different species on hard coal, but very few bacteria and fungi were found to survive on coal producing high molecular weight materials. Again, the degradation process is extremely slow.

The porosity of coal should also be considered in studying rates and mechanisms of coal biotransformations. A certain degree of porosity is associated with almost all coals ranging from micropores (ca. 30 Å) to macropores (300 Å or larger), as shown by X-ray scattering data and other means [19]. Higher rank coals tend to have smaller pores, thus many extracellular enzymes probably would not penetrate micropores since even small enzymes such as chymotrypsin (MW =

22,600; 45 Å x 35 Å x 38 Å) exceed this size. Therefore any process that would increase coal porosity would make more surface area available for enzymatic attack which in turn results in faster and efficient bioprocesses.

Pretreatment of lignite before bioliquefaction, to increase its susceptibility has been attempted, with inconclusive results [20]. The possibility of using an aqueous alkaline oxidation before bioconversion has been proposed.

TABLE 1. Liquid products from microbial interaction with coal in tests at Oak Ridge National Laboratory, USA [18]

Type of coal	Organism	Liquid products
Mississippi lignite	*P .waksmanii* ML20 *Candida* sp ML13 *T .versicolor* ATCC 12679 *Aspergillus* sp	clear, amber and black; inconsistent and generally in moderate amounts
Texas lignite	*Candida* sp *Aspergillus* sp *Paecilomyces* sp *Sporothrix* sp	clear; trace to moderate amounts
Vermont lignite	*Candida* sp ML13 *Sporothrix* sp	brown to black; trace amounts
North Dakota I lignite	*Paecilomyces* sp *P .waksmanii* ML20 *Candida* sp ML13 *Aspergillus* sp *Paecilomyces* sp *Sporothrix* sp	clear; trace amounts clear; moderate amounts
North Dakota II leonardite	*Candida* sp ML13 *Sporothrix* sp *T .versicolor* ATCC 12679 *Aspergillus* sp *Paecilomyces* sp *P .monticola* 11538	brown to black; moderate to profuse amounts; some clear initially, turning brown to black with time (2 to 3 weeks)
Wyodak subbituminous	*T .versicolor* ATCC 12679 *Candida* sp ML13 *Sporothrix* sp *Aspergillus* sp	clear and black; trace amounts clear; trace amounts

Difficulties associated with measuring extremely low rate biodegradation of coal have retarded the progress in this field. Recent advancements have been made in molecular characterization of coal and associated minerals by diffused reflectance Fourier transform infrared analysis [1]. By this methodology it would be possible to study biotransformations of coal. For this purpose several biodegradation of organic compounds as models for microbial processing have been investigated in order to elucidate the mechanisms of breakdown of coal as well as to determine the enzymes responsible for the action. Olson and Brickman gave an excellent review on lignin biodegradation, decarboxylation, decomposition of organic sulfur compounds and organic N biotransformations[1].

TABLE 2. Analyses of the lignite samples used in tests at Oak Ridge National Laboratory, USA [18].

Proximate, as received	Mississippi, %	North Dakota I, %
Moisture	42.0	34.2
Ash	13.6	8.9
Volatile matter	28.1	29.0
Fixed carbon	16.3	27.9
Ultimate (dry)		
C	54.5	62.3
H	5.0	4.4
N	0.8	0.1
S	1.9	1.3
O	14.5	18.3
Cl	0.02	0.01
Sulfur forms (dry)		
pyritic	0.7	0.2
organic	0.4	1.0
sulfate	0.04	0.1

3. Possible Bioreactor Configurations

Even though the scientific basis and the feasibility of microbial coal liquefaction have not been established yet, it is possible to consider some preliminary process concepts which may well influence the direction of further studies [18]. Of these, three possible techniques are:
1. A solid phase fermentation in a fixed bed bioreactor where stationary particles of coal would be exposed to microorganisms and humid air. The liquid product and residual solids are collected at the bottom of the reactor;
2. A fluidized bed bioreactor where suspended coal particles are contacted with a circulating aqueous phase that include the liquid product and carries the nutrients and suspended microorganisms;
3. The rotating bioreactor proposed by Vaseen [21].

Pretreatment of lignite before bioliquefaction, to increase its susceptibility has been attempted, with inconclusive results [20]. The possibility of using an aqueous alkaline oxidation before bioconversion has been proposed.

The proposed schematic views for first two reactors are given in Figure 1. There is no doubt that the systems proposed are prototypes and can not be designed unless a complete understanding of biodegradation process is made clear.

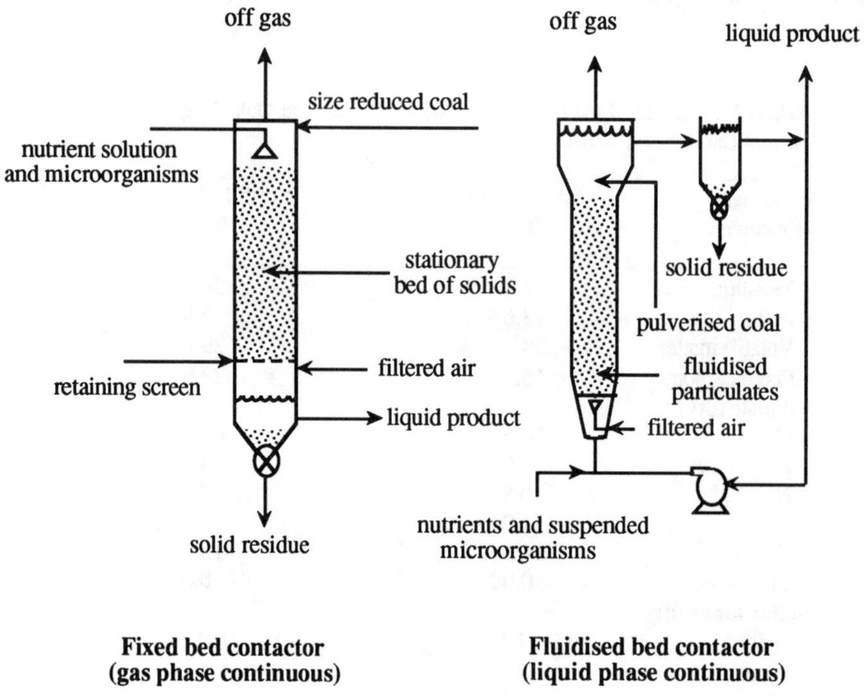

Figure 1. Possible bioreactor configurations for microbial coal liquefaction [18].

4. Bioprocessing of Coal Breakdown Products

Much work has been directed towards identification of microorganisms capable of degrading lignite and its end products. The next phase of the research would look at further bioprocesses to transform the degradation products into useful and valuable end materials [20].

In addition, chemical oxidation by potassium permanganate and sodium hypochlorite is being undertaken, followed by identification and measurement of the acids formed with the carbon dioxide. These acidic degradation products will then be considered as a feedstock for subsequent bioprocesses to convert them to usable products. There is also interest in microbial upgrading of coal derived synthesis gas, for example microbial methanation [8].

Figure 2 shows the chemical steps for a proposed lignite refinery based on bioprocesses which would generate methane from Texas lignite [22].

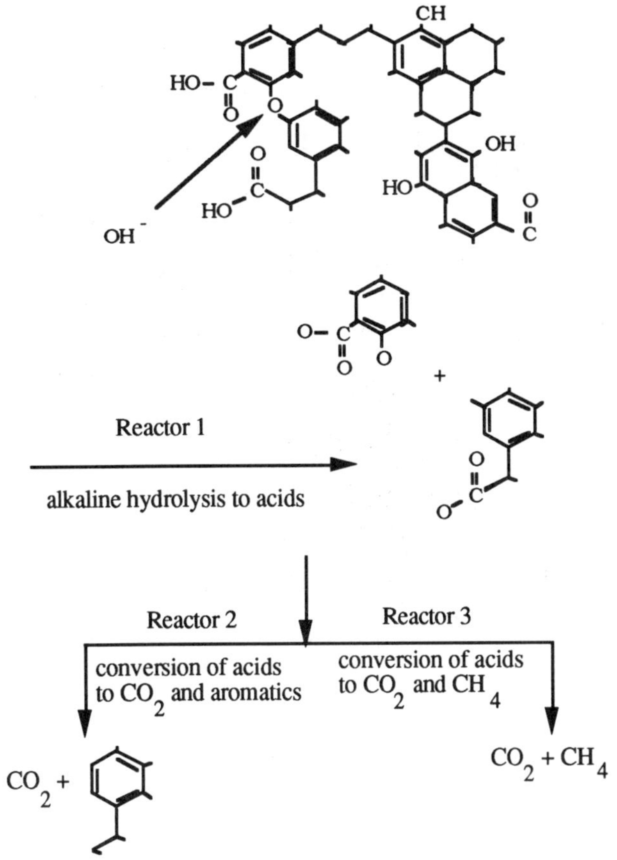

Reactor 1

alkaline hydrolysis to acids

Reactor 2

conversion of acids to CO_2 and aromatics

Reactor 3

conversion of acids to CO_2 and CH_4

CO_2 +

$CO_2 + CH_4$

Figure 2. The chemical steps involved for the proposed lignite refinery [22]

A conceptual design of this lignite refinery was also made as seen schematically in Figure 3 [22]. The objective is to find a clean heat source for power generation through an aqueous alkali oxidation of lignite to simple water soluble aromatic compounds. This can take place at around 300°C and 5 MPa, which is considerably milder than the conditions required for conventional gasification. Conversion of 50% of the acids to CO_2 and immiscible aromatics takes place in a second reactor. The product from the second reactor is separated to give an aromatic stream of benzene, toluene and xylene, and a bottom product. Due to substantial volumes involved, it is proposed to use underground salt caverns as the reactor vessels for the bioprocess and suitable microorganisms which can effect the methanation are being sought. The remaining acids then undergo anaerobic digestion which produces methane and carbon dioxide in the third reactor.

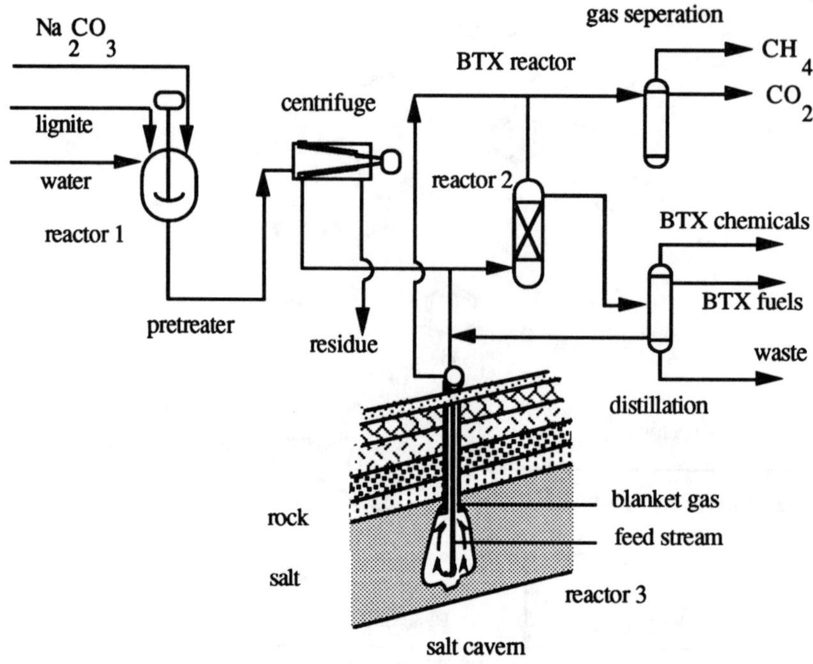

Figure 3. Simplified process diagram for lignite biogasification [22]

5. Methane Production from Lignite

There has been a considerable amount of work on the bioconversion of peat and lignin into methane [23]. Peat contains biochemically altered lignocellulose while lignite has seen additional physical and chemical actions associated with the coalification process. It is thought that lignite itself is likely to resist rapid microbial degradation and that a physical or chemical pretreatment of the feed stock into a biodegradable form would be needed. Alkaline oxidation and wet air oxidation of lignite should give similar products to those obtained from a pretreatment of peat, and the biogasification of these products has been shown to be technically feasible.

The feasibility of peat biogasification was demonstrated in the early 1920s but the conversion rates and methane yields were low [24]. Various pretreatments to convert the polymer fractions to water soluble compounds can improve conversion rates and yields, the efficiency of carbon conversion to methane can be as high as 25-30% [25].

The route that might be used for possible peat biogasification is illustrated in Figure 4. Because of the basic similarities between peat and lignite, a parallel process configuration is likely to be possible for lignite as depicted in Figure 3.

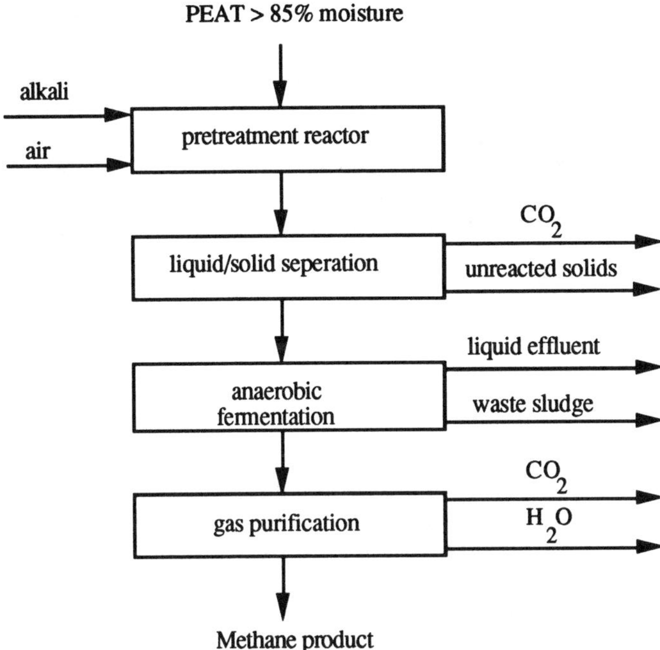

Figure 4. Flowchart for peat biogasification [23]

6. Conclusion

The conversion of coal to liquid and gas products by microorganisms is attractive since bioprocess might involve milder and more selective reaction conditions that could manipulate part of the coal molecules to the best advantage. Understanding of bioprocess reaction mechanisms might even lead to new approaches for conventional chemistry as applied to coal, resulting from insights into the molecular structure of coal.

Preliminary work has been done on the fungal and bacterial degradation of lignite and bacteria have been identified which decompose hard coals. The work is not likely to result in a useful economic process in less than a decade and, more likely, it will be 15 to 20 years before such a process is feasible, even given favorable progress with development work. Given good progress with development work, commercialization of this route might be possible within next ten years. Research has been continued on seeking new and mutant microorganisms for different biotransformations and on biodegradation of organic compounds as models for microbial studies such as destructions of lignin, organic sulfur compounds, carboxyl groups and organic nitrogen bonds.

It is far too early to consider the possible economics of either bioliquefaction or bioconversion of coal into commercial products although possible applications have been speculated immensely. In fact, before economical considerations there will be certain scale up problems for the real process due to relatively complex interactions between different bioprocess variables and conditions which may not be necessarily anticipated from small scale work. There are certain characteristics of bioprocesses implying that there are possibly to be more problems in scaling up mainly due to delicate structure of microorganisms and so of enzymes than there would be an equivalent chemical process. Moreover, the separation and purification of biostreams are less systematized operations with expensive and complex equipment than their conventional chemical engineering equivalent. On the other hand, bioprocesses can be vulnerable to the presence of toxins, which may be well present in coal, and are often unexceptionably slow reactions thus requiring very large bioreactor vessels. For the last, the cost of effluent treatment from these processes may be disproportionate and may need new technologies.

In essence, further development work is essential to explore the potential of bioconversion processes and assess their economics.

7. References

1. Olson, G. J. and Brinckman, F. E. (1986) "Bioprocessing of coal", Fuel, 65, 1638-1646.
2. Olson, G. J., Brinckman, F. E. and Iverson, W. P. (1986) "Processing coal with microorganisms", EPRI Report No: AP-4472, Electric Power Research Institute, Palo Alto, Ca, USA.
3. Davidson, R. M. (1986) "Nuclear magnetic resonance studies of coal" , ICTIS/ IS / TR32, IEA Coal Research Report, London, UK.
4. Fakoussa, R. and Trüper, H. G. (1983) "Coal as a microbial substrate under aerobic conditions", Translation from Kolloquium in der Bergbau-Forschung GmbH Biotechnologie im Steinkohlenbergbau, Essen, pp: 41-50.
5. Wilson, B. W., Lewis, Stewart, L. I., Bean, R. M., Chess, Pyne, J., Cohen, Aranson (1985) "Microbial processing of fuels", in US DOE Coal Liquefaction Contractors Review Meeting, Pittsburg, USA, November 19-21, IV-89/98.
6. Shinn (1985) "The structure of coal and its liquefaction products: A reactive model", in International Conference on Coal Science, Sydney, Australia, October 28-31, pp. 738-741.
7. GRI Report (1983) "Microbiological and chemical studies on the aerobic conversion of coal-derived chemicals to methane", PB-83-218131 Contract Summary Report, Gas Research Institute, Chicago, IL, USA, p.23.
8. Bird, K. and Isaacson, R. (1984) "Biotechnology and the gas industry" in International Gas Research Institute Conference, September, pp:521-526.
9. Gorbaty, M., Wright, F., Lyon, R., Schlosberg, R., Baset, Z., Liotta, R., Silbernagel, B. and Naskora, D. (1979) "Coal science: Basic research opportunities", Science 206, 1029-1034.
10. Scott, C. D., Strandberg, G. W. and Lewis, S. N. (1986) "Microbial solubilization of coal", Biotechnology Progress, 2, 131.
11. Wilson, B. W., Bean, R. M., Franz, J. A., Thomas, B. C., Cohen, M. S, Aaronson, H. and Gray, E. T. Jr. (1987) "Microbial conversion of low rank coal: Characterization of biodegraded product", Energy Fuels, 1, 80.
12. Cohen, M. S. and Gabriele, P. D., (1982) "Degradation of coal by the fungi *Polyporus versicolor* and *Poria monticola* ", Applied and Environmental Microbiology, 44, 23-27.
13. Scott, C. D. (1986) "Microbial coal liquefaction", in Workshop on Biological Treatment of Coals, Washington, USA, July 23-25.

14. Krucher, R. V., Turovskii, A. A., Dzumedzei, M. V., Pavlyuk, M. I. and Khmelnitskaya, D. L. (1977) "Cultivation of *Candida tropicalis* on coal substrates", Microbiologiya, 46, 583-585.
15. Hayatsu, R., Winans, R. E., McBeth, R. L., Scott, R. G., Moore, L. P. and Studier, M. H. (1981) "Structural characterization of coal: Lignin like polymers in coal", in M. L. Gorbazy and Quchi (Eds.), Coal Structure, Am. Chem. Soc., Washington; D.C. pp: 133-149
16. Tien, M. and Kirk, T. K., (1984) "Lignin degrading enzyme from *Phanerochaete chrysosporium*:: Purification characterization and catalytic properties of a unique H_20_2 requiring oxygenase", Proc. Nat. Acad. Sci. , USA, 81, 2280.
17. Ward, H. B. (1985) "Lignite degrading fungi isolated from a weathered outcrop", Systematic and Applied Microbiology, 2, 236-238.
18. Scott, C. D. and Strandberg, G. W. (1985) "Microbial coal liquefaction", in Direct Liquefaction Contractors Review Meeting, Pittsburg, USA, November 19-21, p IV-65/87.
19. Kalliat, M., Kwak, C. Y. and Schmidt, P. W. (1981) "Small angle X-ray investigation into the porosity of coal", in Blaustein, B. D., Bachrath, B. C. and Friedman, S. (Eds.), New Approaches in Coal Chemistry, Am. Chem. Soc., Washington; D. C., pp: 3-22.
20. Yen, T.F. (1986) "Microbial actions on lignite" in Workshop on Biological Treatment of coals, Washington, July 23-25.
21. Vasseen, V. A. (1985) "Commercial microbial desulphurization of coal" in First International Conference on Processing and Utilization of High Sulfur coals, Columbus, Ohio, USA, Elsevier Science Publishers, Amsterdam, The Netherlands, pp. 699-715.
22. Leushner, A. P., Trantolo, D. J., Kern, E. E. and Wise, D. L., (1985) "Biogasification of Texas lignite" in 13 th Biennial Lignite Symposium on Technology and use of low rank coals, Bismarck, USA, May 20-23.
23. NZERDC (1985) "Biological production of methane from southland lignite" Report no: 124, New Zealand Energy Research Committee by Steven, Fitzmaurice and Partners and the Cawthron Institute. ISNN 0110-1692, p.24.
24. Melin, E., Norrbin, S. and Oden, S (1926) "Researches on the methane formation on peat", Ingeniorsvetenskaps-akademiens, Stocholm, Handlingar, 53, 5-42.
25. Leavy, P. F., Sanderson, J. E., De Riel, S. R. and Wise, D. L. (1982) "Development of bio-chemical process for the production of alcohol fuel from peat", Report no: 2201 for US DOE.DE-82016692 DOE/ER/10914-1 Dynatech R/D Co., Cambridge, Ma, USA, 93 pp.

DIRECT ASSAULTS ON WELL BEING "SELECTED CONTAMINANTS IN MARINE ECOSYSTEMS AND HUMAN ACTIVITIES AFFECTING THE ENVIRONMENT"

Turgut İ. BALKAŞ
Middle East Technical University
Environmental Engineering Department
06531 Ankara Turkey

ABSTRACT. In this paper an attempt has been made to review the levels of some of the most important inorganic and organic pollutants e.g., mercury, cadmium, lead and some important organic pollutants in open oceans, coastal zones and semi-closed seas as well as the human activities affecting the environment. According to the available pollution data reliable geographical trends of pollutants are extremely difficult to detect, given that so many intrinsic and external factors can effect measured concentrations, but the open seas is still relatively clean. Low levels of cadmium, lead and synthetic organic hydrocarbons, through detectable, are biologically insignificant. In contrast to the open oceans, the margins of the seas are affected by man around coastal zones, semi-enclosed seas, bays, estuaries. Habitats are being lost due to construction of harbours, and industrial installations, development of touristic facilities, and the growth of settlements and cities.

1. INTRODUCTION

Relative concentrations of different ecological compartments, e.g., water column are valuable in developing models of transfer from the source to target, in establishing spatial trends and possibly identifying a mechanism for toxic action. While recognition that marine pollution arises from the actions of humans, it is becoming clear that it can not be attributed solely to activities performed directly in the seas. The human activities involves a diversity of operations along the coastline, as well as the manipulation of the hydrological cycle and various land-use practices often carried out far inland; they include offshore activities such as waste disposal and marine transportation, and also the exploitation of marine resources, living and

Y. Yürüm (ed.), Clean Utilization of Coal, 109–120.

non-living. Since these activities are often associated with the production of toxic and chemical wastes, it is useful to give the emphasis to examination of distribution of chemical pollutants in the marine environment. The validity of many environmental measurements are questionable (Gill and Fitzgerald, 1984), thus there are difficulties during detection of changes and long-term trends in time and space. The biological impact of man's activities is considered, particularly the effects of wastewater discharge on human health, and the changes of inshore ecosystems caused by nutrient inputs. The increasing loss of natural coastal habitat around the world is well documented, with special reference to wetlands. Longer term problems include the possibility of subtle effects of persistent low levels of pollutants, as well as the effects of an increased ultraviolet flux due to the depletion of the stratospheric ozone layer, and the consequences of increase in the "greenhouse" gases which are expected to produce a rise in sea level and a change in climate patterns with unknown effects on various types of ecosystems (WMO, 1989). The increasing world population, its prefential settlement in the coastal zone and the resulting industrialization of that area will only exacerbate the problems of at the margins of the semi closed seas, in contrast the open oceans. These problems arise at a time of increasing environmental awareness, and it is important that broad and balanced view be taken of how to protect the environment.

2. Environmental Levels of Pollutants

2.1. Substances such as sewage, which have an adverse effect on the oxygen content and cause eutrophication

The term sewage is here used to refer to human excreta and various types of domestic wastes with or without industrial additions. It may be taken that, on the average, human beings produce 100 g of raw sewage per person per day, and that in sewerage systems the flow is 180 liters per person per day. The general effects of sewage at the marine environment may be discussed in terms of constituents of the input as inert solids, nutrient and BOD materials, toxic substances and living organisms. Inert solids, particularly inorganic materials and grit will eventually settle on to the bottom, in a manner which will be determined by the hydrographic regime and by the rate and volume of the input. Nutrients such as nitrate and phosphate stimulate plant growth, their presence in excess can result in massive blooms of algae which alter the chemistry of the water by their life processes and cause oxygen deficiency when they die and decay. Toxic substances such as synthetic organics, metals and oil may be associated with sewage in substantial amounts particularly when industrial wastes are present, and enhanced concentrations of these substances in water column and sediment can lead to accumulation in plant and animals, with potentially toxic effects. The sewage may contain a variety of living organisms, microbes or metazoans. Thus can be public health threat, by direct infection or indirectly via the food chain. Examples to the analysis of sewage sludge discharged to marine environment through direct or outfall discharge is given in Table 1.

Table 1.
(i) Average Composition of Sludge Discharged (McIntyre, 1977)

Constituent	Range Values (percentage of dried solids)
Volatile matter	65-76
Nitrogen (as N)	2.3-3.1
Phosphorus (as P_2O_5)	1.3-1.4
Fat and grease	10.7-20.9
Silicon (as SiO_2)	13.5-19.6
Heavy Metals	
Lead	500-630
Copper	320-740
Cadmium	4-14
Iron	11,400-22,900
Chromium	100-1300
Manganese	300-700
Nickel	70-100
Zinc	1,200-1,900

(ii) Average concentrations of constituents discharge from a submarine outfall (Bascom, 1980)

Constituent	Average Value ($mg\ l^{-1}$ dry weight)
Total solids	8,700
Oil and grease	710
Ammonium nitrogen	290
Total phosphorus	240
Phenols	0.41
Cyanide (CN)	0.37
Silver	0.72
Arsenic	0.22
Cadmium	0.96

Table1 (continued)

Constituent	Average Value (mg l^{-1} dry weight)
Chromium	10
Copper	13.2
Mercury	0.25
Nickel	3.4
Lead	2.42
Zinc	21.0
Selenium	0.63
	$\mu g\ l^{-1}$
DDT	3.7
PCB	20.4

Detecting, evaluating, and monitoring the effects of sewage disposal present several problems. It is often difficult or impossible to isolate effects due to such disposal from others resulting land run-off, river outflow or other dumping.

It is worth considering the ultimate fate of sewage at sea. This clearly depend on number of factors, including its decomposition and dispersion in the water column, its rate of settlement and degradation on the sea bottom, all of which will determine its eventual rate of accumulation in the sediment. Duedall et al. (1975) have pointed out that many cases although total organic carbon values are high, they are not anomalous compared with some areas outside the dump sites, and the C:N ratios are low indicating nitrogen is not limiting. Addition these studies by Grunseich and Duedal (1978) show that the availability of oxygen is the major factor controlling the decomposition rate of sewage, and they estimate that the lower limit for the residence time of sewage sludge in many locations is about 20 weeks. From the information presented here, it would appear that sewage disposal in a properly selected off-shore marine areas would normally produce unacceptable effects where as the major concern comes from coastal and shoreline or in shallow water discharges, where the near-field environmental processes do not provide adequate dilution.

2.2. Heavy metals

Oceanic pollution by metals has been primarily observed in nearshore and coastal waters (Segar and Davis, 1984) as a consequence of river, industrial and domestic sewage discharges and direct dumping of wastes. More recent data for coastal and near-shore water are presented in Table 2.

Table 2. Concentrations (ngl^{-1}) of Hg, Cd and Pb in Coastal Waters.

Location	Hg	Cd	Pb
New York Bight	10-90	3-270	-
Monterey Bay	-	9-23	7.6
Pudget Sound	0.2-1.0	70-100	20-110
Gulf of Finland	-	36-5	324-24
Northern Baltic	-	39-13	189-24
Mediterranean (Spain)	-	76-9	-
Mediterranean (Rhone Delta)	-	11-20	77
Mediterranean (Italy)	6.3-2.0	-	-
North Sea (Kattegat)	12-5	22.5	50

As with information for open waters, the majority of the reliable coastal data come from the northern hemisphere. Recent work has clearly shown that the mercury levels are not necessarily elevated in coastal waters assumed to be contaminated. For example, levels in Eastern Mediterranean Turkish coast line (0.2-1.0 ngl^{-1}) (Balkaş, 1983) are similar those measured in the open oceans (Bloom and Crecelius, 1983). However, as expected in some areas polluted estuaries and bays, large increases in mercury noted. One example is the İzmit Bay where total concentrations at the outlet are high as 35 ngl^{-1} (Tuğrul and Balkaş, 1985). Cadmium shows distinct concentration gradients decreasing from coastal to off-shore waters. A typical example is evident at the outlet of the Elbe River in the Wadden Sea where concentrations drop from more than 500 ngl^{-1} in the estuaries mixing zone to less then 100 ngl^{-1} in the Wadden Sea (Duinker et al., 1982). As for the open ocean, the major source of lead to coastal water is atmospheric input; therefore, outside of the immediate influence of dense population centers, the river input and industrial zones, lead concentrations in coastal waters are not unlike those in the open seas. However, in coastal zones adjacent to populated areas, where most measurements have been made, waters are highly polluted with lead. An example are water from the California Bight where levels ranging from 25 to 150 ng l^{-1} are attributed to sewage and rain storm run-off (Patterson and Glover, 1976).

The most recent studies on heavy metals in open ocean waters stress the importance of ultraclean sampling techniques for obtaining reliable data (Gill and Fitzgerald, 1988). In this respect, reliable data are extremely limited as can be seen from Table 3 belongs mainly to Atlantic and Pacific Oceans. The data in Table 3 indicate that in the open ocean waters inorganic mercury concentrations are often less than 2 ngl^{-1} where as cadmium concentrations in the surface waters can vary widely between 0.2-0.6 ngl^{-1}. Lead is one element whose concentrations in open oceanic waters indicate anthropogenic input from chimney emissions and combustion of leaded petrol (Schaule and Patterson, 1981). It is also one of the most difficult elements to sample and measure without introducing external combustion; thus reliable concentrations in marine waters are limited. Recent data indicate that because of industrial lead emission to the atmosphere, lead concentrations in surface waters of North Pacific and North Atlantic have been estimated to be eight to twenty times as high as those in the South Pacific and are twice as high as natural concentration (Schaule and Patterson, 1981). Furthermore, lead isotope measurements in the Northern Pacific have established that lead concentrations in the surface waters 2,000 km off North American coast originate from aeolian inputs of Asian lead, while those closer to shore are derived from North American lead (Fleagal and Patterson, 1986).

Table 3. Concentrations (ngl^{-1}) of Hg, Cd and Pb in open ocean waters

Ocean	Hg	Cd	Pb
Northwest Atlantic	0.7-0.2	2.5-0.9	33-8.1
Northeast Atlantic	1.2-0.6	11.2-4.8	33
Northeast Pacific	0.37-0.13	4.5-0.16	15-5
Northwest Pacific	5.0-0.5	14	-
Southwest Pacific	0.42	15	4.6-3.5
Arctic	2.3	14.5-8.1	14.8-3.5
Antarctic	-	54-14	-
Mediterranean	0.5-2.5	14-8	150-30
Indian	4.4-1.6	15-0.25	30
Bering	7.0-3.3	-	-
China Seas	5.7-2.3	-	-

2.3. Chlorinated Hydrocarbons

Investigations of the environmental behaviours of the chlorinated compounds have provided a series of scientifically unexpected surprises. Perhaps the most striking result was the discovery that the principal mode of transport of DDT and its metabolites and the PCB's from the land to the oceans appears to be in the vapour phase. The vapour pressures of these substances are in the order 1.5×10^{-7} mm of mercury for PCB's. Such low vapour pressure initially appeared irreconcilable with the widespread occurrences of these compounds. As in the case of heavy metals reliable data for chlorinated hydrocarbons very limited. Comparison of the data must be made with caution since the techniques used by the various investigators in most cases have not been intercalibrated, and with very low concentrations reported (ppt), there is a possibility that some of the data are unreliable due to sample contamination or other analytical errors. DDT residues are most frequently reported compounds in coastal waters, with concentrations generally below 5 ngl^{-1}. Notable exceptions are the high levels reported near Marseille and Blanca Bay, Argentina (>100 ngl^{-1}).

Examination of organochlorine compounds in open ocean water far from any point source lead to a better estimation of global contamination. Average PCB concentrations around 1 ngl^{-1} in open oceanic waters. This values are generally lower than those reported for coastal waters. On the basis of available information, there is clear evidence that PCB levels in surface waters of the northern hemisphere are higher than those in the southern ocean which typically have concentrations of 0.035-0.072 ngl^{-1}.

3. Human Activities Affecting the Marine Environment

3.1. Disposal of plastic litter

It is scientifically well established fact that plastics do not degrade as rapidly at sea as on land (Andrady, 1988). Ultra-violet light from the sun causes most plastic to embrittle and eventually disintegrate. However, sea water filters the ultra-violet light, thus weakling its effect. Weather also keeps the plastic litter cool and reduce thermal degradation. If algae or other marine organisms cover the plastic litter surface, this further slows the rate of degradation. If added weight of the fouling organisms causes the plastic to sink, the degradation will be even slower because of cooler water. The quantities of plastic litter dump to marine environment are well documented. A Mediterranean study revealed that 60-70 % of surveyed debris was plastic (Morris, 1980). The benthic sediment survey off coast of Unite Kingdom showed, 2,000 pieces of plastic per square meter, and beach surveys in New Zealand even greater concentration of plastic on its beaches (Coleman and Wehle, 1984). A potentially significant problem may be created by the increasing quantities of small plastic particles in the ocean. Pellets or spherules of raw plastic used in productions, and insulation and packing material, enter to seas from land based sources such as outfalls, from rivers, and from ships. The major threat of this type of pollution is ingestion, by fish and sea-birds, which can effect the animals feeding and digestive processes. There are numerous studies of turtles, whales and other marine mammals which were apparently killed by ingesting plastic bags, sheeting, and containers. The turtles may mistake it for one of their normal food resources (e.g. jelly fish). In a recent study, plastic particles were found in the digestive tracts of 25 % of the world's sea-bird species. In 1975, U.S. National

Academy of Science estimated that 6.4 million metric tonnes of litter were being discarded annually by shipping activities only and the recent figure must be very much higher than 1975 estimate.

3.2. Development of coastal areas

The coast line is a complex region comprising bay, estuaries, and large semi-enclosed areas where human population and industrial development are concentrated. It is also obvious that, with some exceptions, most coastal development and urbanization occurs around principal coastal cities, ports, and harbours. The nature of ports varies; there are large ports with various of development, much which serves international commerce, and there are smaller ports which tend to function in local commerce. All ports however, are generally dependent on a wide range of services and support facilities. Highways and roads, railways carry commerce and commodities to and from ports. All kind of electric producing power plants, nuclear or fossil-fuel are almost invariably located on harbours, local rivers, or nearby coastal waters. In harbours the accumulation of leaking or released cargoes, discharged wastes, and eroded sands and silts require regular dredging with results that dredge spoils must be deposited in other habitat. When dredge materials settles after sea disposal, it normally returns to an anaerobic and near neutral pH conditions. On the other hand, if this dredge material were placed in an upland contaminant area where oxidizing conditions could occur for a year or more, and if dredge material where high in total sulphide (common in many fine Mediterranean and Black Sea sediments), the pH could become acidic and result in significant release of some pollutant to water soluble phase. Therefore, careful selection of dumping sites, selection of disposal mode and understanding of the long term implication are very important. Similar to other regions, siltation due to erosion, transport of mine tailings and dumping of dredge materials are issues important in the Mediterranean Sea. Several recent literatures commented on how the changes in land use patterns were affecting coastal water quality. For instance near the Black Sea agricultural lands were rapidly being industrialized and urban centres concomitantly grew in size. This was true in Bulgaria, Romania, and Turkey (Balkaş, et al., 1990). Even the finest and most valuable Turkish tobacco growing areas and coastal waters are now industrialized and affected by industrial and urban pollution. The Brundtland report (Brundtland, 1987) emphasized sustainable development. Industrial operations must be conducted in a far more efficient manner in terms of resource use, generating less pollution and wastes, using renewable rather then non-renewable resources, and using less energy. Parallel to this, the scientific community has began to measure the chemical and physical effects resulting from development and to use the data so obtained to asses the events caused by several stresses. A factor most important in managing development, economic cost, is now being dealt with more systematic, realistic manner. Using spills of oil or other hazardous materials as a starting point one can show the importance of the use of economic analyses in valuing natural resource damages.

3.3. Sea Level Rise and Climate Change

Human activities are changing the composition of the atmosphere. In particular the burning of fossil-fuel, which helps meet the world's energy needs, emits carbon dioxide, one of the so-called "green house gases" into atmosphere. The increase of atmospheric carbon dioxide and other green house gases has raised concern about the possibility of induced climate variations during next decades and about their associated impacts.

Substantial scientific research is now being directed towards assessing the extent and nature of these changes. Information collected indicates that the global amounts of carbon dioxide, chlorofluorocarbons, methane, and nitrous oxide have been increasing. These gases are all transparent to incoming short-wave radiation, but absorb and emit long-wave radiation, and increased concentrations can lead to a warming of the Earth's surface and of the lower atmosphere. Also it is necessary to study separately the effects of earth of these gases in order to estimate their relative contributions to the warming at any given time. So far, the main effort has been directed to carbon dioxide. However, because the atmospheric concentrations of the other greenhouse gases are increasing much faster, their relative impact on climate change will soon approach that of carbon dioxide levels to occur if these reserves continue to be exploited in the near future. Atmospheric carbon dioxide has been rising at the rate of about one parts per million per volume (ppmv) per year starting with 315 ppmv in 1958 when good measurements became available. Uncertainty regarding the future increase of atmospheric carbon dioxide centers not only on how much carbon dioxide gas released to atmosphere (about 5 Gt C y^{-1}) and released by biomass destruction-primarily deforestation. At present the atmospheric load of carbon dioxide is increasing by about 2.5 Gt C y^{-1}. It has generally been agreed that neither regional patterns of climate would affect ecosystems and human activities can yet be predicted. It is important to note that climate models indicate that the increase in global mean equilibrium surface temperature due to increases in carbon dioxide and other green house gases equivalent to a doubling of the atmosphere carbon dioxide concentration is likely to be in the range of 1.5 to 4.5 °C. However, the observed increase in global mean temperature of 0.3-0.7 °C over the last hundred years can not be ascribed in a rigorous manner to increasing concentrations of carbon dioxide and other green house gases. The global sea level rise is estimated to have risen some 12 cm. during the present century. On the basis of the observed changes since beginning of this century, it is forecast that a global warming of 1.5 to 4.4 °C will cause the mean sea level rise by 20 to 140 cm.. The major contributing factor to such rise would be the thermal expansion of ocean water. The range of estimates is an indication of their uncertainties. Half of the world's population dwells in coastal regions which are already under great demographic pressure, and exposed to pollution, flooding land subsidence and compaction, and to the effects of upland water diversion. Arise in sea level would have its most sever effects in low-lying coastal regions, beaches and wetlands. In developed countries protection for some regions will be possible, whereas in developing countries without adequate technical and capital resources it may not be. A number Pacific and Indian islands with a maximum altitude of few meters are especially vulnerable and could become unhabitable after rises of the sea level that could hardly be noticed elsewhere.

4. Pollution Control and Prevention

From above discussions it is obvious that pollution should be prevented or controlled, but it is not always clear how much effert and financial resources are justified, what mechanisms are most effective, and what time scale is appropriate. Problem arise in particular when:
- there is an evident cost of the proposed action;
- there is insufficient basic scientific knowledge of pollutants, target or dose-response relationship;
- there is restriction of sources to take a preferred option;
- there is little experience of consequences of alternative options such as; production and disposal and;
- there is inadequate technology available to achieve satisfactory control and prevention.

None of the less common among the modern society, particularly when there is an underlying conviction that something must be done, has led to effective action being taken. However, debate continues between those calling for a total ban of the offending activity, or zero discharge regardless of costs, and those seeking to improve the control of discharge to the environment of potentially harmful substances. The pragmatist would be favour a solution that is effective even if not absolute, rather than one that in practice is difficult to implement (Balkaş et al., 1991).

Over the past two decades, there has been a gradual development of practical control strategies at national and international level. The preferred approach varies with circumstances of the case, and each has its advantages and disadvantages. This approaches, however, have many aspects in common and all have the objectives of reducing pollution in an effective and economic way. Here one can use the concept of environmental capacity. The environmental assimilative capacity is defined as a property of the environment which measures its ability to accommodate a particular activity, or rate of activity without unacceptable impacts. The capacity is finite and quantifiable for any particular site, expected pollutant input and activity range, on the basis of physical and chemical and biological characteristics.

References

Andray A. (1988) The use of enhanced degredable plastics, Proceedings of North Pacific Rim Fisherman Conference on Marine Debris., Hawaii, October, 1987.

Balkaş T.İ. et al. (1982) "Trace metal levels in fish and crustace from Northeastern Mediterranean coastal waters", Marine Environmental Research,4,6.

Balkaş T.İ. et al. (1983) "The methylation of tin in the marine environment", Marine Pollution Bulletin 8, 14.

Balkaş T.İ. et al. (1990) State of Marine Environment in the Black Sea Region, UNEP Regional Seas Reports and Studies, No.24.

Balkaş T.İ., Chung C., and Yetiş Ű. (1991) " Integration of Environmental Consideration to Costal Zone Management", OECD Publication, Paris.

Bascom W. (1980) The effects of sewage disposal in Santa Monica Bay, Coastal Research Project Bienneal Reports, California.

Bloom N.S. and Crecelius (1987) "Determination of mercury in sea water at subnanogram levels, Marine Chemistry, 21.

Brundtland G.H. (1987) Our Common Future, United Nation World Commission on Environment and Development.

Coleman F.C. and Wehle D.H.S. (1984), Plastic pollution, A Worldwide Problem Parks, 9.

Duedal I.W. et al. (1975) "Fate of waste water sludge in New York Bight", Journal of Water Control Federation.

Duinker J.C. et al. (1982) "Processes affecting the behaviour of metals and organochlorine compounds during estuarine mixing", Neth. Journal Sea Research, 15.

Flegal A.R. and Patterson C.C. (1986) " Vertical profiles in the Central Pacific Ocean", Nature, 321.

Gill G.A. and Fitzgerald W.F. (1984) "Mercury sampling of open ocean waters at the picomolar level", Deep Sea Resarch, 32.

Grunseich S.S. and Duedall I.W. (1978) "The decomposition of sewage sludge in sea water", Water Research.

McIntyre A.D. (1977) A review of the effects of the disposal of sewage sludge to sea, Tech. Note No.6. Water Engineering I. Research and Development Division, Department of Environment, London, U.K.

Morris R.J., (1980) "Floating plastic debris in the Mediterranean", Marine Pollution Bulletin, 1.

Patterson C.C. and Glover B.(1976), "Analysis of lead in polluted coastal seawaters", Marine Chemistry, 4.

Schaule B.K. and C.C. Patterson (1981) " Lead concentrations in the Northeast Pasific, Evidence for global antropogenic perturbations", Earth Planet, Sc. Letters, 54, 1891.

Segar D.A. and Dawis P.G. (1984) "Contamination of populated estuaries and adjacent coastal oceans", A Global Review NOAA Technical Memo, NOS OMA 11, Rockville Maryland, USA.

Tuğrul S. and Balkaş T.İ., (1986) İzmit Bay Case Study "The Role of the Oceans as a Waste Disposation", D: Reidal Publishing Comp.

WMO (1989), Global atmospheric background monitoring for selected environmental parameters, BAPMon Data for 1989 Volume I, Atmospheric Aerosols Optical Depths.

UNDERSTANDING of ENVIRONMENTAL DISRUPTION "NATURAL AND POLLUTANT FLUXES, SIMPLE MASS BALANCE MODELS"

Turgut I. BALKAŞ
Middle East Technical University
Environmental Engineering Department
06531 Ankara Turkey

ABSTRACT. An understanding of environmental disruption and the present distribution and abundance of pollutants and contaminants in the various compartments of the environment in particular in the hydrosphere, and prediction of the future distributions and abundance demands a knowledge of the routes, receiving media and the reaction of the substances concerned. In this paper an attempt will be made to elaborate the environmentally important processes such as the major transport paths, fluxes, and residence time of pollutants.

1. Introduction

In our time man's fingerprint is found everywhere, including the hydrosphere, biosphere and the atmosphere. Wide range of activities on land contribute to the release of pollutants to the environment either directly or carried by rivers and the atmosphere. Chemical contamination and litter can be observed from the poles to the tropics, from land surface to the limits of the lower atmosphere and from the coastal zones to abyssal oceanic depths.

Many thousands of substances enter the environment as a consequence of material usage and energy production by human society. Some, such as tributhyltin, DDT, pesticides and dieldrin and artificial radio-active materials produced in nuclear reactors, are alien to the natural environment. Others already exist in the natural environment but their concentrations are altered by man's activities. Only a few of these many compounds are likely to produce unwanted consequences such as changing climate and sea water temperature and the structure of marine communities of organisms or the loss or restricted use of nonliving resources.

Y. Yürüm (ed.), Clean Utilization of Coal, 121–130.

Throughout history, the quality and quantity of water available to man have been vital factors in determining his well-being. Whole civilizations have disappeared because of water shortage resulting from changes in climate. Ambitious programmes of dam construction have reduced flood damage, though not without undesirable side effects. However, problems with water remain, and in some respect are becoming more serious. He is reminded of the problems of water quality by the new "No Swimming" sign. The world's water supply may be found in five places in the hydrologic cycle. A large portion of the earth's water supply is found in the oceans. Another fraction is present as water vapour in the atmosphere. Some water is contained in the solid state as ice and snow in snowpack, glaciers, and in the polar ice caps. Surface water is found in lakes, rivers, streams and reservoirs. Ground water is the water found underground.

2. Understanding of Environmental Disruption: Pollution Dynamics

The time scale of contaminants in a particular environment such as sea water or other geological domains may be estimated by means of simple mathematical models in which the geological domains are depicted as a vast reservoir for materials mobilized by winds, rivers, glaciers, and large outfalls. Substances taking part in these transport processes are usually considered to have time invariant concentrations in the receiving environments such as the oceans and atmosphere as well as the biosphere (Chester, 1990). In the case of the oceans the amount of the material introduced to the waters in any given time period is compensated by the loss of like amounts undergoing destruction through such processes as microbial activity, radio-active decay or precipitating to the sediments. Thus, the time a contaminant spends in a reservoir such as ocean waters, the atmosphere or biosphere is given by the concentration in the reservoir divided by the total flux in or out of the reservoir. This time period is called the residence time. The simplest schematization is given in Figure 1 where the world oceans are considered as a single reservoir for continentally derived materials.

In more complicated models world oceans are divided into parts, such as the South Atlantic, the North Pacific Indian Ocean, Antarctic, etc. Further each basin is also subdivided into two components, a mixed layer and a deep layer. The mixed layer corresponds to the zone in which the primary production of organic matter takes place through photosynthesis. In all of these models the assumption is made that there is a complete mixing of the substances within a reservoir in a relatively short time period compared to the residence time.

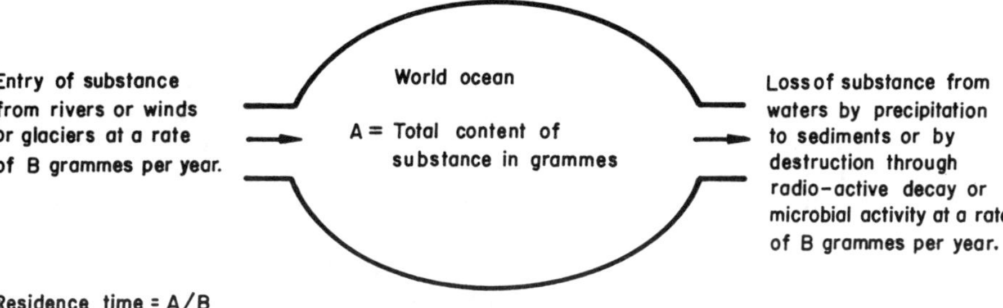

Figure 1. Schematization of Steady State World Oceans

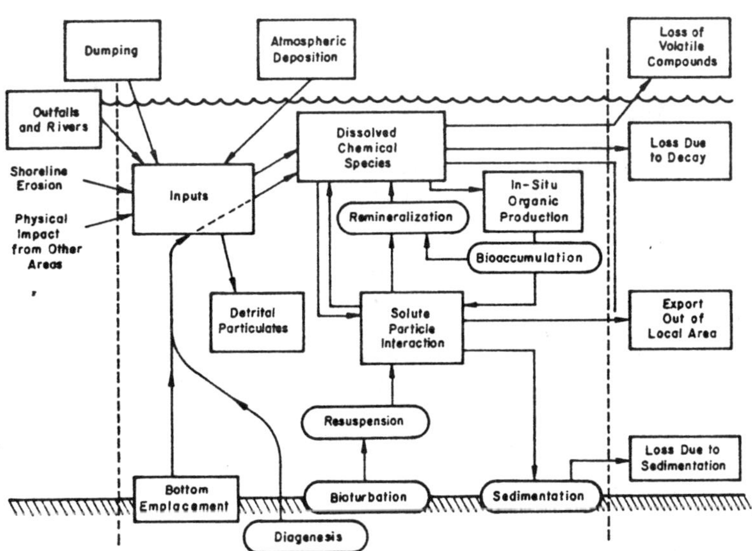

Figure 2. Possible Transformations of Pollutants in the Water Column

3. Environmental Models and Mass Balances

Any contaminant introduced into a sea water column will undergo a series of chemical reactions or transformations that will not only govern its transport through the water column but also affect its toxicity to marine organisms. Figure 2 summarizes the possible chemical and related process or transformations in which a contaminant may be involved from its time of introduction until it is removed from the system. Models of this type provide a dynamic description of a pollutant's dispersion in the environment. They serve as a basis for the development of monitoring strategies. Several characteristics of these models are important in pollution assessment studies. First of all, the residence time of a chemical is inversely related to its reactivity in the reservoir, so that persistent chemicals will be removed more quickly either by breakdown or by precipitation to sediments. Secondly, the time required for a substance to reach a steady- state concentration in a reservoir is given by a period of four residence time. Conversely for a contaminant whose supply to a reservoir is cut-off after a steady state condition is reached, it will take a period of four residence time for its concentration to fall off to less than one percent of its steady state value. A combination of these two observations may allow the prediction of contaminant levels in a reservoir where the residence time of a naturally occurring substance or of a previously studied pollutant with similar chemical characteristics is known (Bender et al, 1979).

The amount of material released to the environment can sometimes be estimated from the production and use data. For example, the gaseous chlorofluorocarbons used primarily as aerosols and refrigerant. Since virtually the total production is released to the atmosphere, knowledge of the production figure gives the atmospheric flux. On the other hand, a knowledge of the annual world production of crude oil does not lead to an estimate of losses to the environment, since the amount lost has not been adequately determined. In addition to the production and use statistics, the geographical sites of production and use are important in the formulation of mass-balance models. For example, nearly all energy production and material utilization takes place in the mid-latitudes of the northern hemisphere. This can be seen through an examination of the geographical distribution of the gross national product. The assumption is made that the energy production and material utilization are directly related to gross national product. One can make the additional assumption that transfer to the environment of material wastes and combustion products from energy production are also proportional to the gross national product, then a latitudinal pollution band can be defined. In certain cases of formulating a global study of the atmospheric dispersion of a synthetic chemical, emphasis should be placed on sample collection from northern rather

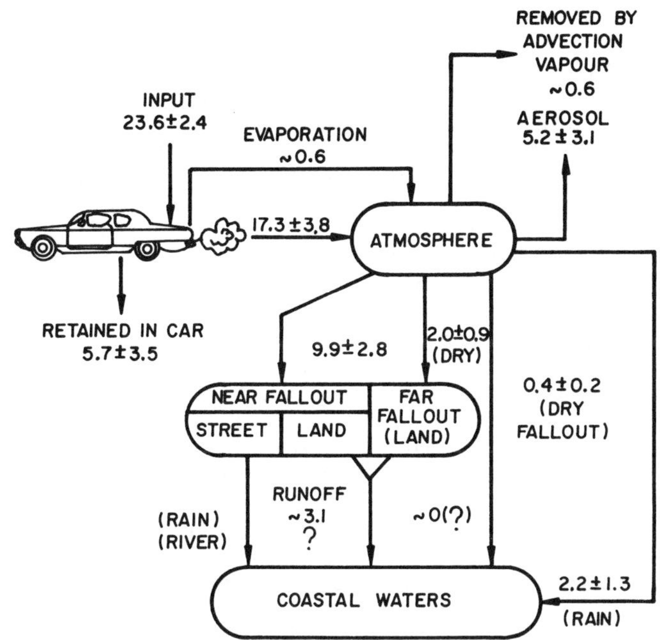

Figure 3. Assessment of the impact of a Pollutant on a Defined Environment "Lead Model"

than the southern hemisphere, especially where steady-state conditions have not been obtained (Global Ocean Flux Study, 1984). The flow of pollutants from land to the coastal zone and finally to the open sea is the main concern of environmental management studies. An example to this is the flow of lead through the Los Angeles area to the nearby coastal and open ocean (Huntzicker et al., 1974). In this model the impact of lead, which originates from traffic, on the coastal zone is investigated. The model is shown in Figure 3 where pollutant lead is assumed to result entirely from the combustion of tetra ethyl lead. According to the investigations in the Los Angeles area 23.6 tonnes were consumed per day during the 1970's. Of this 17.9 tonnes emitted from cars as exhausts to the atmosphere. 17.3 tonnes of this amount was released as particles and the rest as vapours (presumably as alkyl leads); 5.7 tonnes were accumulated in car engines, exhaust systems, etc. The analysis of aerosols exhausted gave the following size distribution:

> 9 μm, 57 percent; 0.3 - 9μm, 27 percent; < 0.3 μm, 16 percent.

However analysis of air samples collected from Los Angeles area showed only 2 percent of the lead particles greater than 9μm. As a result, 9.9 tonnes of was assumed to have been deposited near the roads (near fallout). Of this amount 8.4 tonnes fell directly upon to the road and 1.5 tonnes on the land next to the road. Far fallout was defined as the lead deposited at locations distant from the source. Carbon monoxide was as a tracer in order to determine the rate of removal of lead from the area. Air born carbon monoxide and lead derives entirely from exhaust emissions, co-vary in the atmospheric concentrations. On this basis it has been estimated that 5.2 tonnes of lead per day were removed by convection. Direct measurements of dry fallout on the coastal zone gave value of 0.4 tonnes per day whereas measurements of rain from the area indicated a daily input to the coastal water of 3.1 tonnes. Thus the total amount of lead entering the coastal waters is about 5.7 tonnes from a daily combustion of nearly 24 tonnes.

4. Pollution Transport and Fluxes

Materials reach various sections of the environment from many sources including inflow from rivers and ice melt, transfer from the air by fallout, volcanic activities, weathering of continental rock and direct releases by man as treated or untreated sewage and dredge dumped materials. Estimates of fluxes of materials through the major sedimentary cycle are presented as Table 1.

Table 1: Fluxes in the Sedimentary Cycle

Material	Geosphere	Receiving	Flux in 10^{14}g/year
Suspended Solids	River	Oceans	180
Dissolved Solids	River	Oceans	39
Glaciers		Oceans	30
Sea Salt		Atmosphere	3
Continental Rock Soil Particle		Atmosphere	< 1-5
Volcanic Debris		Stratosphere	0.036
Volcanic Debris		Atmosphere	1.5
Wastes of Society (excluding fossil fuels)		Atmosphere, Hydrosphere and Lithosphere	30
Wastes Dumped from Ships		Oceans	1.4
Industrial Particulates total		Atmosphere	0.54
Carbon, Fly Ash from fossil fuel combustion		Atmosphere	0.25

As can be seen from Table 1 the main mover of materials from land to seas is the system of rivers. Transportation of natural materials is about ten times that of glaciers and about hundred times that of atmospheric transport. Much of the river particulate load is deposited on the shelf region of the seas. Most of the large rivers flow into the Atlantic and Pacific Oceans, compared to the Atlantic the Pacific Ocean receiving only a modest amount of river-carried pollutants and contaminants. The Amazon river accounts for more than ten percent of the total and drains into the equatorial Atlantic where rivers such as the Yellow river enters marginal basins where much of its particulate load is trapped. Many other rivers enter marginal seas, bays or estuaries before entry to oceanic coastal waters. Many rivers can no longer be considered to be in their natural states with regard to their particulate loads. Dissolved materials and solids containing contaminants and pollutants held in suspension enters rivers in the course of weathering and erosion of rocks and soils and as a result of human activity. The tendency of man to settle on river banks, defensive advantages, transportation routes and assuring himself of a steady water supply, has inevitably led to input of man-made materials and wastes into the river waters. The growth of cities, industries and agriculture has gradually increased this human input, thereby also increasing to some extent, inputs of various materials, the amount carried through the atmosphere is relatively small and the contribution through the direct water run-off from the coastal zone and from ground water is unknown (GESAMP,1982).

The mean discharge of the major rivers of the world are listed in Table 2 (River Inputs to Ocean Systems, 1980).

Table 2. Major Rivers of the World.

River	Mean Discharge m^3s^{-1}
1. Amazon	175,000
2. Congo	39,600
3. Orinoco	33,950
4. Yangtse Kiang	22,000
5. Brahamaputra	19,200
6. Missisipi	17,800
7. Yenissei	17,800
8. Lena	16,300
9. Parana	14,900
10. Mekong	14,600
11. Saint Lawrance	14,160
12. Irrawady	13,450
13. Si Kiang	12,500
14. Ob	12,200
23. Danube	6,530
38. Rhine	2,535
60. Tigrus+Euphrates	1,450
200. Grey	294

As can be see from Table 2 these rivers drain about 40 % of the global land surface and in terms of water discharge, the 10 biggest rivers account for 73 % of the total discharge. The mean discharge values of these rivers can also be affected by diversions of river flow for development purposes, to make artificial lakes, and industry and power plant cooling.

Diversions of these kind have enormous effects on the mass balance of materials delivered to the seas and, possibly, they could change the local climate of the areas, the water balance of which has been perturbed (e.g. Nile, Volga, Don, Ingulala, Colarada, and the Tigrus-Euphrates river system). Recent environmental changes in the Black Sea may be a good example of this observation. The manipulation of the Black Sea rivers especially of the Dneper, has not only decreased discharges to the Black Sea but has also significantly altered seasonal flow patterns. The short and intense spring flood now extends over a longer period but has a lower intensity (Balkaş et al, 1990). As a result there is now a nearly permanent stratification in shallow coastal areas and estuaries that strongly precludes downward transport of dissolved oxygen. Following the significant reduction in the discharges of the rivers mass mortalities of major species within the subhalocline waters occurred in large water masses (Tolmazin, 1985). The area affected extended from the mouth of the Dnestr to Danube delta.

The principal wind systems tend to carry pollutants and contaminants along the lines of latitude, although they do meander to the north and to the south from year to year. The amount of time involved in atmospheric transport of materials may be gained from the observation that radio-active contaminants introduced to the troposphere by a Chinese nuclear test at Lop nor in May 1965 encircled the earth in about three weeks with an average velocity of 16 m/s (Cooper and Kruroda, 1966). The three principal wind systems responsible for the transport of various materials including pollutants to the environment are:

(i) The polar easterlies are near surface winds which occur at 70° - 90°N and 65° - 90°S. They decrease in intensity with height and at about 3 kilometres they often reverse their directions.

(ii) The trades which blow between 30°N and 30°S decreasing in intensity with increasing latitude. They blow from east to west. Reversals in direction can take place in the upper troposphere, except near the equator where mid-latitudinal westerlies, the jet streams, prevail between 30°S and 70°N and between 30°N and 65°S, often with greater intensity in the upper troposphere than in the lower troposphere. Wind speeds of over 100 m/s are common in their central parts.

In addition to these three systems, there are the continental monsoons whose flow are determined by continent-ocean temperature difference. They can reverse their direction between summer and winter. As can be easily concluded from the above discussions the amount of material including a variety of pollutants moved about the various section of the environment by the activities of man is about one-tenth of that involved in the major weathering cycle. The rivers carry, primarily to the continental shelf regions, around 20 billion tonnes of dissolved and solid material annually. Glaciers transport around 3 billion tonnes per year of crustal materials to the oceans, primarily in the Arctic region. The wind carries between 0.1 and 0.5 billion tonnes of rock flour and soil materials, and types of contaminants per year excluding the combustion products of fossil fuels. The discards of society are about 4 billion tonnes per year or more than 1 cubic kilometre. A substantial amounts of these discards eventually reaches to land and ocean systems.

Naturally occurring solids are dispersed to the atmosphere as a result of volcanic activity, wind action upon exposed crustal materials, and from the injection of salt from the oceans. The fluxes of sea salt and of rock-soil materials are 0.3 and 0.5 million tonnes per year respectively. These figures are higher by about a factor of ten than man's emissions from industrial activity which are 0.012 million tonnes per year for small particles subject to long-term transport and from fossil combustion resulting in 0.025 million per year. These fine particles originate principally from the production of iron, steel, cement and crushed stone. The oxides of silicon, iron, calcium, aluminum, magnesium and calcium carbonate are dominant materials.

5. References

Balkaş T.İ., et al., State of the Marine Environment of the Black Sea Region, UNEP Regional Seas Reports and Studies, No. 124, 1990.

Bender M. et al. (1979) Proceedings of a Workshop on "Assimilative Capacity of U.S Coastal Water for Pollution, US Department of Commerce.

Chester Roy. (1990) Marine Geochemistry, Academic Division of Unwin Hymann Ltd., London.

Cooper W.W., Kurudo P.K. (1964) " Global Circulation of Nuclear Debries from the May 14 1965 Nuclear Explotion", J. Geophy., 71, 5471.

Global Ocean Flux Study. (1984) Proceeding of a Workshop, National Academy Press, Washington D.C.

Huntzicker J. et al. (1975) "Material Balance for Automobile Emitted Lead in the Los Angeles Basin", Environmental Science and Technology, 9,448.

The Health of the Ocean, GESAMP Review Reports, UNEP, 1982.

Tolmazin D., Changing Coastal Oceanography of the Black Sea, Prog. Oceanogrp., 15, 1985.

River Input to Ocean Systems. (1981) Proceeding of a SCOR/ACMRR/ECOR/IAHS/UNESCO/CMG/IABO/IAPSO Rewiew and Workshop, UNEP and UNESCO Publication.

ASSESSMENT OF ENVIRONMENTAL DAMAGE BY FOSSIL FUELS

F. Barbir and T. N. Veziroğlu
Clean Energy Research Institute
University of Miami
Coral Gables, FL 33124
U.S.A.

ABSTRACT. The objective of this study is to identify the negative effects of the fossil fuels use and to evaluate their economic significance. An economic value of the damage for each of the analyzed effects has been estimated in US dollars per unit energy of the fuel used ($/GJ). This external costs of fossil fuel use should be added to their existing market price, and such real costs should be compared with the real costs of other, environmentally acceptable, energy alternatives, such as hydrogen.

1. INTRODUCTION

Energy is the basis of the economy and for the economic development. Most of the world energy demand is met by fossil fuels today. However, technologies for fossil fuel extraction, transportation, processing, and particularly their end use (combustion) have harmful impacts on the environment, which cause direct and indirect negative effects on the economy.

Excavation of coal devastates the land, which has to be reclaimed, and is out of use for several years. During the extraction, transportation and storage of oil and gas, spills and leakages occur, which cause water and air pollution. Refining processes also have an environmental impact.

Most of the fossil fuel environmental impact occurs during the end use. The end use for all fossil fuels is combustion, irrespective of the final purpose (i.e., heating, electricity production or motive power for transportation). The main constituents of fossil fuels are carbon and hydrogen, but also some other ingredients, which are originally in the fuel (e.g., sulfur), or are added during refining (e.g., lead, alcohols). Combustion of the fossil fuels produces various gases (CO_x, SO_x, NO_x, CH), soot and ash, droplets of tar, and other organic compounds, which are all released into the atmosphere and cause air pollution. Air pollution may be defined as the presence of some gases and particulates,which are not a natural constituent of the atmosphere, or even presence of the natural constituents but in an abnormal concentra-

Y. Yürüm (ed.), Clean Utilization of Coal, 131–152.

tion. Air pollution causes damage to human health, animals, crops and structures, and reduces visibility.

Once in the atmosphere, triggered by sunlight or by mixing with water and other atmospheric compounds, the primary pollutants may undergo chemical reactions, change their form and become secondary pollutants, like ozone, aerosols, peroxyacyl nitrates, various acids, etc.

Precipitation of sulfur and nitrogen oxides, which have dissolved in clouds and in rain droplets to form sulfuric and nitric acids is called acid rain; but also acid dew, acid fog, and acid snow have been recorded. Carbon dioxide in equilibrium with water produces weak carbonic acid, but below pH 5, carbonic acid has no further acidifying effect [1]. Acid deposition (wet or dry) causes soil and water acidification, resulting in damages to the aquatic and terrestrial ecosystems, affecting humans, animals, vegetation and structures.

The remaining products of combustion in the atmosphere, mainly carbon dioxide, together with other so called greenhouse gases (methane, nitrogen oxides and chlorofluorcarbons), result in thermal changes by absorbing the infrared energy the Earth radiates back into the atmosphere, causing global temperature increase. The effects of the temperature increase are melting of the ice caps, sea level rise and climate changes, which include heat waves, droughts, floods, stronger storms, more and bigger wildfires, etc.

The cost of the above described negative effects are not included in the market price of fossil fuels. These costs are paid by the society, and/or eventually will be paid by the society, since in the long term any disturbed ecosystem will affect the human society, its environment and its economy.

The objective of this study is to identify the negative effects of the fossil fuel use and to evaluate their economic significance. The damage estimates have been based on the reported data by various researchers. In cases where no estimates exist, an attempt to evaluate economic effects has been made. In these estimates the following approaches or methods were used, depending on the character of the damage and data availability:

(a) direct damage estimate in proportion to the physical effect,
(b) expenditure estimate of the damage abatement, either as a remedy or as a prevention,
(c) "supplement" method, or the estimate of the costs of implementing the substitutions for the damaged category,
(d) "property value" method, or the estimate of the change of the property value for a given change in environmental quality.

The damages were estimated as the annual costs in US dollars for the year for which data were available, and normalized to a reference year (1990), using implicit deflators. Then damage per unit energy ($/GJ) has been calculated for each type of damage and for each type of fossil fuel consumed. The extension of such damages represents the external cost of fossil fuel use and should be added to the existing market prices of fossil fuels. In this way, the real costs of fossil fuels (which are also called societal costs, because they represent the real price which the society pays for the use of fossil fuels) have been calculated.

2. REPORTED DAMAGES AND THEIR ESTIMATES

Reported environmental damages on various components of the biosphere will now be discussed and evaluated.

2.1. Effects on Humans

It is very difficult to measure health effects of air pollution, because in an exposed population extreme consequences such as clinical disease and death occur only among a small proportion of those exposed, whereas a much larger number of people may have less severe symptoms or physiological changes that can only be measured by specific tests. Different effects can also occur in an individual patient at different times, and it often takes time for a chemical pollutant to accumulate in the body to the extent that the more severe effects occur. In the meantime, more subtle effects may have started but may go unnoticed. Each pollutant (CO, SO_x, NO_x, HC, particulates) has harmful effects on the human organism, particularly on respiratory organs, causing discomfort, disease and even death. These effects can be even more significant because of synergistic effects between various pollutants and the secondary pollutants they generate (e.g., ozone, acid aerosols, acid fog, acid rain).

Many epidemiological studies show higher rates of premature deaths in polluted areas. According to the American Lung Association about 115 million Americans (which is almost 50% of the population) continue to be exposed to air pollution levels exceeding federal health standards [2]. It has been assessed that sulfate and other small particle levels were causing 50,000 premature deaths each year in North America [2]. American Cancer Society and the Chemical Manufacturers Association estimate that 2% of all cancer deaths are caused by pollution [3].

An analysis of the statistical data for the period from 1960 to 1985 shows significant increase in death rates for deaths caused by malignancies of respiratory and intrathoracic organs and by chronic obstructive pulmonary diseases. Although total death rate per 100,000 population decreased from 954.7 in 1960 to 873.9 in 1985, the death rate due to above mentioned diseases increased from 32.1 to 84.6 per 100,000 population in the same period [4]. Taking into consideration that medical services have improved, one may assume that increasing air pollution, and its cumulative effects, are the only plausible reason for such increased death rate. That results in 125,000 premature deaths in the U.S. in 1985 due to air pollution. Assuming the cost of a life as projected lost wages (i.e., $300,000 to $500,000 [3]), the damage to the society becomes as high as $50 billion per year.

However, the premature death is only the most drastic effect of air pollution on humans. The expenditures for healing of the fatal diseases of respiratory organs as well as other non-fatal diseases, poisoning, eye and skin irritation, even skin cancer, must also be accounted for. This could be taken as at least 10-12 percent of the total health expenditures [4], which were $422.6 billion in the U.S. in 1985 [5], including private and public expenditures, health services, medical supplies, medical research and medical facilities construction, resulting in another $50 billion damage per year.

The least measurable harmful effects of air pollution on humans are sicknesses not reported, less severe symptoms or physiological changes, or just discomfort, which certainly affect the working efficiency. Even a reduction of 2% of the working efficiency [4], which can be estimated as 2% loss of income, accounts for $47.4 billion (wages and salaries were $2,370 billion in the U.S. in 1985 [5]).

Because of the foregoing discussion, the total harmful effect of air pollution on humans in U.S.A. has been estimated to be $50 + $50 + $47.4 = $147.4 billion in 1985 $.

2.2. Effects on Animals

The various pollution elements emitted into the atmosphere certainly affect animals in a manner similar to their effects on human beings. However, so far no losses or reduction of livestock have been recorded or estimated.

Damage on farm livestock caused by air pollution could be estimated as 10 percent of the farm income on all livestock and animal products [4], which was $70 billion in the U.S. in 1985 [5]; hence the damage comes to $7 billion.

2.3. Effects on Farm Produce, Plants and Forests

Air pollution and acid precipitation contribute to plant injury and cause significant crop losses. Plants can be affected either directly, by exposure to sulfur oxide, nitrogen oxides and ozone, or indirectly by soil acidification. Ozone fumigation studies have consistently reported reduced yield from the exposure to ozone for all crops. Response models indicate that the ambient levels of ozone across the most of the agricultural regions of the U.S. are high enough to have measurable impact on crop yield. The annual effects on crop production from ambient ozone have been estimated as the 5-10% loss of the actual yield [6]. Other studies also showed considerable productivity losses even without visible symptoms of damage. Test plots in North Carolina, New York and California have shown yield losses from 15% to greater than 50% on crops common to each region [7]. The available information from scientific research has established no measurable and consistent crop yield response from the direct effects of acid precipitation at ambient levels (pH 3.8-5.0); however, there is a threshold (about pH 3.0), below which acid rain and acid fog cause measurable damage to some crops [6].

The value of the annual crop yield may be estimated as the total cash receipts from farm produce marketing, which for the year 1985 in the U.S. were $74.4 billion [5]. Thus, 10% of the crop yield reduction is quite reasonable estimate for direct air pollution damage on the agricultural crops (mostly due to ozone), which becomes $7.4 billion in 1985. An additional loss of 5% may be attributed to the damage by acid precipitation, which comes to $3.7 billion.

The effect of pollution on forests is also quite destructive, although it takes longer time than that for crops to show significant effects. A comprehensive survey and analysis of forest decline in Europe has been carried out recently [8], and the results show that 15 percent of the total growing stock suffered visible moderate, severe or deadly damages. Decline was in average 6.2 times bigger than annual fellings in 1986. Damage on forests has also been reported in the U.S., Canada and China [7,9,10].

The damage on forests may be expressed as an economic value considering the forest usage for timber production and for recreation. An estimate of 10 percent reduction of the annual timber production in Europe in 1983 accounts for $6.6 billion. The recreational value of the U.S. forests may be represented by 225 million visitor-days in 1986 [6], and an estimate of 5 percent reduction results in $1.5 billion loss in 1986.

2.4. Effects on Aquatic Ecosystems

The necessity for oil transportation is connected with risks of spills and leakages, which can have a significant impact on beaches and aquatic ecosystems. The most recent is the disaster in Alaska's Prince William Sound, when 240,000 barrels (35,000 tons) of oil spilled from the tanker *Exxon Valdez*. The slick spread over nearly 900 square miles, and hundreds of miles of shoreline were covered with oil, in places as much as 6 inches deep. Thousands of birds, marine mammals and fish were killed, and the total ecosystem will suffer for at least a decade or two. Projected commercial losses are more than $100 million [11], but the full environmental damage may never be known. In 1970's about 60 tanker accidents have been reported annually, with average of 280,000 tons of oil spilled in and around the U.S. coastal waters per year [12]. Although these isolated spills are large, tanker accidents are not even the major ocean pollution source as far as oil transportation is concerned. They contribute about the same amount globally as coastal refineries. The most of the oil spilled into the ocean comes from the routine tanker operations. It has been estimated that 3 million tons (21 million barrels) of petroleum hydrocarbons are introduced into the ocean annually in the U.S. [13], which includes offshore production, tanker routine operations, dry docking, terminal operations, tanker accidents, bilges bunkering and coastal refinery operations. Extrapolating the estimates for *Exxon Valdez* accident, the total damage of oil spills in the U.S. coastal waters is calculated as $9 billion annually in 1989 $.

Another source of damage to fresh water sources is the leaks from underground storage tanks and pipelines. It is estimated that there are at least one million underground tanks in the United States, and if only 5% of them are leaking, which is a conservative estimate according to the EPA [14], that accounts for 50,000 leaking tanks each holding up to 10,000 gallons. One such incident caused damage amounting to $25 million (1982) [4]. If only 5% of these 50,000 tanks affect freshwater sources, annual damage may be estimated as $12.5 billion.

Acid precipitation due to air pollution can cause acidification of the lakes and fresh water sources. Acidification means lowering the pH value below the natural level for a particular lake. Some of the aquatic ecosystems are very sensitive to the pH level changes, and acidification results in the reduction and loss of fish populations, particularly salmon and trout, as well as aquatic flora. It is reported that acid precipitation has already caused widespread acidification of many aquatic ecosystems in the North-Eastern United States, Canada, Norway, Sweden and the United Kingdom [15]. In Sweden some 4,000 lakes have become fishless, while a total of 14,000 lakes have been acidified and experienced fisheries damage [4]. In southern Norway 1750 out of 5000 lakes have lost fish populations, and 900 others are seriously affected [16]. In southern Ontario, Canada, 24% of lakes surveyed had no fish at all [16]. In the United States more than 200 lakes in the elevated parts of the Adirondacks in New York State, and some lakes and ponds elsewhere are believed to have lost fish due to acid deposition [10,16-18].

One of the suggested - and already applied - techniques of inhibiting the aquatic effects of acidification is adding lime into water. The annual cost of liming is estimated to reach $40 million in 1985 in Sweden, and $500 million in Northeast U.S. [4,19].

The loss of fish production in the U.S. has been estimated as $1.3 billion (1984) per annum [4]. The recreational losses, as a result of up to 10 percent reduction in fishable area and catch rate, has been estimated to be $100 million per year [6].

2.5. Effects on Buildings and Structures

Ferrous metals, zinc, copper, masonry and some paints appear most likely to be degraded during atmospheric exposure [20]. Sulfur oxides act synergistically with oxidants and moisture to form sulfuric acid, which causes corrosion of metals, damage to electrical contacts and erosion of building stone strength by conversion of calcium carbonate to the soluble sulfates. Nitrogen oxides also cause fading of dyes and paints.

While the estimation of repair and replacement costs for contemporary buildings may be straightforward, the cost of damage to cultural properties is much more difficult to asses. Damage to sculptures, and historic monuments and structures diminishes their aesthetic importance far in excess of the amount of materials damage. Only to restore the cathedral in Cologne, West Germany, a sum of $2 million annually is required [4,17], and around the world there are many cathedrals and other historical buildings. Ancient Roman monuments require more than $200 million annually for restoration [17]. It has been estimated that worldwide there is about 10,000 historical buildings and structures, equivalent to that of Cologne Cathedral, which require $20 billion (1984) annually for restoration, because of atmospheric degradation [4].

A Dutch study in 1980 estimated that damage by acid deposition to monuments, libraries and archives in the Netherlands is costing $10-15 million per annum [15].

Degradation also occurs in contemporary public and private buildings and homes, although it is not so evident as it is on the ancient buildings, but quite certainly it shortens their lifetime and results in higher maintenance expenses. For 40 million homes and buildings exposed to air pollution and acid precipitation in the U.S., additional annual maintenance expenses are estimated to be $10 billion dollars.

Maintenance expenses may be extremely high for the bridges and other steel constructions. It has been estimated that damage on paint in the U.S. was $35 billion in 1970 [17]. In addition, damage to the zinc-coated transmission towers and industrial structures (including pipelines for industrial water supply, which corrode because of increased water acidity) must be taken into account. The OECD study of 1981 estimated damage for 12 European countries to galvanized steel and its paint coatings to be as high as $14 billion [16].

Particulate pollutants soil and/or damage the surfaces on which they are deposited, e.g., clothes, cars, windows, etc., and create considerable additional costs for cleaning and/or replacement, which have been estimated to be about $4 billion in the U.S. (1989) [21].

2.6. Other Air Pollution Costs

Visibility is included as an element of public welfare in the Clean Air Act (1970). Reduced visibility impacts human welfare through changes in transportation safety, property values and aesthetics. Because visibility is not directly valued in existing economic market, methods of harm estimate require the establishment of a hypothetical market or identification of complementarity between visibility range and existing market. The former method utilizes surveys in which participants indicate their willingness to pay to obtain higher air quality. The latter method infers a relationship between visibility range and residential property markets, or visibility range and the cost of move to alternative sites with better visibility. Environmental Protection Agency contract study (1984) quotes an annual benefit of $4 billion for visibility improvements in the Eastern United States [17].

Since the Clean Air Act of 1970, a significant decrease in air pollutant emissions has been achieved. Emissions of SO_2 in the U.S. have decreased by 30%, since reaching the peak in 1973, although coal consumption has increased. But this action required considerable efforts and expenditures. These expenditures should also be added to the damage estimate, because they may be considered as negative effect of the fossil fuel utilization. The expenditures for air pollution abatement and control since 1970 were more than $300 billion, and only in 1985 more than $33 billion [5].

2.7. Effect of Strip Mining

Surface coal mining results in extensive disturbance of the land surface. Such a situation can cause serious environmental problems unless the mined land is carefully reclaimed. For many years, reclamation of mine sites was not required by law in the United States. As a result, dangerous high walls were left exposed, trees and other vegetation were buried by waste material and dumped down the slopes, topsoil was buried or washed away, landslides formed on the unstable hillsides, slopes eroded rapidly because of lack of vegetation, bodies of polluted water formed in mine pits, and streams and rivers were frequently polluted by acidic mine drainage. Used land loses its value for crops, farming, recreation and everything else. Surface Mining Control and Reclamation Act of 1977 established two major programs in the United States:

(a) A reclamation program for abandoned mine lands, to reclaim land and water resources adversely affected by pre-Act coal mining.
(b) An environmental protection program, to establish standards and procedures for approving permits and inspecting ongoing surface coal mining and reclamation operations.

In the period 1978-86 more than 25,000 permits were issued for 3.4 million acres for surface coal mining. In the same period reclamation bonds for about 1.4 million acres were released [22]. This means that the rate of disturbance is much higher than the rate of reclamation; thus more effort and more expenses will be needed in the future. The government costs of reclaiming surface mines were estimated to be $19 billion in the period of 1980-89, or $1.9 billion annually, and approximately the same amount is estimated to be spent by industry [4]. This includes the expenses for abandoned coal mines, as well as for the new ones. But it does not include discrepancy between used and reclaimed land. This is at the present rate about 200,000 acres per year, which would cost $1.8 billion to reclaim. Also, during the land use for surface mining and during reclamation period as well, land loses its original value for agriculture, grazing and forestry. This damage has been estimated to be $1 billion annually.

2.8. Greenhouse Effect

As CO_2 and the other greenhouse gases accumulate in the atmosphere, the temperature of the Earth's surface rises. Measurements have shown that the Earth's surface temperature has increased by between 0.5 and 0.7^{o}C since 1860. If the present rates of growth in emissions continue, by the year 2030 the mean global temperature will increase at least 3°C and perhaps as much as 6°C [23]. Such an increase in surface temperature would have catastrophic consequences, such as climatic changes and sea level rise. Global warming will increase the frequencies of extreme weather events such as heat waves, droughts, floods, storms, etc. A warmer earth will have less vigorous ocean currents, less vigorous continental weather systems, more intense tropical storms, and altered jet streams and monsoonal circulation in the tropics. A warmer world will be a wetter world, too. However the distribution of precipitation, both geographically and by seasons, will change; precipitation should increase in low latitude regions (0 to 30 degrees) and during the winter in high latitudes, while mid-latitudes, particularly in continental interiors, may have reduced summer precipitation [23]. Many of the world's major agricultural regions, including the U.S. grain belt, may experience considerably reduced summer soil moisture because of earlier snowmelt, reduced precipitation, and increased water loss through evaporation and transpiration. Run-Off and aquifer recharge rates should decrease by even more than the soil moisture loss rate [24]. Climatic zones, and consequently the ecosystems, should shift as temperature continues to rise. Each 1°C increase would shift temperate zones by about 100 miles north and south [23].

Global warming will cause the water in the world's oceans to expand. However the temperature of all the oceans' various layers will not change at the same rate. The temperature of the surface waters will respond most quickly, increasing with atmospheric temperature increases, though with certain delay, and the transfer of heat downward will be slower. In addition, ice caps and mountain glaciers will start to melt releasing huge amounts of water and contributing to the sea level rise. Because polar temperatures are expected to rise more than the global average, effects of warming on increased melting will be great. It has been estimated that future sea level rise is likely to be between 26-39 cm by 2025 and 91-136 cm by 2075; however, a rise as great as 211 cm cannot be ruled out [25-30]. When melting of the ice caps is complete, the sea level would be 7-8 m higher than it is today.

On the basis of admittedly incomplete evidence, it appears that there is a reasonable chance of negative impacts of climate changes on the following: grain, vegetable and animal agriculture; forest-related industries; activities dependent upon in-stream river flows such as electricity production, urban water uses and barge transport; and many forms of recreation. Global warming may lead to changes in morbidity and increases in mortality, especially for the elderly people during the summers. Water scarcity, particularly in the Western U.S., as well as increasing fire incidence would be of special concern.

Sea level rise will inundate populated and unpopulated coastal areas, flood harbors, cities and agricultural land, erode beaches and cliffs, possibly break up barrier islands, lead to the intrusion of saltwater into drinking water supplies, and destroy coastal wetlands and the fisheries which depend upon them.

The number of hurricanes will probably increase, as will their destructive force, leading to additional economic losses. The ecological or environmental resources that would be affected cut across the entire spectrum of environmental concerns: endangered species, habitat destruction, soil erosion and desertification, air quality, groundwater quality, wilderness and other unique areas, natural parks, forestry and fisheries.

It has been reported that six warmest years in this century were in 1980's [31]. Significant increases in mid-latitude precipitation have occurred over the last 30 to 40 years [32]. Catastrophic droughts and floods in 1988 struck different regions all over the world. Hurricanes Gilbert in 1988 and Hugo in 1989 were the most severe storms ever recorded [33]. The forest fires in the U.S. in 1988, when more than 4 million acres of forest was burned, were the worst wildfires of the century [34]. The 1988 drought in the U.S. was extremely extensive, and baked the soil from Georgia to California, reducing the country's grain harvest by 31% and killing thousands of heads of livestock [35]. Other effects of this drought included the following: water rationing, stranded river barges, reduced electric power generating capacity, and decreased water supplies and quality. During October 1987 an immense chunk (98 miles long, 25 miles wide and 750 feet thick) of Antarctica's Ross Ice Shelf broke off and splashed into the sea [29,36].

An attempt to estimate the monetary value for each of possible negative results of the greenhouse effect will be presented herebelow. Only few estimates of certain economic effects of the climate changes and sea level rise have been done so far, and some were limited to regional damages. However, the effects of climate changes and sea level rise have global character. For this reason, available estimates have been extended.

2.8.1. Human Lives: Human illness and mortality are linked to weather patterns in many ways. Contagious diseases such as influenza and pneumonia, and allergic diseases such as asthma, are directly influenced by weather. It has been reported that weather (temperature, humidity, pressure, etc.) has an impact on daily mortality. During the heat wave of late July 1980 in New York City, mortality was over 50% above normal on the day with the highest temperature [37]. Studies have shown that mortality increases when temperature goes over a certain limit (for New York it was 92°F). At the present, for an average summer season only 5.7% of the total days exceed the threshold temperature, but for the projected 7°F global temperature increase in the next century, over one third of the summer days would exceed the threshold, which would increase daily mortality by 63% during the summer season [37].

Assuming reasonable adaptation to the increased temperature, at least 1 million people would die due to summer heat waves all over the world, which may be estimated as $40 billion average annual loss.

2.8.2. Agricultural Losses: Warming will cause intraregional shifts in productivity. For the faster rates of temperature changes, agricultural adaptations may be out of step in time with effects of climatic change, generating erratic reductions in food availability. Agricultural productivity in the lower latitude zones will be negatively affected because of the increased evapo-transpiration, while the higher latitude zones (but smaller land areas) would benefit from the longer growing season. However, agriculture is dependent on the availability of fertile soils, availability of water for irrigation, changes in soil moisture, etc. Therefore, agriculture in semi-arid areas in the mid-latitudes will probably be adversely affected by warming. It has been reported that the 1988 drought in the U.S. reduced grain harvest by 31% [35], which has been estimated as $25 billion damage. It will be assumed that such a drought could occur each year, but at different regions of the world, resulting in $25 billion annual damage worldwide.

2.8.3. Forestry Losses. The possible responses of mid-latitude forests to climatic changes have been assessed, taking into consideration the rate of seed migration, changes in reproductive success with changes in temperature and climatic stress on standing trees. The reproductive success of many tree species would be reduced by warming, and both the tree and the plant mortality would increase. There is no threshold below which effects do not occur. Major effects on forests were estimated to begin around the year 2000, with forest dieback starting between 2000 and 2050 for the upper limit of the anticipated rate of warming, 0.8-1.0^{o}C per decade, and around 2100 for the lower bound of the temperature scenario, 0.06-0.07^{o}C per decade [36]. The net effect of the warming would be a reduction in the area and the standing stock of carbon in forests. The seriousness of changes depends on the rate of change of temperature.

In warmer and drier climates the probability of wildfires would increase, which could further contribute to the forest decays. Wildfires in the U.S. consumed about 4.2 million acres (nearly 6,600 square miles) of forest in 1988 [34].

Possible annual damage to the forests worldwide is estimated to be in the same range as already estimated damage due to acid rain and air pollution, which comes to $25 billion.

2.8.4. Livestock Losses: Climate changes, particularly intensive droughts, may affect livestock and wild animals in a manner similar to the effects on agricultural crops and forests, because the abruptness of the change would probably surpass the capacity of many species to adjust to higher temperatures and altered climatic conditions. Water shortage and loss of wildlife habitat would lead to the extinction of many species. Even domestic animals may be affected by droughts. As the consequence of several consecutive years of drought in New South Wales in Australia, the number of sheep declined from a peak of about 73 million in the 1970s to about 43 million in 1983 [39]. The 1988 drought in the U.S. killed thousands heads of livestock [35].

2.8.5. Water Shortage: A U.S. National Academy of Sciences study has shown that annual flow in several Western and Southwestern U.S. rivers would be expected to decrease by about 40% to 75% with a temperature increase of only 2°C and a reduction in annual precipitation of only 10% [23,40]. Such an effect would cause shortages in urban and agricultural water supply, reduction in hydro and thermal power production, problems in barge transport, etc. (It should be noted here that the shortage of irrigation water and its consequences have already been included as agricultural losses).

Global warming of only 1.5°F could cut hydropower output by 10% [41]. Severe drought can reduce cooling water availability, and consequently power production by up to 20% [41]. Additional losses (1-2%) would occur by reducing plant efficiency because of the ambient temperature increases. The demand for summer cooling would increase by 7-11% and demand for winter heating would decrease by 5%, hence annual demand would increase by approximately 2-6% [42]. In addition, during the drought, electricity demand for groundwater pumping could increase by 20%. To meet such peak demand and to overcome production losses, additional capacity would have to be built, or electricity would have to be purchased from other regions. The average annual cost of the reduction of power production, including both hydro and thermal power, and of the increase in electricity demand has been estimated for the U.S. to be $8.5 billion.

During the last drought in summer 1988, the Mississippi River had its lowest level since record keeping began in 1871. The barge traffic was stopped, and the barge loads had to be reduced by 15-20% [43]. It resulted in the lost of revenue for the barge operators of $750 million [44].

2.8.6. Floods: Year 1988 will not be remembered only because of the drought in the United States. On the other side of the world, catastrophic floods occurred, first in Sudan and then in Bangladesh, two of the poorer countries in the world. For the second straight year, unusually heavy flood waters surged out of the Himalayan foothills and flooded almost three fourths of Bangladesh. According to the official figures some 800 people died, but other estimates increased this figure to 1500, and more than 100,000 persons were infected with waterborne diseases [45]. In addition 25 million people lost their homes, five million acres of rice growing land were inundated with water, and 43,000 miles of roads and bridges were damaged. A month later heavy floods hit four states in Northwest India, causing at least 400 deaths, tens of thousands of farm animals perished and food stocks were destroyed. Climatologists linked these floods to a sudden cooling of the oceans and with a climatic phenomenon known as El Nino [46].

It is difficult to estimate possible annual damage, but when the climate disruption becomes more noticeable, floods might be expected to occur more often and in the places where normally floods do not occur. Therefore $10 billion annually may be a conservative estimate for the worldwide damage.

2.8.7. Hurricanes, Storms: A change in sea surface temperature of 1°C will change the minimum sustainable pressure in hurricanes by 15 to 20 milibars, and an increase of a few degrees could cause a substantial increase in occurrence and severity of hurricanes. Hurricane Gilbert in September 1988 in the Caribbean was the most severe storm in the history of barometric pressure readings, millions of people were left homeless, several hundreds were killed, and property damage was estimated to be more than $10 billion [33].

2.8.8. Sea Level Rise: For a temperature rise in the range of 1.5 to 4.5 °C, published estimates of sea level rise generally range from 0.5 to 2.0 meters by 2100 [25-30].

Sea level rise will have a number of major effects:

(a) About 70% of the world's beaches are already eroding because of combination of sea level rise and human intervention, and future sea level rise and subsidence because of river discharge modifications are expected to force land use changes in coastal areas [28,38]. Dykes and other diversions will be required to counteract this process.

(b) In places where inland migration is not possible, sea level rise will mean a loss of wetlands.

(c) An increase in sea level would increase the frequency and severity of flooding and damage to coastal structures, port facilities and water management systems.

The direct physical effects of sea level rise will have a major influence on the use of coastal zones. Economic cost of sea level rise can be estimated on two different ways, (1) as the direct damage or (2) as the cost of anticipating sea level rise and preparing for it [47]. A study for Charleston, South Carolina, estimated the total economic damage in 2075 to be $2.5 billion (1980) and the value of anticipatory preparations as $1.4 billion. A similar study for Galveston, Texas, estimated the total economic damage to be $1.84 billion, and the value of anticipatory preparations as $1.1 billion (1980) [47].

Preliminary estimates by U.S. Environmental Protection Agency [42] suggest that the cumulative capital cost of protecting currently developed areas in the U.S. would be $73 to $111 billion in 1988 dollars for 1 meter rise. The cost of holding back the sea was based on (a) the quantity of sand necessary to elevate the beaches and coastal barrier islands, (b) rebuilding roads and elevating structures, and (c) constructing levees and bulkheads to protect the developed lowlands. On an annual basis this would be about $2 billion per year. Even with these precautions, some 7,000 square miles of dryland and most of the coastal wetlands, would be lost [42].

The Shell Oil Company is planning to raise the height of the natural gas platforms, to be built in the North Sea, by one meter, because of the expected sea level rise, and will spend an extra $20-40 million above the estimated costs [48]. This is probably the first example of a major engineering project taking into account the potential global warming induced sea level rise.

However, global warming and resulting sea level rise is a global problem and effects should be estimated for the whole world, not only for some regions or countries. Awad and Veziroglu [4] estimated the cost of constructing a dyke along the coastlines of the continents, and found it to be as high as $52 billion per year (1984), assuming annual sea level rise of 1 cm, and taking into account water pumping, reconstruction of bridges, roads and locks.

2.9. Effect on Biological Diversity

Some of the environmental impacts, that have a direct measurable economic value, have already been described. The consequences of fossil fuel consumption, land devastation, air and water pollution, acid rains, greenhouse effect and climatic changes have already caused severe economic damages. However, some of the ecosystems just do not have a direct economic value; but it does not mean that they are worthless and that they should be ignored.

Unlike most other impacts, loss of species and reduced biological diversity are irreversible. The effect of the human impact on species and ecosystems varies, with some species benefiting and others facing extinction. The uncertainties about the rate of the impact, particularly that of the global warming, individual species response, and interspecies dynamics, make the effect on biological diversity difficult to assess. However, it is clear that air and water pollution, acid rains and climatic changes will alter competitive outcomes and destabilize natural ecosystems in unpredictable ways. In many cases, the indirect effects, such as changes in habitat, in food availability, in ecological chains, may have a greater impact than the direct physiological effects [42]. Therefore, as a conservative estimate, the damage is estimated to be twice as much as the measurable and already calculated damage to the various ecosystems (plants, forests, animals, aquatic ecosystems, etc.).

The above discussed environmental damages and their estimates are summarized in Table I. It should be noted that the damages are given for different years and for different regions.

TABLE I. Environmental Damage Estimates

Type of Damage	References	Region	Year	Damage Estimate $\$10^9$/yr
Effects on humans:				
Premature deaths	[2,3,5]	U.S.A.	1985	50.0
Medical expenses	[4]	U.S.A.	1985	50.0
Loss of working efficiency	[4,5]	U.S.A.	1985	47.4
Effects on animals:				
Loss of domestic live stock	[4,5]	U.S.A.	1985	7.0
Loss of wildlife		U.S.A.	1985	14.0
Effects on plants and forests:				
Crop yield reduction - ozone	[5-7]	U.S.A.	1985	7.4
Crop yield reduction - acid rain	[6,7]	U.S.A.	1985	3.7
Effect on wild flora (plants)		U.S.A.	1985	22.2
Forest decline (economic value)	[6-9]	Europe	1983	6.6
Forest decline (effect on biological diversity)		Europe	1983	13.2
Loss of recreational value	[6]	U.S.A.	1986	1.5
Effects on aquatic ecosystems:				
Oil spills	[11-13]	U.S.A.	1989	9.0
Underwater tanks leakages	[4,14]	U.S.A.	1982	12.5
Liming the lakes	[4,19]	NE U.S.A.	1985	0.5
Loss of fish population	[4,15-17]	U.S.A.	1984	1.3
Loss of recreational value	[6]	U.S.A.	1985	0.1
Effect on biological diversity		U.S.A.	1985	5.0
Effects on man-made structures:				
Historical buildings and monuments degradation	[4,16,17,20]	World	1984	20.0
Buildings and houses detriment	[4]	U.S.A.	1984	10.0
Steel constructions corrosion	[16,17]	W. Europe	1981	14.0
Soiling of clothes, cars, etc.	[21]	U.S.A.	1989	4.0
Other air pollution costs				
Visibility reduction	[17]	E. U.S.A.	1984	4.0
Air pollution abatement costs	[5]	U.S.A.	1985	33.0
Effect of strip mining				
Land reclamation & loss of value	[4,22]	U.S.A.	1984	6.6
Effects of global warming and climate changes				
Heat waves - effects on humans	[37,42]	World	1989	40.0
Droughts -				
Agricultural losses	[35,49,50]	World	1989	25.0
Livestock losses	[35,39]	World	1989	20.0
Forests losses	[42,51]	World	1989	25.0
Wild flora and fauna losses		World	1989	40.0
Water shortage and power production problems	[23,40-44]	U.S.A.	1989	9.4
Floods	[45,46]	World	1989	10.0
Storms, hurricanes	[33]	World	1989	10.0
Sea level rise	[4,25-30,47]	World	1984	52.0

3. DATA EVALUATION

Fossil fuels in general consist of coal, petroleum and natural gas. Their contributions to environmental damage are different, e.g., they are not equal to their energy contents). Hence, in order to apportion a given type of damage amongst the fossil fuels, a pollution factor (P_{in}) will be used. It can be defined as a factor which is proportional to the environmental damage produced for a given fuel "i" and for a given type of damage "n". In the absence of more definitive data, in general, CO_2 emissions will be used to obtain the pollution factors, i.e., pollution factor for each fuel will be assumed to be proportional to CO_2 emissions per unit of energy. In case of oil spills and leakages, the pollution factors for coal and natural gas should be equal to zero, since they do not contribute to the damage in question. Similarly, for strip mining, the pollution factors for petroleum and natural gas are zero. Fossil fuel CO_2 emissions and the estimated pollution factors are presented in Table II.

TABLE II. CO_2 Emissions and Pollution Factors

	Fossil Fuels		
	Coal	Petroleum	N. Gas
CO_2 Emissions, kg/GJ	85.5	69.4	52.0
	Pollution Factor (P_{in})		
Damage Type (n)	Coal (i=1)	Petroleum (i=2)	N. Gas (i=3)
Those Due to Emissions	1.00	0.81	0.60
Oil Spills, Leakages	0	1.00	0
Strip Mining	1.00	0	0

As stated earlier, Table I presents the various environmental damages for different regions and for different years. The results have been divided by the fossil fuel consumption (in energy units) of the region in question. In order to bring them to the same base, year 1990 has been selected as the year of reference, and all the damage estimates have been recalculated for the year 1990 by using appropriate inflation coefficients.

The following relationship can then be used to obtain the environmental damage for a given fuel and for a given type of damage, C_{in} (the subscript "n" refers to the type of damage, and the subscripts 1, 2 and 3 refer to coal, petroleum and natural gas respectively):

$$C_{in} = \frac{D_n \cdot K_{in}}{F_i} \qquad (1)$$

where D_n is estimated damage, F_i is consumption of fossil fuel "i" for the given region and the given year, and K_{in} is a factor to apportion damage, D_n, to particular fossil fuel "i", and is defined with a help of previously defined pollution factor P_{in} as follows:

$$K_{in} = \frac{F_i \cdot P_{in}}{\sum_{i=1}^{3} (F_i \cdot P_{in})} \qquad (2)$$

4. RESULTS AND DISCUSSION

Using the appropriate data and Equations (1) and (2), the unit damages for coal (C_{1n}), petroleum (C_{2n}) and natural gas (C_{3n}) have been calculated for each of the discussed environmental damage for the year 1990. The results are presented in Table III and summarized in Table IV.

It can be seen from Table IV that the environmental damage due to fossil fuel consumption is \$9.82 per GJ of coal consumed, \$8.47 per GJ of petroleum consumed, and \$5.60 per GJ of natural gas consumed. The weighted mean damage for the world is \$8.14 per GJ of fossil fuel consumption (all values are in 1990 US \$). These damage costs are not included in the prices of fossil fuels, but they are paid for (or will be paid for) by the people directly or indirectly through taxes, health expenditures, insurance premiums, and through reduced quality of living. In other words, today fossil fuels are heavily subsidized. If the respective environmental damages were included in the fossil fuel prices, it would force the earlier introduction of cleaner fuels, such as hydrogen, with many benefits to economy and environment.

In order to see the world-wide dimensions of the fossil fuel environmental damage, Table V has been prepared. It can bee seen that 40% of the total damage is caused by coal while coal consumption is 33% of the total fossil fuel consumption. On the other side, only 18% of the damage is caused by natural gas which has a market share of 27%. It is clear that increasing the natural gas consumption at the expense of coal and petroleum, will environmentally be beneficial. However, natural gas is not a completely clean fuel either, and its reserves will be depleted sooner or later as the other fossil fuels. Therefore, a development of a clean energy system based on renewable energy sources and coupled with clean and efficient energy carrier(s) is urgently required.

It can also be seen from Table V that the worldwide environmental damage caused by the fossil fuels is 1990 \$2,360 billion, or equal approximately to 14% of the Gross World Product. This is a very large figure. Conversion to a cleaner fuel, such as for example hydrogen, could enable the world to save this enormous sum and perhaps use it to further improve the quality of life worldwide.

TABLE III. Environmental Damage Caused by Each of Fossil Fuels

Type of Damage (n)	Environmental Damage (1990 $/GJ)		
	Coal (C_{1n})	Petroleum (C_{2n})	N. Gas (C_{3n})
Effect on humans			
Premature deaths	1.18	0.96	0.71
Medical expenses	1.18	0.96	0.71
Loss of working efficiency	1.12	0.91	0.67
Effect on animals			
Loss of domestic live stock	0.17	0.14	0.10
Loss of wildlife	0.34	0.28	0.20
Effect on plants and forests			
Crop yield reduction - ozone	0.17	0.14	0.10
Crop yield reduction - acid rain	0.09	0.07	0.05
Effect on wild flora (plants)	0.52	0.42	0.31
Forest decline (economic value)	0.18	0.15	0.11
Forest decline (effect on biological diversity)	0.36	0.29	0.22
Loss of recreational value	0.03	0.02	0.02
Effect on aquatic ecosystems			
Oil spills	-	0.30	-
Underwater tanks leakages	-	0.61	-
Liming the lakes	0.03	0.02	0.02
Loss of fish population	0.03	0.02	0.02
Effect on biological diversity	0.12	0.10	0.07
Effect on man-made structures			
Historical buildings and monuments degradation	0.12	0.10	0.07
Buildings and houses detriment	0.25	0.20	0.15
Steel constructions corrosion	0.67	0.54	0.40
Soiling of clothes, cars, etc.	0.08	0.06	0.05
Other air pollution costs			
Visibility reduction	0.20	0.16	0.12
Air pollution abatement costs	0.78	0.63	0.47
Effect of strip mining	0.49	-	-
Effect of climatic changes			
Heat waves - effects on humans	0.18	0.15	0.11
Droughts -			
Agricultural losses	0.11	0.09	0.07
Livestock losses	0.09	0.07	0.05
Forests losses	0.11	0.09	0.07
Wild flora and fauna losses	0.63	0.51	0.38
Water shortage and power production problem	0.17	0.14	0.10
Floods	0.05	0.04	0.03
Storms, hurricanes	0.05	0.04	0.03
Effect of sea level rise	0.32	0.26	0.19
Totals	9.82	8.47	5.60

TABLE IV. Environmental Damage Caused by Fossil Fuels (Summary)

Type of Damage (n)	Environmental Damage (1990 $/GJ)		
	Coal (C_{1n})	Petroleum (C_{2n})	N. Gas (C_{3n})
Effect on humans	3.48	2.83	2.09
Effect on animals	0.51	0.42	0.30
Effect on plants and forests	1.35	1.09	0.81
Effect on aquatic ecosystems	0.18	1.05	0.11
Effect on man-made structures	1.12	0.90	0.67
Other air pollution costs	0.98	0.79	0.59
Effect of strip mining	0.49	-	-
Effect of climatic changes	1.39	1.13	0.84
Effect of sea level rise	0.32	0.26	0.19
Totals	9.82	8.47	5.60

TABLE V. World-Wide Environmental Damage by Fossil Fuels for 1990

Fossil Fuel Consumption	(10^{18} J/yr)
World coal consumption	96
World petroleum consumption	117
World natural gas consumption	77
World fossil fuel consumption	290
Environmental Damage Estimate	(1990 billion $)
Damage due to coal	940
Damage due to petroleum	990
Damage due to natural gas	430
Total environmental damage	2,360
Demographic and Economic Data	
World population (billion)	5.2
Damage per capita	$460
World GWP (billion $)	17,000
GWP per capita	$3,270
Damage/GNP	0.14

5. CONCLUSION

Environmental damage caused by each of the fossil fuels has been estimated on an item-by-item basis. It has been shown that worldwide this damage adds up to a very large figure of 1990 $2,360 billion per year, or $460 per capita per year. This is what the society pays in addition to the market prices for using fossil fuels. Internalizing these costs, i.e., adding them to the fossil fuels market prices, would allow the market to move toward wise and beneficial long-term energy-use decisions.

Environmental damage is not the only external cost of fossil fuels. In a complete economic study, one must take into account military costs for protecting oil and gas supplies, full costs of international conflicts and other external economic effects (such as employment, balance of trade, depletion of non-renewable resources, etc.).

With estimated environmental damage in mind, it becomes clear that a speedy and planned conversion to cleaner fuels, such as natural gas and ultimately hydrogen, would benefit the world both economically and environmentally.

6. REFERENCES

1. Record, F.A., D.V. Bubenick, and R.J. Kindya, (1982) *Acid Rain - Information Book*, Noyes Data Corp., Park Ridge.
2. Luoma, J.R., (1982) The Human Cost of Acid Rain, *Audubon*, Vol. 90, No. 6, pp. 16-23.
3. Zweig, R.M., (1982) Hydrogen - Prime Candidate for Solving Air Pollution Problems, in T.N. Veziroglu, W.D. Van Vorst and J.H. Kelley (eds.), *Hydrogen Energy Progress IV*, Proceedings of the 4th World Hydrogen Energy Conference, Pergamon Press, New York, Vol.4, pp. 1789-1808.
4. Awad, A.H., and T. N. Veziroglu, (1984) Hydrogen Versus Synthetic Fossil Fuels, *Int. J. Hydrogen Energy*, Vol.9, No5, pp. 355-366.
5. *Statistical Abstract of the U.S.*, (1988) United States Department of Commerce, 108th edition, Washington D.C.
6. The National Acid Precipitation Assessment Program, (1988) Vol. IV, *Effects of Acidic Deposition*, Washington D.C.
7. Borman, F.H., (1985) Air Pollution and Forests: An Ecosystem Perspective, *Bioscience*, Vol. 35, No. 7, pp. 434-440.
8. Nilsson, S., and P. Duinker, (1987) A Synthesis of Survey Results: The Extent of Forest Decline in Europe, *Environment*, Vol. 29, No. 9, pp. 4-11.
9. Postel, S., (1984) Air Pollution, Acid Rain, and the Future of Forests, *Worldwatch Paper* 58, Worldwatch Institute, Washington D.C.
10. Johnson, A.H., (1986) Acid Deposition: Trends, Relationships, and Effects, *Environment*, Vol. 28, No. 4, pp. 7-11 and 34-39.
11. Marshall, E., (1989) Valdez: The Predicted Oil Spill, *Science*, Vol.244. pp.20-21.

12. *Environmental Quality*, (1980) 11th Annual Report of the Council on Environmental Quality, U.S. Government printing Office, Washington, D.C.
13. Fowler, J.M., (1984) *Energy and the Environment*, 2nd ed., McGraw-Hill, New York.
14. The Leaking Underground Storage Tank Trust Fund, (1987) *EPA Journal*, Vol.13, No.1.
15. Schindler, D.W., (1988) Effects of Acid Rain on Freshwater Ecosystem, *Science*, Vol. 239, pp.149-157.
16. *Acid Rain, A Review of the Phenomenon in the EEC and Europe*, (1983) A Report prepared for the Commission of the European Communities, Directorate General for Environment, Consumer Protection and Nuclear Safety by Environmental Resources Ltd., Unipub, New York.
17. *Acid Rain: A Survey of Data and Current Analyses*, (1984) A Report prepared by the Congressional Research Service, U.S. Government Printing Office, Washington D.C.
18. Crocker, T.D., (1982) What Economics Can Currently Say About the Benefits of Acid Deposition Control, presented at the Conference on "Adjusting to Regulatory, Pricing, and Marketing Realities, Williamsburg, Virginia.
19. Sheppard, C.R., (1986) Liming Program Restores Lakes, *American Mining Congress Journal*, Vol.72, p.9
20. Graedel, T.E.,and R. McGill, (1986) Degradation of Materials in the Atmosphere, *Environ. Sci. Technol*. Vol. 20, No. 11, pp.1093-1099.
21. Barbir, F., (1989) Effects and Damage of Using the Fossil Fuels, Special Problem Report, MEN 692, Clean Energy Research Institute, University of Miami, Coral Gables, FL.
22. Surface Coal Mining Reclamation: 10 Years of Progress, 1977-1987, (1987) United States Department of Interior, Office of Surface Mining Reclamation and Enforcement, Washington, D.C.
23. Abrahamson, D.E., (1989) Global Warming: The Issue, Impacts, Responses, in D. E. Abrahamson (ed.) *The Challenge of Global Warming*, Island Press, Washington D.C., pp. 3-34.
24. Manabe, S., (1989) Changes in Soil Moisture, in D. E. Abrahamson (ed.) *The Challenge of Global Warming*, Island Press, Washington D.C., pp.146-150.
25. Hoffman, J.S., (1984) Estimates of Future Sea Level Rise, in M. C. Barth and J. G. Titus (eds.), *Greenhouse Effect and Sea Level Rise*, Van Nostrand Reinhold Co. New York.
26. Monastersky, R., (1987) Rising Sea Levels: Predictions and Plans, *Science News*, Vol. 132, pp.326,
27. Titus, J.G., et al., (1987) Greenhouse Effect, Sea Level Rise, And Coastal Drainage Systems, *Journal of Water Resources Planning and Management*, Vol. 113, No.3, pp.216-227.
28. Titus, J.G., (1989) The Causes and Effects of Sea Level Rise, in D. E. Abrahamson (ed.) *The Challenge of Global Warming*, Island Press, Washington D.C., pp.161-195.
29. Wolkomir, R., (1988) The Greenhouse Revolution, *Oceans*, Vol. 16, No. 4, pp.17-20.

30. Dean, R.G., (1986) The Ultimate High Tide, *Civil Engineering*, Vol. 56, pp.64-66.
31. Houghton, R.A., and G. M. Woodwell, (1989) Global Climatic Change, *Scientific American*, Vol. 260, No. 4, pp.36-44.
32. Bradley, R.S., et al., (1987) Precipitation Fluctuations over Northern Hemisphere Land Areas Since Mid-19th Century, *Science*, Vol.237, pp.171-175.
33. Severe Hurricanes Expected to Occur More Frequently, (1988) in *Climate Alert*, Vol.1, No.4, pp.6
34. 1988 U.S. Wildfires Burn 6,600 Square Miles, (1988) in *Climate Alert*, Vol.1, No.4, pp.4
35. Sancton, T.A., (1989) What on Earth Are We Doing? (Destruction of the Earth's environment), *Time*, Vol.133, pp.24-30,
36. Owen, O.S., (1989) The Feat Is On, THe Greenhouse Effect and the Earth's Future, *Futurist*, Vol. 25, No. 5, pp.34-40 (1989)
37. Kalkstein, L.S.,et al., (1986) The Impact of Human Induced Climatic Warming Upon Human Mortality: A New York City Case Study, in J. G. Titus, (ed.) *Effects of Changes in Stratospheric Ozone and Global Climate*, Vol.3: Climate Change, EPA/UNEP, pp.275-293.
38. Jaeger, J., (1989) Developing Policies for Responding to Climate Change, in D. E. Abrahamson (ed.) *The Challenge of Global Warming*, Island Press, Washington D.C., pp.96-112.
39. Wilhite, D.A., (1986) Drought Policy Implications of CO_2-Induced Climatic Change in the United States and Australia, in J. G. Titus, (ed.), *Effects of Changes in Stratospheric Ozone and Global Climate*, Vol.3: Climate Change, EPA/UNEP pp.73-88.
40. Revelle, R.R., and P. E. Waggoner, (1989) Effects of Climatic Change on Water Supplies in The Western United States, in D. E. Abrahamson (ed.) *The Challenge of Global Warming*, Island Press, Washington D.C., pp.151-160
41. Palmer, R.N., and J. R. Lund, (1986) Drought and Power Production, Vol.112, *Journal of Water Resources Planning and Management*, No.10, pp.469-484.
42. Smith, B.J., and D. A. Tirpak (eds.), (1988) The Potential Effects of Global Climate Change on the United States (draft), U.S. Environmental Protection Agency, Office of Policy, Planning and Evaluation and Office of Research and Development, Washington, D.C.
43. Wilson, L.J., and E. Britton, (1988) Barges Just Keep Rolling Along, *Chemical Week*, Vol.143, August 17, pp.9-10.
44. Vogel, T., (1988) Barges up a Creek: The Drought is Still Snarling Traffic on the Mississippi, *Business Week*, August 8, pp.36.
45. Bangladesh: The Water This Time, (1989), *Newsweek*, September 19,
46. Pearce, F., (1988) Cool Oceans Caused Floods in Bangladesh and Sudan, *New Scientist*, Vol.119, September 8, pp.31.
47. Gibbs, M.J., (1984) Economic Analysis of Sea Level Rise: Methods and Results, in M. C. Barth, J. G. Titus (eds.), *Greenhouse Effect and Sea Level Rise*, Van Nostrand Reinhold Co., New York.
48. Shuldiner, H., (1989) Global Warming, in *Popular Science*, December, p.8.

49. Trenberth, K.E., G. W. Branstator, and P. A. Arkin, (1988) Origins of the 1988 North American Drought, *Science*, Vol.242, pp.1640-1644.
50. Kunkel, K.E., (1988) A Drought's Unyielding Cycle, *Natural History*, Vol.98, No.1, pp.48-49.
51. Brown, L.R., C. Flavin, and E. C. Wolf, (1988) Earth's Vital Signs, *Futurist*, Vol. 22, No. 4, pp.13-20.
52. Cannon, J.S., (1987) Controlling Acid Rain, A New View of Responsibility, An Inform Report, Inform, New York.
53. Brown, L.R., et al., (1991) *State of the World*, A Worldwatch Institute Report, W.W. Norton & Co., New York.
54. German Bundestag (ed.), (1989) *Protecting the Earth's Atmosphere - An International Challenge*, German Bundestag, Bonn.
55. Ince, M., (1990), *The Rising Seas*, Earthscan Publications, London
56. Ginsberg, I.W., and J.A. Angelo, Jr., (eds.), (1990) *Earth Observations and Global Change Decision Making, 1989: A National Partnership*, Krieger Publishing, Malabar, FL.
57. *Energy Statistic Yearbook 1987*, (1988) United Nations, New York.
58. *Annual Energy Review 1989*, (1990) Energy Information Administration, Washington, D.C.
59. The World Bank, *World Development Report 1989*, (1989) Published for the World Bank by Oxford University Press, New York.

PRESENT PRACTICE IN PHYSICAL COAL CLEANING OPERATIONS

Y. YÜRÜM
Department of Chemistry
Hacettepe University
Beytepe
Ankara 06532
Türkiye

ABSTRACT. Most of the coals require some preparation before use which may range from simply crushing to provide a size suitable for certain types of boilers to extensive size reduction and cleaning to remove sulfur and ash forming mineral matter. There are three basic kinds of cleaning; a. conventional physical methods, b. advanced physical operations and c. chemical coal cleaning processes, each with a wide range of capabilities. This chapter will examine the physical cleaning operations and the advanced physical methods.

1. Introduction

Coal is an abundant, relatively cheap and widely distributed fuel. Coal is also known to be a polluter therefore has major barriers affecting its utilization. Coal utilization gives rise to air and water pollution, corrosion and fouling of boilers and heat exchangers. The major enviromental and operational problems in coal utilization have been due to the ash forming mineral material and sulfur constituents present in the feed coal. When low rank coals are used instead of bituminous coals these difficulties are intensified. The importance of cleaning coals was first recognized in the metallurgical industry. Therefore, coal cleaning operations have been firstly used almost exclusively in coke manufacturing by the metallurgical industry (Decker and Hoffmann, 1963). Over recent years, more advanced forms of coal cleaning have started to evolve. Coal burning industries in the United States have already spent more than $250 billion to control emissions since 1970 and as a result, sulfur dioxide emissions have dropped dramatically, although much of the drop is also attributed to the use of lower sulfur coals (Haggin, 1991). Technology for clean utilization of coals can be divided into three groups: a. **preparation of coal before its utilization**, b. **developing new and efficient utilization technologies** and c. **cleaning of the products of coal utilization**. All of these subjects will be investigated throughout the present book, but within this chapter only physical preparation operations will be reviewed. Basically, the purpose in all of the cleaning operations is to improve the quality of raw coal by separating varying amounts of sulfur, ash and other wastes before its use. The resulting product after cleaning is a higher quality fuel for utilization and is easier to be used in plant equipment. Whether or how extensively a coal will be cleaned before use depends entirely on the purposes for which it is mined and on the provisions it must meet in the consuming market. However, some decrease in the mineral matter content of the raw coal is generally required if the coal is intended to be utilized in coke production or if it contains relatively large amounts of inorganic minerals of sulfur. In the former situation, cleaning serves to improve the caking properties of the coal and in the latter, it can make the coal an environmentally acceptable fuel where its use would otherwise be prevented by law. For some applications, coal cleaning is essential. One such application is for coal used in the preparation of slurry fuels, that is, coal-oil mixtures and coal-water mixtures, for use in generation of steam in boilers designed

Y. Yürüm (ed.), Clean Utilization of Coal, 153–172.

for oil firing. These boilers cannot handle the large amounts of ash forming mineral matter present in run-of-mine coal and therefore mineral matter of the coal to be used should be separated.

Coal cleaning offers several important economic benefits for coal users namely industrial boilers and coke making industry. These benefits may include:

a. **Reduced transportation costs**. If the coal must be transported to distant markets, removal of inorganic matter brings economic benefits through lowering of the effective transportation costs per ton of usable coal.

b. **Reduced operating an maintenance costs**. Although cleaned coal is more expensive than the raw coal, use of the clean coal saves money by lowering boiler maintenance and other plant operating costs and increasing boiler efficiency. Boiler performance which experiences from slagging, fouling and corrosion due to ash and sulfur is extensively improved by the cleaning operations. Fouling which is intensified by the presence of sodium or potassium in the fly ash, reduces the rate of heat transfer and contributes to boiler tube corrosion. Coal washing will not only lessen these problems but also reduce the cost of ash disposal and flue gas desulfurization. Excessive ash causes erosion of some parts of the boiler and induced draft fans; plugs the air preheaters and overloads the electrostatic precipitators (Cole, 1979). Consequently, ash and sulfur reduce peak boiler capacity, lead to outages, reduce availability and effectuate increased operating and maintenance costs.

c. **Reduced equipment costs**. Cleaning may reduce or eliminate flue gas desulfurization and scrubber requirements and costs depending on the sulfur content of the coal used and local emission standards. By using a coal with a reduced sulfur content less lime or other material is required for absorbing sulfur oxides in the scrubbers and less waste is generated. Therefore, the absorbent preparation and waste disposal sections of the system can be reduced insize. Moreover, in some cases, cleaner coal may reduce scrubbing requirements sufficiently so that part of the flue gas can be allowed to bypass the scrubbing system.

d. **More reliable pollution control**. Increasing coal use creates some potentially serious environmental problems. Compared with the other fossil fuels, oil and natural gas, coal is an unclean fuel. In its raw form, coal contains many impurities, such as rocks that line coal seams, organic and inorganic sulfur constituents, mineral matter and other wastes. When burned, many of these contaminants can be transformed into sulfur dioxide, particulates and other air pollutants. Power plants emit millions of tons of sulfur dioxide. These emissions contribute to the air pollution load that has been linked to respiratory and heart problems and increasingly to the formation of acid precipitation which is affecting hundreds of rivers and lakes, destroying fish and aquatic life all over the world. Coal cleaning can help plants to meet sulfur dioxide standards established by clean air requirements. The level of reduction in sulfur dioxide emissions to the atmosphere and flue gas desulfurization technologies used to control these emissions are currently the subjects of international pollution control research in many countries.

e. **Higher quality coal**. In the past five decades the quality of coal being produced has steadily deteriorated. One reason for this is that automatic mining has replaced mechanical mining and loading methods which allowed coal to be selected more carefully. A second reason is that many higher quality seams have been used up, as a result greater amounts of waste are present in the raw coal, especially from underground mines. The metallurgical coke producing industry has long recognized the benefits of clean coal and most of the coal used in this industry is cleaned by removing ash forming minerals from coking coal. Coke manufacturing requires coals of uniform consistency with a low percentage of sulfur and mineral matter. Excessively high amounts of sulfur in metallurgical cokes give rise to brittleness in the iron and also renders the final product highly vulnerable to corrosion (Mc Gannon, 1971; Evans, 1981). Chemical industry has become increasingly heavy user of cleaned coal for steam generation. In addition to removing a portion of the mineral matter and mining wastes, cleaning can provide coals of improved quality, higher heating value and great consistency. Cleaning also offers the possibility of enlarging the amount of coal appropriate for burning by converting a coal of low quality into one of higher quality.

f. **Future applications**. In the future, coal gasification and liquefaction processes should benefit from coal cleaning operations. Withdrawal part of the sulfur and mineral matter from the feedstock reduces the load of these components which has to be handled by the conversion plant with a consequent saving in capital, operating and maintenance costs (Simbeck, 1981; Killmeyer, 1982). Coal cleaning needs to be carefully integrated and matched to each major conversion process because the requirements vary considerably. Whereas some conversion processes can benefit from using coal with the least amount of mineral matter, others can gain using coal containing certain minerals. An example of the latter is the direct liquefaction process in which iron pyrites seem to catalyze the hydrogenation of coal (Killmeyer, 1982). High conversions of about 97% in the liquefaction of coals with reduced mineral matter are observed, this was due to increased rejection of less reactive components of coal and leaving pyrites selectively by oil agglomeration operation (Gollakota et al. 1989). For liquefaction processes, thus, it would be advantageous to leave the pyrites in the coal while removing other minerals. For some gasification processes it may be advantageous to desulfurize the feedstock with a hot alkaline solution. This treatment would not only reduce the sulfur content but also reduce any caking tendency of the coal and add some alkali to catalyze the gasification process. The future application of various coal conversion methods could also benefit from the prior separation of the various maceral groups which compose the carbonecous part of the coal. By separating these components, each could be directed into the application for which it is best suited. For example, the more reactive components could be utilized for hydrogenation, the less reactive ones for combustion. Although some success has been achieved in separating the carbonaceous components of coal by using conventional physical separation methods (Oder et al., 1981; Pommier et al., 1981), the commercial application of this concept probably expects further innovation in separation technology.

Most of the coal need some preparation before use which may range from size reduction providing a size suitable for certain types of boilers, to extensive size reduction and cleaning to remove sulfur and ash forming mineral matter. There are four basic kinds of coal preparation methods: i. **conventional physical methods**, ii. **developing physical methods**, iii. **chemical coal cleaning processes** and iv. **biological coal cleaning processes**, each with a wide range of capabilities. This chapter will examine the conventional physical cleaning methods and the advanced physical methods. Chemical and biological coal cleaning will be presented in a different chapters.

i. At present, coal cleaning is achieved through **physical separation** of particles low in mineral content from particles high in mineral content. Physical cleaning is limited to mostly washing to separate pyritic sulfur and mineral impurities. Conventional physical methods, the most generally known and used, reduce the size of coal pieces and separate the coal from its noncoal components by exploiting the large density differences between the organic coal substance and mineral matter, and surface properties between coal and minerals. Clay minerals, iron sulfides, carbonates and silica make up most of the mineral matter in coal (Gluskoter, 1975; Harchand et al. 1989). With the utilization of these methods removal of an average of 25% sulfur and 50% of the mineral matter of a coal can be accomplished (Gibbs and Hill, 1978). How easily or completely a coal can be cleaned by physical methods depends on the distribution of the mineral matter in the carbonaceous material. In autochthonous coals, which developed at the growth site of the parent plants, inorganic material is for the most part concentrated along the bedding planes and in major fractures and such material can normally be quite easily removed. In allochthonous coals, which formed from transported plant debris, much of the mineral matter exists as colloidally dispersed form throughout the coal substance and therefore it is virtually impossible to remove such mineral matter, utilizing the differences in density by physical methods. Cokes produced from coals in which pyrites are finely disseminated within the vitrinite exhibit high sulfur contents (Ndaji and Imobighe, 1989). Thus, the kind of physical cleaning method for a definite coal should be specifically determined according to the type of the coal.

ii. **The advanced physical operations** attempt to clean coal more efficiently and at a lower cost than the conventional methods.

iii. **Chemical coal cleaning processes**, which can be viewed as the true state of the art, attempts to disintegrate the molecular structure of coal to remove almost all of the organic sulfur and a considerable part of the total sulfur, while the most sophisticated conventional methods are known to remove only up to 50% of the total sulfur and none of the organic sulfur. Coal is cleaned to remove the troublesome contaminants present as discrete mineral phases, but part of the sulfur and practically all of the nitrogen in coal are an integral part of the organic structure of the coal. The organic sulfur is likely to be present as alkyl or thiols, sulfides and disulfides and as heterocyclic compounds of the tiophene type (Given and Wyss, 1961; van Krevelen, 1961; Attar and Corcovan, 1977; Attar, 1978; Yürüm and Tuğluhan, 1990; Yürüm et al. 1990). X-ray photo electron spectroscopy and sulfur x-ray absorption near edge structure techniques were also used for the determination and quantification of organic sulfur forms in coals (Boudou et al. 1987; Kelemen et al. 1990). Nitrogen seems to occur in heterocyclic rings of the pyridine or pyrrolic type (van Krevelen, 1961; Whitehurst, 1978). The organic frame work may also contain minerals of certain trace elements such as germanium, beryllium and boron (Glustoker, 1975). Additionally, sodium and other alkali metal ions may be held in ion exchange connection by the carboxylic and phenolic groups present in subbituminous coal and lignite (Sondreal et al., 1977; Paulson and Fowkes, 1968). There have been some encouraging results in chemical coal cleaning and it is hoped that that up to 90% of organic sulfur will be removable by appropriate chemical cleaning technology (Haggin, 1991).

iv. **Biological coal cleaning**, might be suitable for small scale operations where the use of coal-water slurries is proposed or for cleaning fines which arise from modern mining operations and coal preparation. Although bioprocess introduces the possibility of organic sulfur removal, whereas most physical processes separate only the inorganic pyrite, too slow desulfurization rates limits the large scale commercial applications (Güllü et al. 1992).

2. Physical Coal Cleaning Operations

Coal preparation technologies have been used for over a century (Hutton and Gould, 1982). In general, coal preparation produces a fuel with uniform quality of greater heating value by removing mineral matter and concentrating organic carbonaceous part of the coal. The coal preparation term is usually used to describe all of the operations that coal undergoes between mining and utilization. These include transportation, storage, crushing, classification by size, separation of contaminants, removal of water and disposal of waste. However, throughout this chapter, beneficiation, cleaning or washing will be used interchangeably with coal preparation and its meaning is narrowed to only those physical operations and chemical processes that upgrade the quality of the fuel by regulating its size and reducing the quantities of ash, sulfur and other impurities found in coal. The chemical coal cleaning technologies promise greater total sulfur removal, although these processes are not technically and economically feasible at the present. Dry physical coal cleaning methods, such as pneumatic cleaning and electrostatic separation, have also been used in the past without great commercial success (Bechtel Corporation, 1977). Further, few of the dry cleaning methods have shown some commercial potential, with the possible exception of the magnetic separation technologies. At the present time, only the physical coal cleaning operations are used commercially.

2.1. WASHABILITY TESTS

The extent to which a coal seam can be beneficiated to reduce its mineral matter is understood by float and sink analysis and the resulting washability data. Coals vary in washability characteristics. Most of the seams can be cleaned by a two stage operation and the possibility of even a third stage might be investigated. Because of large density differences between the organic part of the coal and mineral matter associated with it, cleaning is normally done by gravity separation methods. While separation would be successful in aotochthonous coals in which inorganic material is concentrated as discrete planes, it might not be possible in allochthounous coals in which the mineral matter is colloidally dispersed in the carbonecous matter of the coal.

As already indicated, most of the commercial coal cleaning methods separate particles on the basis of differences in specific gravity. To forecast the response of a particular coal to such cleaning methods needs conducting a float-sink or washability test. To perform such a test, the coal is first crushed and then screened to obtain particles within a certain size range. A quantity of the sized material is placed in a container which the bottom part is made of a fine screen and the container is immersed successively into a series of liquid baths of incresing specific gravity. The particles which float in each bath are collected separately, dried, weighed and analyzed for mineral matter and sulfur. The particles which sink even in the liquid of the highest specific gravity are separated and analyzed. These particles represent the ultimate sink fraction. The results of such a test are recorded and the specific gravity of the liquids which floated the fractions with different percentages of sulfur and mineral matter is determined. The results of such an analysis by Wheelock and Markuszewski, (1984) is presented in Table 1. These results indicate that 64.1% of a sample of Iowa coal floated at 1.30 specific gravity and an additional 6.2% floated at a specific gravity of 1.35 and so on, with the mineral matter and sulfur contents of each succeeding fraction being larger than the contents of the preceding fraction. If a specific gravity of 1.50 was selected as the basis for cleaning this sample of Iowa coal, 84.6% of the material would be regained in the floating product under ideal conditions and the product would contain 9.5% mineral matter and 2.72% total sulfur. These values correspond to a decrease in mineral content of 38% and in sulfur content of 48%.

TABLE 1. Washability analysis of a bituminous coal from Iowa, [after Wheelock and Markuszewski (1984)].

Specific gravity of the bath	Individual fractions, %			Cumulative float, %		
	Wt	Minerals	S	Wt	Minerals	S
Float 1.30	64.1	8.4	2.22	64.1	8.4	2.22
1.30 to 1.35	6.2	8.8	2.97	70.3	8.4	2.29
1.35 to 1.40	8.1	13.1	3.90	78.4	8.9	2.45
1.40 to 1.45	4.0	16.5	5.15	82.4	9.3	2.58
1.45 to 1.50	2.2	19.4	7.89	84.6	9.5	2.72
1.50 to 1.55	1.4	21.8	9.07	86.0	9.7	2.82
1.55 to 1.60	1.0	25.2	10.49	87.0	9.9	2.91
Sink 1.60	13.0	52.7	20.77	100.0	15.4	5.23

The results of a washability test which is obtained in a laboratory is not applicable directly in industry. The separation which would be achieved industrially would be lower depending on the type of equipment used and the separation efficiency. One indication of the difficulty of the separation is the percentage of near gravity material. For the example given in Table 1, 8.6% of the material is within the range of specific gravity from 1.40 to 1.60. This indicates that the separation would be a difficult one for a jig but better separation would be achieved in an efficient cleaning device (Corriveau and Schapiro, 1979).

The separation of coal components given in Table 1, is based on specific gravity and mineral matter content. Other properties of the sample are distributed as follows.

Calorific value: Highest calorific values are found in the lowest density fractions, this trend declines as density increases. If separations are carried out in liquids of specific gravity greater than 2.0 in the case of high sulfur coals the calorific value of the fraction may rise slightly in the highest density fractions.

Volatile matter: The volatile matter of the fractions tends to increase in the low density fractions and generally decreases with increasing density. There may be a small increase in the volatiles produced from inorganic decomposition such as carbonates, in the higher density fractions.

Sulfur: With the exception of few uncommon coals, percentage of sulfur tends to be the lowest in the low density fractions and increases as density increases. High sulfur content in low density fractions is associated with organic sulfur.

Maceral composition: The concentration of reactive macerals tends to be higher in the lower density fractions and the content of less reactive macerals increases in the higher density fractions.

2.2. EQUIPMENT USED IN PHYSICAL COAL CLEANING

Coal preparation plants contain a variety of systems, ranging from those designed simply to remove coarse residue from raw coal to detailed systems designed to remove the maximum amount of pyritic sulfur and mineral matter. Thus, not to contain a specific cleaning process, but rather a number of different operations applied sequentially or in various combinations is wiser and more economical for a coal cleaning plant. Gibbs and Hill (1978) state that a modern coal cleaning plant is a continuum of technologies rather than one distinct technology.

Coal to be beneficiated is usually separated into three size fractions: coarse, intermediate and fine. Coal preparation for more extensive cleaning involves crushing the raw coal feed to smaller size fractions and cleaning the coal particles of smaller and smaller size fractions (EPRI, 1979). Additionally, each fraction can be treated with more specific and sophisticated methods and equipment.

The six major operations to obtain coal for better use are crushing, sizing, cleaning, dewatering, clarifying and drying (Decker and Hoffmann, 1963; Gibbs and Hill, 1978). The equipment available to achieve these operations is produced by different manufacturers and each unit has its own specific applications. These different operations and variations are described below. Most industrial coal cleaning plants employ some version of the basic operations shown in Figure 1.

2.2.1. *Crushing*. Crushing, or comminution, liberates the larger impurities present in raw coal and gives a product of more uniform size, easing transportation and proper combustion. Over the years, there has been a great tendency in the amount of coal which is crushed after mining (Decker and Hoffmann, 1963) and there has also been a general increase in the size of crushing and grinding equipment, which has made it possible for big tonnages of coal to be cleaned.

As Figure 1 shows, large size, raw coal is first crushed to satisfy handling, processing and end use requirements. Crushing generally generates a wide distribution of particle sizes which may range from a top size of 4-15 cm down to fine dust depending on the breakability of the material and type of equipment used in size reduction. Crushing frees mineral matter or at least particles having a very high content of mineral matter. Liberation of the mineral matter is essential if coal and mineral matter are to be physically separated. However, excessive size reduction is undesirable since ultrafine particles are difficult to deal with, clean and dehumidify. Thus, in many large plants rotary breakers are used for crushing because they produce relatively few fines. Large quantities of finer coals are being used for briquetting, though high percentages of dust in the coal render the briquetting operation difficulty and usually require that more binders be added (Schinzel, 1981). Favorable degrees of comminution with the production of little dust have been achieved with roll breakers (Winterhalder, 1975).

2.2.2. *Sizing*. Sizing breaks up feed coal into various fractions suitable for washing by different cleaning operations. Sizing can be achieved by wet or dry screening equipment, which use variations in the flow rates of different size particles through a fluid which is usually water or air. In a general cleaning plant the comminuted coal may be separated into coarse (+9.5 mm), fine (-9.5 mm to +0.6 mm) and ultrafine (-0.6 mm) sizes. These separations can be done by different types of screens. Vibrating deck screens are usually used for separating the rough solids from the finer particles. Water sprays are applied to the screens to prewet the coal and to suppres dust. For smaller particles of coal, sieve bends and classifying cyclones are used (Decker and Hoffman, 1963; Bechtel Corporation, 1977; Gibbs and Hill, 1978). Finer sizes suspended in water are separated by vibrating screens, classifying hydrocyclones or sieve bends.

2.2.3. *Cleaning*. The specific gravity of clean coal is smaller than that of mineral matter, pyritic sulfur or the other impurities found in raw coal. Most coals have a specific gravity that ranges from 1.12 to 1.70. Mineral matter and other contaminants have a wide range of specific gravities: pyritic sulfur 4.6-5.2; gypsum, kaolinite and calcite 2.3, 2.6, 2.7; and sandstone, clay and shale 2.6 (Decker and Hoffman, 1963; Gibbs and Hill, 1978). This difference in specific gravity is utilized by almost in all of the commercial wet physical coal cleaning technologies to separate larger size coal particles (larger than 0.5 mm) from their accompanied impurities.

Coal cleaning technologies depend on differences of specific gravities, however, are usually not effective in cleaning raw fine coal particles smaller than 0.5 mm in size. When the size of coal is reduced, the surface area of the resulting fine coal particles is increased and due to the increase in surface area, the time needed for separation is longer for fine particles. The pyrite in coal is often found in small and highly distributed particles requiring extensive size reduction of the raw coal to ease its removal.

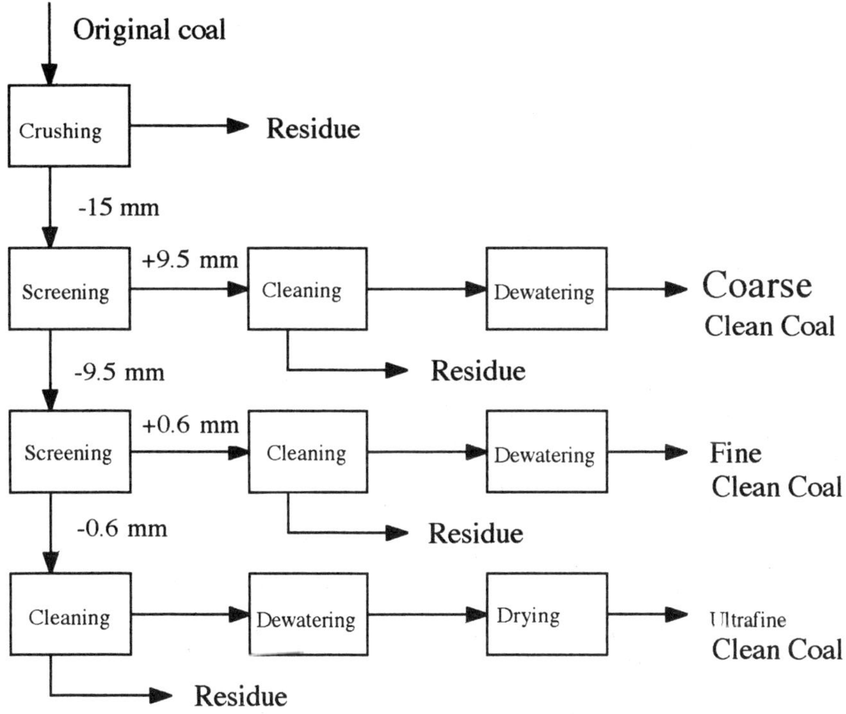

Figure 1. Basic operations in coal preparation, [after Wheelock and Markuszewski (1984)].

Five different types of equipment are generally used to clean coal:

Jigs are one of the oldest and still the most common equipment used for washing, separating and concentrating mineral wastes. Jigs were first introduced in Pennsylvania in 1873 for washing coal, are used today to treat minerals with a wide range of specific gravities, including gold (which has a specific gravity of 19.0). Jigging separates raw coal in water by pulsations that rearrange the particles so that the lighter coal particles move to the top of the bed while the heaviest impurity particles settle on the bottom and the two products can then be collected separately (Decker and Hoffman, 1963).

Dense media vessels utilize liquids that have higher specific gravities than water to separate coal from its contaminants. Particles of coal as small as 6 mm in size can be treated with these equipments. Finely ground magnetite mixed with water is used to make a slurry of the specific gravity required for the separation of coal and residue. The specific gravity of the separating medium can be adjusted by changing the concentration of the magnetite to achieve the desired result (Bechtel Corporation, 1977). The specific gravity can be changed to sustain a constant product, even with variation in the amount of impurities accomodated in the run-of-mine coal.

The utilization of dense media vessels is second only to the use of the jig. However, it is gaining acceptance because very productive separations between coal and residue can be obtained by its use. While there was only one commercially operated coal plant in the U.S. used a heavy medium process in 1948, by 1963, at least 100 plants in the U.S. were using this equipment to treat 49 million tons of coal annually (Decker and Hoffman, 1963). This equipment is extensively utilized all over the world (Hutton and Gould, 1982).

Magnetite recovery is important for the economics of the operation; since nearly 500 g of magnetite is used for each ton of raw coal to be cleaned. The amount of magnetite used depends largely on the amonunt of the coal cleaned; the smaller the size of the particles to be cleaned, the greater the amount of magnetite consumed.

Dense medium cyclones are used to achieve a better separation of intermedieate sized run-of-mine coal of -12.7 mm to +0.5 mm. In these equipment, the gravity separation of a dense media mixture is intensified by centrifugal force (Decker and Hoffman, 1963; Bechtel Corporation, 1977).

Hydrocyclones or water-only cyclones also enhance separation by centrifugal force, but they do not utilize a dense-media mixture. They are now being used more often in modern coal preparation plants as rougher separating devices that decrease the load on down stream cleaning equipment. They are also used to beneficiate very small floation size coal (-0.6 mm) (Bechtel Corporation, 1977; Gibbs and Hill, 1978). Hydrocyclone equipment is advantageous because it can clean large tonnages of raw coal at a relatively low cost and it needs relatively less space and maintenance (Gibbs and Hill, 1978). However, the power and water requirements of hydrocyclones are high (Bechtel Corporation, 1977).

The froth flotation operation is utilized to clean fine coal particles of less than 0.5 mm in size. Technologies used to remove the impurities combined with fine coal particles take benefit of the surface properties of coal that make it hydrophobic. The most important of these operations is froth flotation in which raw fine coal particles are fed into a water bath, through which air is bubbled. The hydrophobic coal particles bond to these air bubbles with the help of flotation reagents and can be separated in the surface froth. The heavier coal impurities, which are not hydrophobic, sink and settle on the bottom of the tank (Gibbs and Hill, 1978).

The increasing demand for coal has forced utilization of low rank coal slimes. Tests of beneficiation of low rank coals can be carried with froth flotation. The success of this method depends on the possibility to form a stable contact between air bubbles and the surface of coal particles. The studies of the detachment force of air bubble from the surface of solids in liquids (Janczuk, 1983) indicated that the stability of the solid-air bubble-liquid system depends on the size of particles and air bubbles, surface tension of liquid, surface free energy of solid and liquid film pressure under the air bubble. Therefore, when surface free energy is known for various ranks of coal, it is possible to understand the stability of the coal particle-air bubble aggregates as well as the efficiency of the flotation operation. These forces calculated theoretically by Janczuk et al. (1988) and correlation between the stability of three phase coal-air bubble-water system and the natural flotability of these coals were conducted. Measurements of the detachment force of an air bubble from the surface of various coal ranks were carried out in water (Janzcuk et al., 1989a). These values were compared with those calculated from the contact angle values measured in coal-air bubble-water system and from surface free energy of the studied coals. It is shown that the detachment force of air bubble from coal surface depend on its rank. taking into consideration the values of the detachment force, the studied coals may be divided into three groups of increasing ranks.

2.2.4. *Dewatering*. Significant amounts of water remain in the coal after wet physical coal cleaning operations. This additional moisture decreases the calorific value of the coal, increases transportation and handling costs by adding to the weight of the coal transport and freezes in winter causing problems in handling and use. To decrease the water content of the coal after primary cleaning operations, coals are first dewatered mechanicaly and then dried thermally in the preparation plants.

Screens are the most commonly used coal dewatering equipment and improved designs in recent years have enhanced their dewatering capabilities. Sieve bends are now used throughout the world for preliminary dewatering before the coal is conveyed to the vibrating screens and centrifuges.

Centrifugal dryers are utilized to dewater coal particles of 38mm and smaller in a spinning centrifuge. The effectiveness of this equipment depends on the centrifugal force developed, the size of the particles and length of time the coal is left in the centrifugal dryer (Decker and Hoffman, 1963; Bechtel Corporation, 1977; Gibbs and Hill, 1978).

The vacuum filter is on of the most common dewatering devices for fine sized coal particles. 90% of vacuum filters are of the continuous vacuum type and these are less expensive to purchase and install per unit of filtration area and require less space than other filters. Solids form a cake on the leaves of this filter and are withdrawn by blasts of air (Bechtel Corporation, 1977).

After dewatering, different sizes of coal are generally recombined but they may also be used separately for different purposes. The coarse product is generally crushed further before shipment. Combining the ultrafine coal with the larger sizes usually eases handling and transportation unless the material is to be transported and used in a slurry. Clean coal may be stocked in open piles, bins or large silos at a preparation plant before shipment. Coals in an open stock piles are liable to oxidation and spontaneous combustion and soluble constituents of the coal may be leached by rainfall, thus care must be taken in the storage.

2.2.5. *Thermal Drying*. Thermal drying equipment separates water in cleaned coals by passing hot gases over them. It can remove The surface moisture from cleaned coal can be removed in this manner although the internal moisture is largely unaffected. Different types of thermal dryers are currently in utilization; fluidized bed, rotary dryer, screen dryer, flash dryer and vertical dryer. The fluidized bed dryer introduced to the coal industry in 1955, represented more than half of the drying equipments in operation (Hutton and Gould, 1982), processing over 72% of the coal to be dried (Deurbrouck and Hucko, 1979).

2.2.6. *Clarifying*. Waste disposal is an important aspect in coal cleaning operations. Dry solid refuse may be stocked in open piles which are compacted and sealed to prevent admission of air and water. Unless such access is prevented, spontaneous combustion of organic matter may occur and iron pyrites are oxidized to acids which are washed subsequently by rainfall causing pollution. In the case of surface mines, the refuse may be returned to the pit and buried. Fine waste particles are usually entrusted to settling pools. After the bulk of the solids settles, the water is regained and returned to the preparation plant. Closed water circuits are necessary for pollution control and modern plants have extensive facilities for water clarification and regaining.

Water clarifiers are utilized in most coal preparation plants to remove nearly all of the solid refuse from process water. Preparation plants have very high recirculating requirements of up to 60,000 liters per minute for a plant cleaning 1,200 tons of coal per hour, 80% of this water is recycled and overall water management is of key importance for the economics of the operation (Spaite, 1979).

Now, all wet coal preparation plants have some type of partially closed water circuit systems. Recirculation and clarification of process waters are critical parts in cleaning plant operation because such treated waters decrease the loss of fine coal in waste, prevent the build up of solids in the water recirculated to the cleaning equipments. Some coal preparation plants discharged a large portion of the polluting water into streams to keep the level of solids within the system low in the past. Currently this is not practiced due to the restrictions brought to reduce pollution.

Gravitational clarifier is the most commonly used equipment for water clarification. It is a large circular pool in which solids settle on the bottom where they can be pumped away, leaving the clarified water to be removed at the top. Chemicals for flocculation are added to the water in the clarifier to increase the rate of clarification by the formation of flocs which settle more easily.

2.3. LEVELS OF COAL PREPARATION

All coals are not good candidates for coal cleaning, owing to variations in their compositions and physical properties. In addition to that all coal preparation operations have several problems and restrictions that may decrease potential usage of the equipment.

Depending upon the physical constituents of the run-of-mine coal and the market demand for the cleaned coal, various preparation plants may employ different combinations of the operations and equipment described above. For the cleaning of finer coal particles and removal of bigger amounts of sulfur, mineral matter and other impurities more rigorous and expensive levels of cleaning may be required (Gibbs and Hill, 1978).

All of the preparation plants may not employ the full spectrum of operations described in Figure 2 and some may employ more extensive and more complicated schemes. The general layout of most prepartion plants can be derived from the basic scheme by adding or taking out different steps. The number and type of steps which are used in any given plant usually determine the level of beneficiation which is accomplished and therefore it becomes possible to classify cleaning plants according to the level of preparation. The following system of classification was proposed by Philips and De Rienzo (1976).

Level A: No preparation.
Level B: Crushing to control top size with limited, removal of coarse refuse and trash.
Level C: Beneficiation of coarse coal. The $<$ 9.5 mm coal is shipped without cleaning.
Level D: Coarse and fine sizes are cleaned. Ultrafine coal is not cleaned and may be shipped as is or discarded. Some thermal drying may be employed.
Level E: All sizes are cleaned. Thermal drying of finer sizes is generally required to limit moisture content.
Level F: Most rigorous degree of coal beneficiation. This involves production of two or more usable products in a multistream coal cleaning system.

A similar system was given by Horsfall (1984) .

Route A: The seam could be sold untreated as a power station fuel for on site power generation. The high ash and low calorific value would not allow long distance transportation of unwashed coal seam to be cost effective, hence its use is on site power generation.
Route B1: The seam could be divested of its highest density component and sold as a medium to high grade steam coal, with about 15% mineral matter and a calorific value in the range 26-28 MJ/kg.

Route B2: The coal seam could be washed in a single stage, removing components over 1.6 specific gravity as discard. The resulting washed coal then be a high grade steam coal with a calorific value in excess of 28 MJ/kg. The penalty of this method is the presence of components of 1.8-2.0 specific gravity in the discards. They contain combustibles and if dumped incautiously will cause spontaneous combustion.
Route C: Removal of the components of specific gravity greater than 1.8 as the discard, as in B2, followed by rewashing the clean coal at a still lower specific gravity to produce a low mineral matter coal of about 7% mineral matter and over 30 MJ/kg and a steam coal of 27-28MJ/kg as a secondary product. If a steam coal of 28 MJ/kg is required, then the washing density of the high density washing stage may have to be reduced to remove material over 1.7 or even over 1.6 specific gravity.

A coal preparation plant can be designed to produce two basic washed products, but by taking into account the washing characteristics of the coal and the flexibility of the beneficiation plant, product qualities may be arranged according to market demands within quite wide limits.

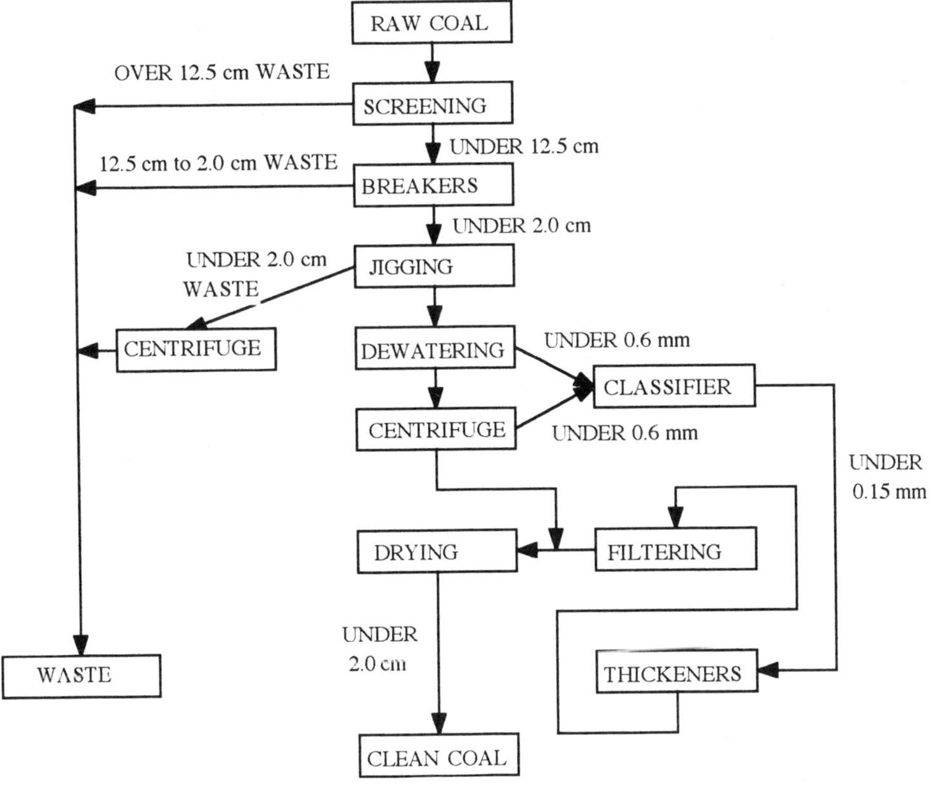

Figure 2. General purpose coal preparation plant.

3. Developing Methods in Physical Coal Cleaning

Physical coal cleaning has been developing for a long time and this development is enduring with minor progresses in equipment and technique occurring every year but major developments take place rarely. Some of the more encouraging and few of the more unusual new developments are presented below. Several of the new techniques have arrived or about to arrive the industrial stage, whereas others are still in an introductory stage.

3.1. FRAGMENTATION BY CHEMICALS

One of the more unconventional developments which has been presented on a bench scale is a technique of comminuting coal by the application of specific chemical agents such as anhydrous ammonia (Quackenbush et al., 1979; Howard and Datta, 1977). When coals are treated with anhydrous ammonia they are pulverized and easily crumbled, particles of the treated coals show cracks and an increase in surface area (Matida et al. 1977). It appears that the bedding planes and interfacial boundaries between coal and mineral matter are separated. Bonds between organic and inorganic material are cleaved selectively, much of the mineral matter is freed without excessive size reduction. The method seems to have an alternative over mechanical crushing where fragmentation tends to be less selective.

Chemical comminution depends on the nature of the coal, the chemical agent used and the technique and conditions of application (Howard and Datta, 1977). Although some coals are reactive than others, the coal properties which are responsible for this difference are not well proven. Coal rank is considered to be a factor, but other properties such as pore structure, swelling tendency and maceral composition are also important. Among various chemicals which fragment coal, anhydrous ammonia is one of the most efficient and it can be used either as liquid or vapor forms. Hydrogen fluoride is a more effective comminuting agent than ammonia, coals which are slowly affected by ammonia appeares to be crumbled instantaneously in hydrogen fluoride (Jensen, 1979). One of the advantages with ammonia is good separation of mineral matter while producing a minimum amount of material below 0.15 mm in size (Howard and Datta, 1977).

Coal does not react chemically with either liquid or gaseous ammonia under the conditions used for comminution since most of the ammonia can be recovered by desorption and washing with water. In addition to fragmentation, other physical properties of coal are not affected by this treatment (Datta et al., 1976; Howard and Datta, 1977).

It has been suggested that solvent swelling of coal is the cauese of the comminuting action of ammonia as well as other agents (Keller and Smith, 1976). The volume of bituminous coal can increase up to 50% or more with the ammonia treatment. Since various constituents of the coal do not swell equally, stresses developed along boundaries between constituents cause comminution where the stresses are concentrated.

A number of experiments in which coal was first fragmented and then subjected to a washability test showed that chemical fragmentation freed more pyrite than mechanical size reduction (Datta et al., 1976).

Plant designs employing chemical fragmentation and cost estimates based on these designs have been reported (Datta et al., 1976; Quackenbush et al., 1979). The general procedure first involves crushing of the coal to 4 cm top size by mechanical means. The crushed coal is placed next in a large vessel which is evacuated and then pressurized with saturated ammonia vapor at ambient temperature. After an experimental time of 1-2 hr, the pressure is released and the coal is transferred from the treatment vessel. The coal is then washed with hot water to separate any traces of ammonia and is subsequently beneficiated by conventional methods.

3.2. SEPARATIONS USING SURFACE PROPERTIES OF COALS

Several new separation techniques utilize the difference in surface characteristics between the organic and inorganic constituents of coal, but among these techniques, froth flotation is the only one in extensive use. Even though froth flotation has been used in coal cleaning plants for a number of years, it has several disadvantages (Burger, 1980). Generally pyrites and some clays tend to float with the coal in a poor separation operation. Also, oxidized and lower rank coals do not react well to froth flotation methods because oxygen containing groups make these coals hydrophilic. Kelemen et al. (1991) used XPS in the study of iron and inorganic sulfur forms in coals. Results indicate the presence of iron oxides or oxyhydroxides. No evidence was found for the presence of pyritic iron at the surface of any coal sample. This shows that the initial surface oxidation of the pyrite is very rapid and unavoidable even in carefully handled samples. The aim of much recent research and development has been to overcome these problems.

The isolation of coal and mineral matter by froth flotation can be improved by careful control of the operating conditions (Miller and Deurbrouck, 1982). The pH of the mixture, the characteristics of the metal ions present and the reagents used can affect the separation. Certain metal ions can react differently at various pH values either to increase or decrease pyrite removal. Kerosene utilized as a collector for coal can also increase the separation of pyrite.

Another encouraging method to separate coal and pyrite is based on the differences in the rate of flotation. The rate of recovery in the floating of a bituminous coal can be 10 to 30 times faster than the rate of recovery of pyrite under mild conditions. Thus, by using such mild conditions and short residence times, much of the coal can be isolated before the slower floating pyrite is seen in the froth. On the other hand, an oily collector and more severe conditions can increase the rate of flotation of pyrite more than that of coal and result in a unsuccesful separation.

For some coals appropriate liberation of the mineral matter may be accomplished only by grinding to such a fine size that common froth flotation is not practical. In this situation it may be possible to use a new flotation technique which uses micrometer size gas bubbles. This method is developed in Germany and has reached to the stage a large pilot plant (Bethe, 1980). Coal samples consisting of -0.030 mm size particles with an average mineral matter content of 30% have been cleaned in this unit to produce a clean product of 12% mineral matter.

Flotation of oxidized coals received new stimulus from the study by Wen and Sun (1981), which suggested that the hydrophobicity of oxidized coal was maximum when the coal was at its isoelectric point and that maximum adsorption of an oily collector by the coal occured when the collector was at its isoelectric point. The isoelectric point of these substances was controlled by pH, concentration of heteropolar reagents and adsorption of metal ions. Through adsorption of ferrous hydroxide and control of pH, the isoelectric point of an oxidized coal can be made to agree with that of an oily collector resulting in optimim flotability. This condition corresponded to a pH of 6.5-7.0 when ferrous hydroxide was used.

Another isolation technique of important concern is one based on the use of fuel oil to agglomerate specifically ultrafine coal particles suspended in water. The oil is mixed with the suspension vigorously making the oil to be dispersed into droplets that become attached to coal particles (Pawlak et al. 1984). The partially oil coated particles join together and form agglomerates larger than 1 mm which can be separated on a screen with the exception of pyrite or marcasite and most of the mineral particles are unaffected by this method. Therefore, selective enrichment of minerals like pyrite and marcasite occurring in coals may be utilized for catalytical purposes if the coal is going to be used in liquefaction (Gollakota et al. 1989).

But, on the other hand, if the coal is going to be used in combustion presence of pyrites, of course, creates problems. Most minerals found in coal are hydrophilic and do not become oil coated, pyrite and marcasite are the exception and some of the earliest oil agglomeration experiments indicated that pyrite particles were easily coated by fuel oil and agglomerated (Perrot and Kinney, 1921). Different techniques have been suggested to solve this problem. One method involves pretreating ultrafine coal with iron oxidizing bacteria which give the surface of the pyrite hydrophilic properties. At first it was claimed that the effect created by bacterial treatment was due to surface oxidation (Capes et al., 1973), but more recent work (Kempton et al., 1980) showed that the effect was due to adsorption of *thiobacillus ferrooxidans* on the surface of the pyrite.

A couple of versions of the oil agglomeration techique have been used in a small number of pilot plants located in different countries. Commercial scale oil agglomeration systems have been used to separate ultrafine coal from samples with a high clay content at two cleaning plants in Germany (Mehrotra and Sastry, 1980). Also big demonstration plants employing the Shell pelletizing separator are being used to treate coals in both Japan and England. This unique equipment developed by Shell International Petroleum Company produces larger and more uniform agglomerates by multistage treatment in a single unit. A large pilot plant employing more conventional mixing and preemulsified oil has been founded in Australia. In addition to these, development and application of oil agglomeration methods are progressing in India (Hutton and Gould, 1982). The oil agglomeration operation which is still in the research and development stage, utilizes the same hydrophobic properties of coal particles, use oils instead of flotation reagents. Nonpolar liquids such as kerosene or diesel oil are used as collectors to enrich coal. These reagents are of hydrocarbon types. Janczuk et al. (1989b) measured contact angles in coal-normal alkanes film-air bubble-water systems. The values of the contact angles were found to be depend on the rank of coal.

3.3. MAGNETIC AND ELECTROSTATIC CLEANING OPERATIONS

Differences in the magnetic properties between the organic matter of coal and pyrites and the other minerals is utilized in separating these components by magnetic methods (Oder, 1976; Liu and Lin, 1976; Maxwell, 1978; Beddow, 1981). On the other hand, since the difference in magnetic susceptibility of the constituents is minute, the separation needs the combination of an intense magnetic field and a large field gradient. This combination has been accomplished in several laboratory devices and in commercial separators developed for the ceramic industry to isolate weakly magnetic impurities. The industrial equipment and some laboratory devices for conducting high gradient magnetic separation (HGMS) use a matrix of magnetizable filamentary material placed in an extensive magnetic field. High field gradients in the neighborhood of the filaments in association with the intense field capture even slightly magnetic particles when a suspension of particles is transported through the matrix. Particles with nonmagnetic properties pass through the matrix unaffected. After the matrix becomes charged with magnetic particles, the field is turned off and the particles are purged from the matrix. Different types of filamentary material can be utilized for the matrix and the best data are obtained when ferromagnetic stainless steel is used as the matrix material because it is both magnetizable and resistant to corrosion.

In industry both batch and continous HGMS units are used (Oder, 1976; Ianicelli, 1976). In the batch equipment the separator matrix is covered by a powerful electromagnet and thus it remains fixed, whereas in one type of continuous unit the matrix material is in a series of canisters which are situated on a carousel, (Hise et al., 1981). As the carousel spins, the canisters are brought subsequently into a zone where there is a very strong magnetic field. While the canisters move through the magnetic field, a suspension of particles to be isolated is conducted through the canister first followed by a purging stream to remove mechanically the captured coal particles. After leaving the magnetic field, the canisters are cleaned to remove the magnetic impurities which have been collected. The canisters then proceed around the circle and are brought back to the magnetic field for recharging.

Laboratory tests have indicated that HGMS can accomplish cleaning efficiencies equal to or greater than froth flotation of pyrite (Hucko and Miller, 1980). For comparison purposes ultrafine coals were first cleaned by conventional froth flotation and the floated coals were then treated either by HGMS or by the method which floats the pyrite. The two methods accomplished comparable pyritic sulfur reductions when applied to 0.50 mm size coals but HGMS achieved much greater pyritic sulfur reductions when applied to 0.15 mm size coals. In addition HGMS could separate other mineral matter to a greater extent.

The cost of magnetic separation could be decreased if the difference in magnetic susceptibility of coal and mineral matter was increased. One of the methods to increase the difference in magnetic susceptibility of coal and mineral matter involves treating coal with microwave energy to heat the pyrite minerals selectively, without affecting the coal (Ergun and Bean, 1968; Bluhm et al., 1980; Zavitsanos et al., 1982). Pyrite can be thermally dissociated and partially converted into pyrrhotite by heating to 450°C. Magnetic susceptibility of pyrrhotite is about four orders of magnitude greater than that of the pyrite, therefore only 1% of the pyrite needs to be converted into pyrrhotite to produce a composite material having a susceptibility which is 100 times larger than that of pyrite. Small amounts of coal or pyrite under an unreactive atmoshere have been heated with microwave energy for short periods (e.g. 30-60 s) and the partial conversion of pyrite to pyrrhotite was observed. Microwave heating alone has converted 40 to 60% of the pyritic sulfur from various bituminous coals in laboratory experiments with emission of gaseous sulfur compounds (H_2S, COS, SO_2). More sulfur was removed when coal heated in this manner has been magnetically separated. Weight losses of about 6% have been observed when specific bituminous coals were treated with microwave energy.

Electrostatic separation techniques seem to be promising dry coal-cleaning methods and they can be used for separating coal macerals. When this method is applied, separation is accomplished by the action of electrical field on small particles holding various electrical charges. Particle charging is achieved by different ways. Particles not conducting electricity (e.g., coal) keep the charge which holds them to the surface of the separator whereas particles which are relatively conducting (e.g., pyrite and other minerals) quickly lose the charge and are separated by the charging surface. Good separation of coal and pyrite was accomplished with electrostatic separation when it was used in a closely sized, mixture of cleaned coal and precleaned pyrite (Abel et al., 1973). Further experiments of the equipment together with a centrifugal air separator succeeded overall pyritic sulfur reductions of 35 to 60% and coal recoveries of 80 to 90%.

Interesting results were observed in bench scale electrostatic separation systems using a developed drum separator and other new aspects (Advanced Energy Dynamics, 1980). This system was utilized to clean several coals at input rates of 5 to 70 kg/h. Coal was first crushed to below 0.074 mm, the size used in boilers fired with pulverized coal. Particles smaller than 0.020 mm were separated from the feed by a classifying cyclone; these particles bypassed the cleaning system. The remaining particles in an inert gas were first removed by a different classifying cyclone into a coarse fraction of sizes larger than 0.053 mm and a fine fraction with particles smaller than 0.053 mm and each fraction was transferred to a discharge ionizer where a 60 Hz alternating current electrical field was used to deagglomerate the particles and by discharging electrical charges on the particles. The particles were next isolated from the gas by a cyclone and cleaned with a revolving drum electrostatic separator (Inculet et al., 1982). The results accomplished with this system indicated that the product obtained from the coarse fraction was cleaner than that obtained from the fine fraction.

3.4. SONIC ENHANCEMENT OF PHYSICAL COAL CLEANING

Sonic energy can be used in coal cleaning,(Celik and Somasundaran, 1980; Bonner, 1989; Buttermore et al., 1990) to disperse fine clay particles particles which create problems during coal preparation (Arnold and Aplan, 1986; Fairbanks et al., 1990). Laboratory tests done by Slomka and Buttermore (1991) have shown substantial benefits resulting from the use of sonic energy to deslime Illinois No.6 coal. The ash forming mineral matter removed as a result of brief sonic treatment ranged from 11.6 to 31.1% accompanied by a minimum loss of combustible material. Coal recoveries up to 99.8% were achieved. Exposure of coal tosonic energy was observed to enhance the selctivity of flotation. Enhanced flotability of the coal can be best achieved by rinsing following sonic treatment, but prior to cleaning by flotation. It was shown that the beneficial effect of sonic treatment is not limited to dispersion of all mineral particles, but also includes enhanced separation of coal-mineral composite particles.

4. References

Abel, W.T., Zulkowski, M., Brady, G.A. and Eckerd, J.W. (1973), 'Removing pyrite from coal by dry-separation methods', U.S. Bureau of Mines Report of Investigations 7732.

Advanced Energy Dynamics, Inc. (1980), 'Pilot plant program for the AED advanced coal cleaning system', Phase II Interim Final Report, Natick, Massachusetts.

Arnold, B.J. and Aplan, F.F. (1986), 'The effect of clay slimes on coal flotation. Part 1. The nature of the clay', Int.J.Miner.Process., **17**, 225-242.

Attar, A. (1978), 'Chemistry, thermodynamics and kinetics of reactions of sulfur in coal-gas reactions', Fuel, **57**, 201-211.

Attar, A. and Corcovan, W.H. (1977), 'Sulfur compounds in coal', Ind.Eng.Chem. Prod.Res.Dev., **16**, 168-170.

Bechtel Corporation (1977), ' Environmental Control Implications of Generating Electric Power from Coal', Technology Status Report, Appendix A (Part 1), Coal Preparation and Cleaning Assessment Study (Preparaed for Argonne National Laboratory; Work sponsored by the Division of Enviromental Control Technology, U.S. Department of Energy, ANI/ECT-3, pp. 222-224.

Beddow, J.K. (1981), 'Dry separations techniques', Chem.Eng., **88**, 70-84.

Bethe, W.P. (1980), 'Impact of changing raw coal properties and market trends on coal preparation', in, Proceedings of the Fifth International Conference on Coal Research, Düsseldorf, Germany, Vol. II, pp. 701-721.

Bluhm, D.D., Fanslow, G.E., Beck-Montgomery, S. and Nelson, S.O. (1980), 'Selective magnetic enhancement of pyrite in coal by dielectric heating at 27 and 2450 MHz', in B.R.Cooper and L.Petrakis (eds.), Chemistry and Physics of Coal Utilization, AIP Conf.Proc. No. 70, Am. Inst. Physics, New York, pp. 417-437.

Bonner, H.E. (1989), 'Ultrasonic's application to fine coal recovery', Annual meeting of Society of Mining Engineers, Nevada, Prepr. No. 89-164.

Buttermore, W.H., Slomka, B.J. and Dawson, M.R. (1990), 'The sonic enhancement of physical coal cleaning', Proc. 6th Annu, Coal Preparation, Utilization and Enviromental Control Contractors Conf., U.S. DOE, Pittsburgh, pp. 283-289.

Boudou, J.P., Boulege, J., Malechaux, L., Nip, M., de Leeuw, J.W. and Boon, J.J. (1987), 'Identification of some sulfur species in a high organic sulfur coal, Fuel, **66**, 1558-1569.

Burger, J.R. (1980), 'Froth flotation is on the rise', Coal Age, **85**, 99-108.

Capes, C.E., Mc Ilhinney, A.E., Sirianni, A.F. and Puddington, I.E. (1973), 'Bacterial oxidation in upgrading pyritic coal', Can. Min. Metall. Bull., **66**, 88-91.

Celik, M.S. and Somasundaran, P. (1980), 'Effects of preteatments on flotation and electrokinetic properties of coal', Colloids and Surfaces, **1**, 121-124.

Cole, R.M. (1979), 'Economics of coal cleaning and flue gas desulfurization for compliance with revised NSPS for uitility boilers', in S.E.Rogers and A.W.Lemmon Jr. (eds.), Proceedings of Symposium on Coal Cleaning to Achieve Energy and Enviromental Goals, Florida, Vol. I, pp. 324-359.

Corriveau, M.P. and Schapiro, N. (1979), 'Projecting data from samples', in J.W.Leonard (ed.), Coal Preparation, Am. Inst. Mining, Metallurgical and Petroleum Engrs., New York, pp.4.30-4.32.

Datta, R.S., Howard, P.H. and Hanchett, A. (1976), 'Feasibility study of precombustion coal cleaning using chemical comminution', Final Report, FE-1777-4, Syracuse Research Corp., Syracuse, New York.

Decker, H.W.Jr. and Hoffmann, J.N. (1963), 'Coal Preparation', Vol. I, The Pennnsylvania State University, University Park, Pennsylvania, pp. 181-295.

Deurbrouck, A.W. and Hucko, R. (1979), 'Coal Cleaning and Desulfurization', Direct Use of Coal, Vol. II, Part A, Working Papers, Appendix IV, Washington D.C., p. 25

EPRI Journal (1979), 'More Coal Per Ton', Electric Power Research Institute, California, p.10.

Ergun, S. and Bean, E.H. (1968), 'Magnetic separations of pyrite from coals', U.S.Bureau of Mines Report of Investigations 7181.

Evans, U.R. (1981), 'Corrosion and oxidation of metals', Edward Arnold, London, p.948.

Fairbanks, H.V., Morton, W.E. and Wallis, J.W. (1990), 'Mineral recovery enhanced by ultrasonic treatment', Annual meeting of Society of Mining Engineers, Nevada, Prepr. No. 90-21.

Gibbs and Hil, Inc. (1978), 'Coal Preparation for Combustion and Conversion', Electric Power Research Institute, EPRI AF-791, Project 466-1, Final Report, California, pp. 1-3.

Given, P.H. and Wyss, W.F. (1961), 'The chemistry of sulfur in coal', Br. Coal Util. Res. Assoc. Mon. Bull., **25**, 165-179.

Gluskoter, H.J. (1975), 'Mineral matter and trace elements in coal', in S.P.Babu (ed.), Trace Elements in Fuel, Adv. in Chem. Series No. 141, Am.Chem.Soc., Washington, D.C., pp. 1-22.

Gollakota, S.V., Lee, J.M. and Davies, O.L. (1989), 'Process optimization of close-coupled integrated two-stage liquefaction by the use of cleaned coal', Fuel Processing Technology, **22**, 205-216.

Güllü, G., Durusoy, T., Özbaş, T., Tanyolaç, A. and Yürüm, Y. (1992), 'Biodesulphurization of coal', in Y.Yürüm (ed.), Clean Utilization of Coal and its Products, NATO ASI Series, Series C, Kluwer Academic Publishers, Dordrecht.

Haggin, J. (1991), 'Coal, in the national energy picture, the Department Energy's clean-coal program is key to future utility of coal', Chemical and Engineering News, **69** (24), 32-35.

Harchand, K.S., Taneja, S.P., Raj, D. and Sharma, P. (1989), 'Mössbauer studies of coal ash', Fuel Processing Technology, **21**, 19-24.

Hise, E.C., Wechsler, I. and Doulin, J.M. (1981), 'The continuous separation of dry crushed coal at one ton per hour by high-gradient magnetic separation, ORNL-5571, Oak Ridge National Laboratory, Tennessee.

Horsfall, D.W. (1984), 'Beneficiation, storage and handling of coal', 9th RSA/ Israel Binational Symposium on Coal Technology, Beer Sheva, Israel, pp. 129-154.

Howard, P.H. and Datta, R.S. (1977), 'Chemical comminution, a process for liberating the mineral matter in coal', in T.D.Wheelock (ed.), Coal Desulfurization: Chemical and Physical Methods, ACS Symposium Series 64, Am. Chem. Soc., Washington, D.C., pp. 58-69.

Hucko, R.E. and Miller, K.J. (1980), 'A technical performance comparison of coal-pyrite flotation and high-gradient magentic separation, RI-PMTC-10(80), U.S. Department of Energy.

Hutton, C.A. and Gould, R.N. (1982), 'Cleaning up Coal', Ballinger, Massachusetts.

Ianicelli, J. (1976), 'New developments in magnetic separation', IEEE Trans. Magn. MAG **12**, 436-443.

Inculet, I.I., Bergougnou, M.A. and Brown, J.D. (1982), 'Electrostatic beneficiation of coal', in Y.A.Liu (ed.), Physical Cleaning of Coal: Present and Developing Methods, Marcel Dekker, Inc., New York, pp. 87-131.

Janczuk, B. (1983), 'Detachment force of air bubble from the solid surface in water', Journal of Colloid Interface Science, **93**, 40-52.

Janczuk, B., Wojcik, W. and Bialopiotrowicz, T. (1988), 'Wettability of coal surface and its surface free energy components, Croat.Chem.Acta, **61**, 51-63.

Janczuk, B., Wojcik, W. and Bialopiotrowicz, T. (1989a), 'Stability of coal-air bubble-water system', Fuel Science and technology International, **7**, 989-1004.

Janczuk, B., Wojcik, W. and Bialopiotrowicz, T. (1989b), 'The influence of n-alkanes on wettability of low coal ranks', Fuel Science and technology International, **7**, 969-987.

Jensen, H.P. (1979), 'Process for comminuting and reducing the sulfur and ash content of coal', U.S.Patent 4,169,710.

Kelemen, S.R., George, G.N. and Gorbaty, M.L. (1990), 'Direct determination and quantification of organic sulfur forms by XPS and K-edge absorption spectroscopy', Fuel Processing Technology, **24**, 425-429.

Kelemen, S.R., Gorbaty, M.L., George, G.N. and Kwiatek, P.J. (1991), 'Surface composition of iron and inorganic sulfur forms in Argonne premium coals by X-ray photoelectron spectroscopy', Energy and Fuels, **5**, 720-723.

Keller, D.V. and Smith, C.D. (1976), 'Spontaneous fracture of coal', Fuel **55**, 273-280.

Kempton, A.G., Moneib, N., McCready, R.G.L. and Capes, C.E. (1980), 'Removal of pyrite from coal by conditioning with *thiobacillus ferrooxidans* followed by oil agglomeration', Hydrometallurgy, **5**, 117-125.

Killmeyer, R.P. (1982), 'Coal preparation for synfuels-where do we stand?', Energy Prog., **2**, 14-17.

Liu, Y.A. and Lin, C.J. (1976), 'Assessment of sulfur and ash removal from coals by magnetic separation', IEEE Trans.Magn., MAG **12**, 538-550.

Matida, M. Nishiyama, Y. and Tamai, Y. (1977), 'Gasification of coals treated with non-aqueous solvents. 1. Liquid ammonia treatment of a bituminous coal.' Fuel, **56**, 177-180.

Maxwell, R. (1978), 'High-gradient magnetic separation in coal desulfurization', in B.R.Cooper (ed.), Scientific Problems of Coal Utilization, DOE Symposium Series 46, U.S. Department of Energy.

Mc Gannon (Ed.) (1971), 'The making, shaping and treating of steel', 9th Ed., United States Steel Corporations, p.1069.

Mehrotra, V.P. and Sastry, K.V.S. (1980), 'Oil agglomeration offers technical and economical advantages', Min. Eng. (N.Y.) **32**, 1230-1235.

Miller, K.J. and Deurbrouck, A.W. (1982), 'Froth flotation to desulfurize coal', in Y.A.Liu (ed.), Marcel Dekker, Inc. New York, pp. 255-291.

Ndaji, F.E. and Imobighe, G.A. (1989), 'Controlling effects of ash, total sulfur and chemical forms of sulfur in coals on the selection of components of coal blends for making metallurgical cokes', Fuel Processing Technology, **21**, 49-61.

Oder, R.R. (1976), 'High gradient magnetic separation theory abd applications', IEEE Trans. Magn., MAG **12**, 428-435.

Oder, R.R., Datta, R.S. and Pommier, L.W. (1981), 'Solid coal refining', Second Symposium on Separation Science and Technology for Energy Applications, Gatlinburg, Tennessee.

Paulson, L.E. and Fowkes, W.W. (1968), 'Changes in ash composition of North Dakota lignite treated by ion exchange', U.S. Bureau of Mines Report of Investigations 7176.
Pawlak, W., Saymocha, K., Janiak, J. and Ignasiak, B. (1984), 'Application of agglomeration for upgrading low rank coal', in Proceedings of 9th Annual EPRI Conference on Clean liquid and Solid Fuels, Palo Alto, CA.

Perrot, G.St.J. and Kinney, S.P. (1921), 'The use of oil in cleaning coal', Chem.Metall.Eng., **25**, 182-188.

Philips, P.J. and DeRienzo, P.P. (1976), 'Steam coal preparation economics', in Papers Presented Before the Second Symposium on Coal Preparation, NCA/BCR Coal Conference and Expo III, National Coal Association, Washington, D.C., pp. 50-63.

Pommier, L.W., Oder, R.R. and Datta, R.S. (1981), 'Separation of coal components to achieve improvements in coal utilization', Second Symposium on Separation Science and Technology for Energy Applications, Gatlinburg, Tennessee.

Quackenbush, V.C., Maddocks, R.R. and Higginson, G.W. (1979), 'Chemical comminution: An improved route to clean coal', Coal Min. Process. **16**, 68-72.

Schinzel, W. (1981), 'Briquetting', in M.A.Elliot (ed.), Chemistry of Coal Utilization, John Wiley and Sons, Inc., New York, pp. 609-664.

Simbeck, D.R. (1981), 'Economic and technical considerations of coal preparation for synfuel feedstocks', 91st National Meeting of Am. Inst. Chem. Engr., Detroit.

Slomka, B.J. and Buttermore, W.H. (1991), 'Sonically enhanced beneficiation of Illinois No.6 coal', Fuel Procesing Technology, **29**, 133-142.

Sondreal, E.A., Tufte, P.H. and Beckering, W. (1977), 'Ash fouling in the combustion of low rank western U.S. coals', Combust. Sci. Technol., **16**, 95-110.

Spaite, P.W. (1979), ' Enviromental Assessment of Coal Cleaning Operations: Technology Overview', U.S. Enviromental Protection Agency: EPA-600/7-79-073, p.33

van Krevelen, D.W. (1961), Coal, Elsevier, New York.

Wen, W.W. and Sun, S.C. (1981), 'An electrokinetic study on the oil flotation of oxidized coal', Separation Sci. Technol., **16**, 1491-1521.

Wheelock, T.D. and Markuszewski, R. (1984), 'Coal preparation and cleaning', in B.R.Cooper and W.A,Ellingson (eds.), The Science and Technology of Coal and Coal Utilization, Plenum, New York, pp. 47-123.

Whitehurst, D.D. (1978), 'A primer on the chemistry and constitution of coal', in J.W.Larsen (ed.), Organic Chemistry of Coal, ACS Series 71, Am.Chem.Soc.,Washington D.C., pp. 1-3.

Winterhalder, J. (1975), Aufbereitungstechnik, **4**, 168-171.

Yürüm, Y. and Tuğluhan, A., (1990), 'Supercritical extraction and desulphurization of Beypazarı lignite by ethyl alcohol/Na0H treatment. 1. Effect of the alcohol/coal ratio and NaOH', Fuel Science and Technol. International, **8**, 87-105.

Yürüm, Y., Özyörük H., Gülce, H. and Tuğluhan, A., (1990) 'Supercritical extraction and desulphurization of Beypazarı lignite by ethyl alcohol/Na0H treatment. 3. Electrochemical characterization of organic sulphur compounds, Fuel Science and Technol. International, **8**, 699-718.

Zavitsanos, P.D., Golden, J.A. and Bleiler, K.W. (1982), 'Coal desulfurization by a microwave process', Technical Progress Report, General Electric Co., Philadelphia, Pennsylvania.

CHEMICAL PROCESSES FOR DEMINERALIZATION AND DESULFURIZATION OF FUELS

HAROLD H. SCHOBERT
Fuel Science Program
Department of Materials Science and Engineering
The Pennsylvania State University
University Park, PA 16802 USA

ABSTRACT: It is desirable to have processes for removal of mineral and sulfur impurities that are as simple, and hence as inexpensive, as possible, to avoid a large incremental increase in cost of fuel. Physical processes (e.g., gravity separation) are ideal from a standpoint of simplicity and low cost. Some fuel impurities, such as organically bound sulfur or ion-exchangeable cations, are not amenable to removal by physical techniques. As environmental regulations governing emissions from fuel use become more stringent, it will be necessary to consider greater and greater degrees of removal of impurities. Both issues require consideration of chemical processes for demineralization and desulfurization. The chemical basis for desulfurization, and, to some extent, demineralization, in reducing, neutral, and oxidizing environments are considered. The emphasis is on fundamentals of chemistry rather than specific processing schemes.

1. Introduction

Why consider chemical processes for the demineralization or desulfurization of fuels? Compared to physical processes, such as gravity separation for example, chemical processes are likely to result in additional unit operations for for conversion of fuel to desired products. The added complexity of a process flowsheet will be reflected in increased capital costs for plant and increased operating costs. Hence there is little, if any, economic incentive for considering or adopting chemical processing schemes. Rather, it is the nature of the fuel, and the impurities in it, that provide the impetus for chemical processing for demineralization and desulfurization.

Although some of the inorganic components of fuels might have beneficial catalytic properties for fuel processing, as discussed elsewhere in this volume [1], in many fuel processing systems they are, at best, inert diluents that must be passed through the system, collected, and appropriately disposed of. At worst, the inorganic components can catalyze undesirable reactions (such as retrogressive char-forming reactions), can poison externally added catalysts, or can react chemically to produce undesirable by-products or emissions, such as hydrogen sulfide. Inorganic materials not recovered from processing streams but remaining in the product could form undesirable amounts of ash when the synthetic fuel is eventually burned. Thus generally (that is, excepting the cases in which inorganic components catalyze desired reactions) the removal of inorganic components is a benefit for fuel processing. Sulfur has of course achieved a notoriety far out of proportion to its actual concentration in most fuels, the notoriety stemming from pollution problems associated with emission of sulfur compounds from fuel conversion and combustion processes. It is almost certain that any change in environmental regulations in the future will be to favor stricter limits on emissions, generating greater concern for the removal of sulfur. Desulfurization of the feedstock prior to conversion to synthetic fuels eliminates a post-conversion desulfurization step for purification of the fuel product.

In any processing it is always desirable to have the simplest and least expensive operations, consistent with desired product quality and yields. In many instances, some demineralization can be effected by grinding to liberate the minerals and then carrying out a separation based on some difference in physical properties of the minerals and carbonaceous part of the fuel (density differences, for example). When sulfur is present as the mineral pyrite, FeS_2, demineralization is also accompanied by desulfurization. However, there are instances in which a simple physical separation is not adequate. Some mineral grains may be so tiny and so thoroughly encapsulated in the surrounding carbonaceous material, that grinding fine enough to effect their liberation is impractical. An extreme example is the alkali and alkaline earth cations held in low-rank coals as

Y. Yürüm (ed.), Clean Utilization of Coal, 173–183.

counterions with the carboxylate groups that are a part of the coal structure. These ions are dispersed on essentially an atomic scale, and no amount of grinding will liberate them. Some sulfur is usually incorporated in the carbonaceous portion of the fuel as so-called organic sulfur, in which the sulfur atoms are chemically bonded to the carbon structure. This form of sulfur can be removed only by chemical processes that cleave C-S bonds. While some measure of demineralization and desulfurization can be achieved for many fuels by simple physical processes, extensive cleaning will require some chemical processes as well.

The discussion that follows focuses on the chemistry of fuel demineralization and desulfurization processes as might be relevant to synthetic fuel production. Demineralization and desulfurization of fuels are also very important for combustion applications (that is, considering combustion as an alternative to conversion to synthetic fuels), but that topic is not relevant to the theme of this Advanced Study Institute. Most of the discussion deals with the demineralization and desulfurization of coals, because it seems likely that coal conversion to synthetic fuels may be closer to commercialization than conversion of other feedstocks. However, the discussion is presented mainly in terms of chemical fundamentals, rather than details of specific process configurations or processing technology. Thus much of what follows should be of general relevance, despite its having sprung from the coal literature. The discussion is organized along the broad types of chemical reactions: oxidizing, neutral, and reducing reactions. This chapter is not by any means a comprehensive survey of all recent literature in the field. Rather, selected examples have been chosen that should be illustrative, and, I hope, interesting.

2. Oxidation Processes

2.1. CHLORINATION REACTIONS

A variety of reaction strategies has been employed for effecting desulfurization by chlorinolysis. The reaction conditions vary greatly, as do the extents of dechlorination. For example, coal can readily be chlorinated by the reaction of a slurry of coal in carbon tetrachloride / water with chlorine at 65-70°C [2]. Reaction of a slurry of moist coal in carbon tetrachloride with chlorine at ambient conditions can remove most of the pyritic sulfur and sulfate sulfur, as well as 30% of the organic sulfur [3]. Subsequent reactions, first in water at 70-80°C followed by dechlorination in steam at 350-400°C, reduce the sulfur content by about 50% [2].

Low-temperature chlorinolysis involves three reaction steps: chlorination, hydrolysis, and dechlorination [3. Chlorination is readily effected by reaction of chlorine gas with a slurry of coal in water, organic solvents inert to chlorine, or mixed aqueous-organic solvents. A potential advantage that might accrue from the use of organic solvents is that some solvents can cause significant swelling and "'loosening" of the coal structure, possibly facilitating mass transport. The reactions that occur during chlorination are

$$2\ FeS_2 + 15\ Cl_2 + 16\ H_2O \rightarrow 2\ FeCl_3 + 4\ H_2SO_4 + 24\ HCl$$

and

$$RSR' + 4\ Cl_2 + 4\ H_2O \rightarrow RCl + R'Cl + H_2SO_4 + 6\ HCl$$

As customary, the symbols R and R' represent generic organic structures, in this case portions of the fuel (i.e., coal) structure. As apparent from these equations, some water is necessary to satisfy reaction stoichiometry, even if the process were run in an organic solvent. In essence, the sulfur in the fuel is converted to soluble sulfates which can be removed by washing with water at ~60°C. Desulfurization accomplished at the price of chlorination of the fuel (represented, in this case, by RCl and R'Cl) would simply trade one evil for another. Combustion of chlorinated fuels would lead to undesirable emissions of HCl; chlorine or chlorides present in fuel conversion processing can cause stress corrosion cracking of process vessels. Consequently, the chlorine-desulfurized fuel must subsequently be dechlorinated. One approach for dechlorination is hydrolysis in steam at 300-500°C:

$$RCl + H_2O \rightarrow ROH + HCl$$

Evidently, chlorine introduced to the coal structure in high temperature chlorination is much more difficult to remove [3].

This approach to desulfurization removes both pyritic and organic sulfur. The reactions proceed at atmospheric pressure and relatively low temperatures. The maximum sulfur removals from Turkish lignites were in the range 50-66% [3]. In principle another advantage of a process such as this is that the necessary chlorine could be regenerated from the HCl (as, for example, by reaction with pyrolusite [4]).

An alternative to the use of gaseous chlorine is sodium hypochlorite. The oxidation of pyrite to sulfate can be written as

$$FeS_2 + 8\ H_2O \rightarrow Fe^{+2} + 2\ SO_4^{-2} + 16\ H^+ + 14\ e^-$$

with a half-cell potential of -0.362 v [5]. Hypochlorite can readily effect this oxidation. The potential for

$$ClO^- + H_2O + 2\ e^- \rightarrow Cl^- + 2\ OH^-$$

is +0.90 v [5]. Nevertheless, tests of three bituminous coals showed that sodium hypochlorite removed only the organic sulfur and sulfate sulfur [5].

Hypochlorite is able to attack a variety of organic sulfur functional groups. The reactions can be represented as

$$RSR' \rightarrow RSO_2R'$$

$$RSSR' \rightarrow RSO_3H + R'SO_3H \rightarrow \rightarrow H_2SO_4$$

$$RSH \rightarrow RSSR \rightarrow RSO_3H + H_2SO_4$$

In some cases the oxidation of sulfur functional groups may not proceed beyond the sulfones. This oxidation does not actually remove the sulfur from the fuel. Then it is necessary to include an additional desulfurization step, such as treatment with hot sodium carbonate [5]. A pre-treatment with aqueous ammonia appears to facilitate desulfurization by masking the reaction sites more accessible to the hypochlorite [5].

A disadvantage of the hypochlorite oxidation is that some chlorination of the coal occurs at the same time. Residual chlorine contents of bituminous coals treated with hypochlorite were in the range of 2.0-2.7% [5]. High-temperature reaction with steam might be required to reduce the chlorine content of the desulfurized coal.

2.2 DIRECT OXIDATION REACTIONS

Reaction of pyrite with oxygen results in formation of iron(III) oxide or sulfate, with sulfur eliminated as sulfur dioxide [6].

$$4\ FeS_2 + 11\ O_2 \rightarrow 2\ Fe_2O_3 + 8\ SO_2$$

$$2\ FeS_2 + 7\ O_2 \rightarrow Fe_2(SO_4)_3 + SO_2$$

The specific reaction products depend on temperature, partial pressure of oxygen, and particle size [6]. Excess oxygen favors formation of the oxide. Formation of the sulfate is favored by relatively low reaction temperatures. Sulfate formation is undesirable. First, formation of the sulfate leaves some of the sulfur in the coal. In addition, the iron(III) sulfate has a larger molar volume than the pyrite; hence formation of the sulfate results in an increased volume of solid, which in turn can lead to closing of the pores and stifling of mass transport. In that case the rate of reaction would be governed by oxygen diffusion through the sulfate or oxide layer on the pyrite.

Oxydesulfurization in the presence of water can lead to the formation of iron(II) sulfate via [7]

$$2\ FeS_2 + 7\ O_2 + 2\ H_2O \rightarrow 2\ FeSO_4 + 2\ H_2SO_4$$

The iron(II) sulfate is not a desirable product, because its retention in the coal would mean that not all of the sulfur had been removed. However, the iron(II) sulfate can be oxidized to iron(III) oxide under these conditions [7]:

$$4\ FeSO_4 + O_2 + 4\ H_2O \rightarrow 2\ Fe_2O_3 + 4\ H_2SO_4$$

Under these conditions only about 20% of the organic sulfur is removed [7]. However, an advantage of the process is that air and water are the only reagents required.

Pyrrhotite, which might form by thermal decomposition of the pyrite, will react with oxygen analogously: [6]

$$2\ FeS + 3\ O_2 \rightarrow 2\ FeO + 2\ SO_2$$

$$FeS + 2\ O_2 \rightarrow FeSO_4$$

In both cases, that is, FeS_2 and FeS, sulfate formation is favored by reaction temperatures at or below 400°C [6]. Sulfate formation leads to pore closing in both cases.

Organic sulfur compounds can also be oxidized to sulfur dioxide. The reactions of the organic functional groups in oxidizing regimes is accompanied by some loss of the carbon. Alkyl sulfides and disulfides react well at relatively low temperatures; the maximum rate of oxidation occurring around 230°C [6]. Aromatic sulfur compounds are less reactive. The maximum rate of oxidative desulfurization of aryl sulfides or thiophenes does not occur until temperatures around 460°C have been reached [6].

Oxydesulfurization at 150°C and 1.4 MPa oxygen in 0.2M aqueous sodium carbonate can effect removal of 40% of the organic sulfur in an hour [8]. Pyrite is converted to hematite at these conditions [9].Carbon-sulfur bond cleavage in sulfides is initiated by an oxidative attack at benzyl carbons. In support of this proposed mechanism, it can be shown that fluorene is oxidized quantitatively to fluorenone [e.g., 9], but dibenzothiophene is unreactive at these conditions. In the presence of base, the benzylic carbon reacts to form an α-carbanion. This ion will then react with oxygen to produce an α-peroxide anion. The key step in effecting C-S bond cleavage is then the decomposition of the α-peroxide anion to form an aldehyde and a sulfenic acid. Neither of these intermediates survives the strong oxidizing conditions, the final product being a carboxylic and a sulfonic acid [8]. For example, the reaction of benzyl phenyl sulfide under these conditions would be expected to produce benzoic and benzene-sulfonic acids.

Under these conditions, thiophenols and disulfides are oxidized at the sulfur atoms [9]. For example, the reactions

$$C_6H_5SH \rightarrow C_6H_5SSC_6H_5$$

and

$$C_6H_5SSC_6H_5 \rightarrow 2\ C_6H_5SO_3^-\ Na^+$$

are envisioned with oxidation in basic conditions [9].

2.3. POTASSIUM PERMANGANATE

Potassium permanganate is a powerful oxidizing agent used to carry out a variety of oxidations in organic chemistry [10]. Prior to the advent of instrumental methods such as nuclear magnetic resonance spectroscopy of solids, potassium permanganate was one of a group of oxidants widely used in coal structural studies. Reaction of coal with permanganate in alkaline solution at ambient conditions can remove 90% of the pyritic sulfur and 50% of the organic sulfur [11].

A problem experienced with permanganate oxidation is the precipitation of manganese salts, which in effect increases the ash content of the coal. Furthermore, some of the manganese compounds may contain sulfur, and their precipitation and incorporation into the coal thus means that not all of the sulfur has been removed. For example, in alkaline solution permanganate is converted to the insoluble manganese dioxide (probably via the manganate):

$$MnO_4^- + 2\,H_2O + 3\,e^- \rightarrow MnO_2 + 4\,OH^-$$

Manganese double sulfates may also form, along with a by-product jarosite:

$$K^+ + 3\,Fe^{+3} + 2\,SO_4^{-2} + 6\,OH^- \rightarrow KFe_3(SO_4)_2(OH)_6$$

There are methods for removing the precipitated manganese compounds, or the jarosite, from the coal. These include washing the coal with dilute acids (preferably nitric or hydrochloric) or leaching the coal with ethylenediamine tetraacetate (EDTA) [11].

2.4. REACTIONS WITH PEROXIDES

Organic sulfur functional groups readily react with peroxides and are converted to sulfones. This reaction will occur even for organic sulfur compounds such as dibenzothiophene which are unreactive toward many other reagents [12]. Peroxides are intermediates in the oxidation of sulfur dioxide, a reaction which is catalyzed by transition metal ions. Since Fe ions would catalyze oxidation of sulfur dioxide, and hence the formation of transient peroxide species, the necessary catalyst for oxidation and peroxide formation could be derived from minerals impurities in fuels, as for example in the oxidation of pyrite.

The net reaction for the oxidation of sulfur dioxide in aqueous systems is [12]

$$2\,SO_2 + 2\,H_2O + O_2 \rightarrow 2\,H_2SO_4$$

The reaction mechanism, including the transient peroxide species that could actively intervene to oxidize organic sulfur is given by the following sequence of reactions [12]. In these equations the symbol • represents a radical, M is a metal, I an inert species and transient species are shown in italic type.

hydrolysis	$SO_2 + H_2O \rightarrow HSO_3^- + H^+$
initiation	$M^{(n+1)+} + HSO_3^- \rightarrow M^{n+} + \bullet HSO_3$
propagation	$\bullet HSO_3 + O_2 \rightarrow \bullet \mathit{HSO_5}$
propagation	$\bullet \mathit{HSO_5} + HSO_3^- \rightarrow \mathit{HSO_5^-} + \bullet HSO_3$
product formation	$\mathit{HSO_5^-} + HSO_3^- \rightarrow 2\,HSO_4^-$
initiator regeneration	$2\,M^{n+} + O_2 \rightarrow 2\,M^{(n+1)+} + \mathit{H_2O_2} / \mathit{HO_2^-}$
termination	$2\,\bullet HSO_3$ or $2\,\bullet HSO_5 \rightarrow$ inactive species

inhibition $\bullet HSO_3$ or $\bullet HSO_5$ or $M^{(n+1)+} + I \rightarrow$ inactive species

These concepts were put into practice for oxidation of Polish bituminous coals at atmospheric pressure in acidic solution (pH < 1), 67°C, with an air flow rate of 1100 dm^3/h and SO_2 added to the reactor at 0.65-0.82 dm^3/h [12]. Under these conditions 73% of the pyrite was oxidized. Dibenzothiophene is oxidized under these reaction conditions, but evidently the fate of organic sulfur in the coal samples was not elucidated [12]. Of course if sulfur groups are converted only to sulfones no net removal of sulfur occurs. The importance of the autooxidation by SO_2 is illustrated by the finding that the reaction run without added SO_2 but at otherwise comparable conditions, effected less than 20% conversion of pyrite [12].

3. Chemical Processes in Neutral Environments

3.1. PYROLYSIS REACTIONS

Pyrite decomposes to pyrrhotite during pyrolysis. The reaction can be represented as [13]

$$2\ FeS_2 \rightarrow [FeS_{1+x}] \rightarrow 2\ FeS + S_2$$

Pyrolytic decomposition of pyrite begins above 400°C [13]. The maximum rate loss for decomposition of pyrite in Polish metaanthracite occurs at 480°C [13] Conversion to pyrrhotite is complete about 800°C [13]. For Turkish lignites, this transformation occurs around 500-550°C [3], a temperature also reported by other investigators [14]. Although the reaction could result in desulfurization, some of the sulfur released from the pyrite might become trapped in the carbonaceous matrix as organic sulfur. In addition, coals with alkaline ashes, or strongly basic inorganic components, could trap some of the released sulfur.

Pyrolysis of organic sulfur groups leads to the formation of H_2S, COS, CS_2, CH_3SH, and SO_2 as gaseous products, along with a variety of thiophenes, benzothiophenes, and dibenzothiophenes in the tars [14]. For example [15]

$$RCH_2CH_2SH \rightarrow RCH{=}CH_2 + H_2S$$

About 25-50% of the sulfur appears in the gaseous products [14]. The formation of carbon disulfide occurs at pyrolysis temperatures above 800°C [14]. This product may arise from carbon reduction of pyrite

$$2\ FeS_2 + (\text{coal-C}) \rightarrow CS_2 + 2\ FeS + (\text{coal})$$

Participation of methane, also formed as a pyrolysis product, is also possible [14]

$$4\ FeS_2 + CH_4 \rightarrow CS_2 + 4FeS + 2\ H_2S$$

Carbon disulfide formation is favored by high temperatures; the higher the temperature of pyrolysis, the greater the CS_2 production. The carbonyl sulfide likely arises from reactions with carbon monoxide, which is also a product of pyrolysis [14]

$$CO + S \rightarrow COS$$

The sulfur source in this case is pyrite. Pyrite can be converted to sulfur by such reactions as [12]

$$FeS_2 + 2\ Fe^{3+} \rightarrow 3\ Fe^{2+} + 2\ S$$

When the amount of pyrite is reduced, COS production decreases. Sulfur dioxide may arise in pyrolysis reactions from the interaction of sulfates with pyrite [14]:

$$CaSO_4 + FeS_2 + H_2O \rightarrow CaO + FeS + 2\ SO_2 + H_2$$

Sulfur formed in the pyrolytic decomposition of pyrite can be converted to hydrogen sulfide by reacting with available hydrogen in the coal [13,14]

$$S + (\text{coal-H}) \rightarrow H_2S + (\text{coal})$$

Other reactions in the coal matrix can involve the formation of H_2S from organic sulfur. For example [15]

$$\text{coal-}C_6H_4SH + CH_3C_6H_4\text{-coal} \rightarrow \text{coal-}C_6H_4CH_2C_6H_4\text{-coal} + H_2S$$

It is important to recognize that gaseous sulfur species liberated in pyrolyses (or, for that matter, any other reactions discussed in this chapter) can react with alkaline species in the coal and become trapped [15]. For example,

$$H_2S + CaCO_3 \rightarrow CaS + H_2O + CO_2$$

The effect of this trapping is to reduce the amount of sulfur actually removed from the coal. The sulfur may be converted to another chemical form, but it is still present.

3.2. REACTION WITH STEAM

Pyrite is converted to pyrrhotite in the presence of steam via the reaction [6]

$$FeS_2 + H_2O + C \rightarrow FeS + H_2S + CO$$

The reaction can occur at 400°C, if the H_2S concentration in the gas phase is held below 8000 ppm [6]. Since pyrrhotite is one of the reaction products, this process does not effect complete desulfurization. Various reactions have been proposed for the subsequent decomposition of pyrrhotite in steam, which would remove the remaining sulfur.

$$FeS + H_2O \rightarrow FeO + H_2S$$

$$3\ FeS + 4\ H_2O \rightarrow Fe_3O_4 + 3\ H_2S + H_2$$

$$2\ FeS + 3\ H_2O \rightarrow Fe_2O_3 + 2\ H_2S + H_2$$

Steam appears to have little affect on organic sulfur functional groups.

3.3. TREATMENT WITH SULFUR DIOXIDE

Liquid sulfur dioxide is an excellent solvent [16]. Of particular relevance to desulfurization and demineralization is its ability to dissolve alkyl, aromatic, and heterocyclic sulfides [17].The key reaction is

$$RSR' + SO_2 \rightarrow RR'S\text{-}SO_2$$

and the products are highly soluble in the solvent. Treatment of lumps of coal with liquid sulfur dioxide resulted in removal of 25-37% of the organic sulfur [17] The reaction is enhanced in part by a physical breakdown of the coal particles by the liquid sulfur dioxide. Grinding to -60 mesh enhanced the removal of organic sulfur, to about 48% [17]. Unfortunately pyrite is inert in this system.

3.4. ION EXCHANGE REMOVAL OF METAL CATIONS

Coals of subbituminous and lower rank contain significant amounts of metal cations in ion-exchangeable form. These ions are associated with the carboxylate functional groups in the coal structure. The principal ions present in this form are sodium, calcium, and magnesium. A portion of the total amount of potassium in low-rank coals is also ion exchangeable. Some coals also appear to have small amounts of other exchangeable cations, most notably iron and aluminum.

Although in principle the exchangeable ions could be removed completely, by washing with dilute acid, for example, much of the interest in ion exchange has focused on the replacement of sodium by calcium. The concern stems from the combustion behavior of low-rank coals, those coals having high concentrations of sodium forming extensive ash deposits on heat transfer surfaces in boilers. Reduction of sodium content reduces the deposition problem; in addition, in some systems calcium appears to be an antagonist to deposit formation.

Since low-rank coals are moderately good ion-exchange systems, the fundamental principles of mass transport in ion exchange [e.g., 18] are generally applicable. Particle size of the coal affects the rate and extent of exchange, since particle size essentially determines the distance through which the exchanging ion must travel in the pore system of the coal [19]. There is a direct correlation between the concentration of the in-coming exchanging ion in solution and the displacement of ions from the coal. For example, in exchanging sodium with calcium, the quantity of calcium present in solution is one of the factors governing the quantity of sodium removed [19]. Furthermore, the calcium content of the coal increased in direct proportion to the decease in the sodium content. Other factors affecting removal of ions include the moisture content of the coal, the liquid/solid ratio, solution temperature, and contact time. The liquid/solid ratio and the concentration of the exchanging ion in solution are the variables that establish the total amount of the exchanging ion that could exchange with a given quantity of coal. For Ca^{+2}/Na^{+} exchange in lignite, the rate of exchange increases with temperature, up to ~50°C. The quantity of sodium exchanged increased with contact time, up to 30 minutes.

4. Reduction Processes

4.1. HYDROGENATION REACTIONS

Pyrite can be effectively removed from coals by reaction with lithium aluminum hydride [2].

Hydrodesulfurization of lignites containing high organic sulfur contents can be effected by reaction with hydrogen in the presence of impregnated catalysts. Impregnation of the lignite with ammonium tetrathiotungstate, ammonium tetrathiomolybdate, or nickel sulfate, followed by reaction with hydrogen at 275-325°C, can eliminate over half the total sulfur [20]. The sulfur is eliminated from the lignite as hydrogen sulfide and as a variety of organic sulfur compounds occurring in by-product liquids. Ideally, if one were interested in the desulfurization of these high organic sulfur coals to produce a low-sulfur char suitable for boiler fuel, it would be desirable to have the sulfur eliminated as H_2S with minimal liquids production, preserving the carbon values in the char. Ammonium tetrathiotungstate catalyst precursor, which presumably decomposes to WS_2 at reaction conditions, provides the best selectivity to H_2S, at least for Turkish lignites. Hydrogen sulfide selectivity increases with increasing reaction temperature.

Hydrodesulfurization of high organic sulfur lignites appears to proceed mainly through a depolymerization of the lignite structure, the depolymerization proceeding through attack on weak C-S bonds. The first product is organosulfur compounds present in a coal-derived liquid. These organosulfur compounds are then hydrotreated on the catalyst in the second step of the mechanism. For example, hydrogenation of thiols can be represented by [15]

$$RSH + H_2 \rightarrow RH + H_2S$$

Of course some H_2S could be produced directly from hydrogenolysis of C-S bonds. In fact, if the lignite is not impregnated with a catalyst precursor, there is little evidence that the desulfurization is sequential. Instead, it appears that hydrogen sulfide is formed by direct hydrogenation of the sulfur functional groups in the lignite. Nevertheless, the selectivity to H_2S is lower in the non-catalyzed reactions than it is for catalytic hydrodesulfurization [20]. Impregnation with a catalyst precursor significantly improves total sulfur removal.

Hydrogen is a moderately expensive reagent, and a requirement for high hydrogen-to-solids ratios in conventional hydrodesulfurization could have a negative effect on process economics. An alternative strategy in hydrodesulfurization is the so-called "convert-remove" approach [21]. The first step involves the reaction of hydrogen with organic sulfur, as shown above. However, the hydrogen sulfide then reacts with hydrogen scavengers *in situ*, as

$$FeO + H_2S \rightarrow FeS + H_2O$$

To this point there is no net removal of sulfur. Rather, the sulfur has been converted from an organic to inorganic form. Several methods can then be used to remove the inorganic sulfur, including oxidation or leaching with acids. Of particular interest is reaction with steam:

$$FeS + H_2O \rightarrow FeO + H_2S$$

because this reaction regenerates the *in situ* sulfur scavengers. A high steam flow rate is required to insure that the H2S is swept out of the reaction vessel, but the actual consumption of steam is low. It appears that repetitive cycling between the "convert" and "remove" steps with short reaction times for each is preferable to a simpler two-step process with one long reaction step for each [21].

Pyrite is readily reduced by hydrogen [6]:

$$FeS_2 + H_2 \rightarrow FeS + H_2S$$

This reaction is proceeds readily at temperatures of 400°C and above. Any remaining pyrrhotite (or pyrrhotite that had been produced by thermal decomposition of pyrite) will also react with hydrogen.

$$FeS + H2 \rightarrow Fe + H2S$$

However, this reaction is not nearly as favorable as the reduction of pyrite to pyrrhotite. At 850°C the equilibrium constant for pyrite reduction is 1000; whereas that for pyrrhotite reduction is only 0.002 [6]. A hydrogen sulfide concentration of 2000 ppm in the gas phase would be sufficient to reverse the reduction of pyrrhotite.

4.2 ALKALI METAL REDUCTIONS

The reaction of alkali metals - lithium, sodium, and potassium - in liquid ammonia solution removed about 40% of the sulfur from coal [2. Details of the alkali metal / liquid ammonia systems and behavior of ammoniated electrons are available in the inorganic chemistry literature [22]. The liquid ammonia reactions are carried out at -55°C. Of the three metals, potassium seems to be the most effective.

Potassium in dimethoxyethane reacts with sulfides and heterocyclic compounds at the C-S bond [2]. Benzyl phenyl sulfide is converted to benzene, toluene, and thiophenol, along with hydrogen sulfide.

The use of an electron transfer agent, particularly naphthalene, in conjunction with alkali metal solutions, facilitates desulfurization. For example, a solution of potassium in tetrahydrofuran (THF) provides some measure of desulfurization, but both the extent and rate of desulfurization can be enhanced by the addition of naphthalene. Potassium in THF cleaves dibenzothiophene to thiophenol and biphenyl. The principal bond cleavage occurs at sp^2 C-S bonds. Under mild conditions (albeit long reaction times, 7 days) virtually complete desulfurization can be achieved by reaction with potassium naphthalenide [2]. The reaction of potassium naphthalenide with pure compounds showed conversion of dibenzothiophene to biphenyl, di-1-naphthyl sulfide to naphthalene and 1-naphthalenethiol, and di-1-octyl sulfide to octanethiol and octylnaphthalenes [2]. The aromatic thiols can be desulfurized further, but C-S bond cleavage in these compounds is slower than the corresponding reactions of sulfides or heterocyclic compounds. Aliphatic thiols like 1-nonanethiol and 1-decanethiol are unreactive. Thus the sp^3 C-S bond is unreactive in this system. Nevertheless, about 50-90% of the organic sulfur can be removed.

5. Summary

A variety of chemical treatments can be considered for demineralization and desulfurization of fuels. Since oxidative, reductive, and neutral processes have been demonstrated at the laboratory scale, and in some cases at pilot plant scale, the potential exists for selecting chemical conditions appropriate for the specific fuel and specific impurity removal desired. Processing economics is an issue not directly related to the chemical problems of demineralization or desulfurization. Nevertheless it must be recognized that chemical processes in general are likely to be more expensive, in terms of both capital and operating costs, than physical processes. For synthetic fuel production, a balance must be struck between the costs and complexity of treatment of the feedstock before the conversion process relative to treatment of the product synthetic fuel downstream of the conversion reaction.

6. Literature Cited

1 Schobert, H.H. (1991) "Catalytic and chemical behavior of coal mineral matter in coal conversion processes", in This Volume

2 Chatterjee, K., Wolny, R., and Stock, L.M. (1990) "Coal desulfurization by single electron transfer reactions", Energy and Fuels 4, 402-406.

3 Ozdemir, M., Bayrakceken, S., Gurses, A., and Gulaboglu, S. (1990) "Desulfurization of two Turkish lignites by chlorinolysis", Fuel Processing Technology 26, 15-23.

4 Wagner, Rudolf (1872) A Handbook of Chemical Technology, D. Appleton and Company, New York.

5 Brubaker, I.M., and Stoicos, T. (1985) "Precombustion coal desulfurization with sodium hypochlorite", in Y.A. Attia (ed.), Processing and Utilization of High Sulfur Coals, Elsevier Science Publishers, Amsterdam, pp. 311-326.

6 Stephenson, M.D., Rostam-Abadi, M., Johnson, L.A., and Kruse, C.W. (1985) "A review of gas-phase desulfurization of char", in Y.A. Attia (ed.), Processing and Utilization of High Sulfur Coals, Elsevier Science Publishers, Amsterdam, pp. 353-372.

7 Akhtar, S.S., and Cliffe, K.R. (1990) "The effect of shear on the oxydesulfurization of coal", in R. Markuszewski and T.D. Wheelock (eds.), Processing and Utilization of High Sulfur Coals III, Elsevier Science Publishers, Amsterdam, pp. 405-413.

8 Chang, L.W., Goure, W.F., Squires, T.G., and Barton, T.J. (1980) "Evaluation of the oxydesulfurization processes for coal. 1. The effect of the Ames process on model organosulfur compounds", Amer. Chem. Soc. Division of Fuel Chemistry Preprints 25(2), 165-170.

9 Squires, T.G., Venier, C.G., Chang, L.W., and Schmidt, T.E. (1981) "Chemical studies of the Ames oxydesulfurization process", Amer. Chem. Soc. Division of Fuel Chemistry Preprints 26(1), 50-59.

10 Vollhardt, K. Peter C. (1987) Organic Chemistry, W.H. Freeman and Company, New York.

11 Attia, Y.A., Lei, W., and Carlton, R.W. (1990) "Removal of reaction products from chemical oxidation of coal using potassium permanganate for desulfurization", in R. Markuszewski and T.D. Wheelock (eds.), Processing and Utilization of High Sulfur Coals III, Elsevier Science Publishers, Amsterdam, pp. 391-404.

12 Bronikowski, T., Pasiuk-Bronikowska, W., and Ulejczyk, M. (1990) "Coal desulfurization facilitated by coupled autooxidation of sulfur dioxide", in R. Markuszewski and T.D. Wheelock (eds.), Processing and Utilization of High Sulfur Coals III, Elsevier Science Publishers, Amsterdam, pp. 425-435.

13 Gryglewicz, G., and Jasienko, S. (1990) "The behavior of pyritic sulfur during pyrolysis of metaanthracite", in R. Markuszewski and T.D. Wheelock (eds.), Processing and Utilization of High Sulfur Coals III, Elsevier Science Publishers, Amsterdam, pp. 437-445.

14 Calkins, W.H. (1987) "Investigation of organic sulfur-containing structures in coal by flash pyrolysis experiments", Energy and Fuels 1, 59-64.

15 Attar, A., and Hendrickson, G.G. (1982) "Functional groups and heteroatoms in coal", in R.A. Meyers (ed.), Coal Structure, Academic Press, New York, Chapter 5.

16 Waddington, T.C. (1965) "Liquid sulfur dioxide", in T.C. Waddington (ed.), Non-Aqueous Solvent Systems, Academic Press, London, Chapter 6.

17 Burow, D.F., and Glavincevski, B.M. (1980) "Removal of organic sulfur from coal: The use of liquid sulfur dioxide", Amer. Chem. Soc. Division of Fuel Chemistry Preprints 25(2), 153-164.

18 Treybal, Robert E. (1968) Mass-Transfer Operations, McGraw-Hill Book Company, New York.

19 Paulson, L.E., and Futch, J.R. (1980) "Removal of sodium from lignite by ion-exchanging with calcium chloride solutions", Amer. Chem. Soc. Division of Fuel Chemistry Preprints,25(1), 224-232.

20 Garcia, A.B., and Schobert, H.H. (1990) "Catalytic hydrodesulfurization of a high organic sulfur Turkish lignite: amount, form, and mechanism of sulfur removal", Fuel Processing Technology 26, 99-109.

21 Tipton, A.B. (1980) "Sulfur removal from coal char using "convert-remove" technology", Amer. Chem. Soc. Division of Fuel Chemistry Preprints 25(4), 143-149.

22 Lagowski, J.J. (1973) Modern Inorganic Chemistry, Marcel Dekker, Inc., New York.

BIODESULPHURIZATION OF COAL

G.GÜLLÜ*, T.DURUSOY, T.ÖZBAŞ, A.TANYOLAÇ, Y.YÜRÜM*
Hacettepe University
Department of Chemistry and*
Department of Chemical Engineering
Beytepe 06532 Ankara, Türkiye

ABSTRACT. A summary of the work done for microbial desulphurization during past ten years was presented along with our study of biodesulphurization by *Sulfolobus solfataricus* for a Turkish lignite. In the text; sulfur compounds in coal, microorganisms for biodesulphurization, microbial action, possible process schemes and economics were outlined. The bioprocess parameters affecting the growth kinetics of *Sulfolobus solfataricus*, a recent strain for the removal of organic sulfur from coal, were investigated and formulated through least square regression methods as a part of experimental design. Preliminary runs with a Turkish lignite have shown 35% reduction in organic sulfur and 50% in the total sulfur content under thermophilic conditions yet to be optimized. New and mutant strains are still in search for higher desulphurization rates and conversion which will make the bioprocess feasible for industrial applications.

1. Introduction

One of the most challenging tasks of our time is to find proper means to efficiently utilize natural energy sources without polluting the environment. Some of the major pollutants are sulfur dioxide, nitrogen oxides, carbon monoxide and hydrocarbons. There is a growing interest using of coal as a clean energy source because its conversion to liquid fuels and gas products is available through liquefaction and gasification processes.

The major problem associated with direct utilization of coal is the emission of sulfur and nitrogen containing gases (mainly sulfur dioxide and nitrogen oxides) into the atmosphere upon combustion of coal [1, 2]. The sulfur dioxide has deleterious effects upon human, animal and plant life meanwhile the sulfur compounds present in coal cause equipment wear and corrosion problems in certain industries, such as the steel industry. Acid mine waters containing sulfuric acid cause serious environmental pollution problems and acid production in acid mine is due to oxidation of reduced sulfur compounds in coal as a result of bacterial action. Reduction of sulfur content of coal before or after combustion is essential to eliminate the pollution problems caused by the sulfur present in coal [3].

The major processes developed for the removal of sulfur from coal can be classified in two categories: (1) precombustion desulphurization of coal and (2) desulphurization after combustion (e. g., flue gas desulphurization). In general, precombustion desulphurization methods offer significant advantages over flue gas desulphurization, since they eliminate both environmental pollution problems and other problems associated with the equipment corrosion [4].

Precombustion coal desulphurization processes can employ physical or chemical methods. Physical methods, such as flotation or magnetic separation of pyrite from coal, are, in general,

Y. Yürüm (ed.), Clean Utilization of Coal, 185–205.

more cost effective than chemical methods [4]. However, these methods are only effective for the removal of pyritic sulfur and result in significant energy loss because they also remove pyrite containing coal particles. The waste product of flotation (i.e., the coal refuse) contains nearly 10-15 % pyritic sulfur and may have a heating value as high as 18,000 kJ/kg. The chemical precombustion desulphurization methods operate at high temperatures (100°C-500°C) and usually at high pressures (e.g., hydrogenation at 4 - 7MPa) and are energy intensive [2, 4] .

One of the major problems associated with chemical desulphurization is the cost of recovery of chemical oxidizing agents (e.g., oxidation by ferric salts, chlorination, ozonation). Other chemical processes involve reaction at high temperatures with the alkali metal carbonates, bicarbonate or hydroxides to produce soluble superheats, much of the organic sulfur can be removed by hot dilute sodium carbonate containing dissolved oxygen under moderate pressure [5,6].

Flue gas desulphurization after combustion of coal requires large volumes of processing water and causes some equipment wear and corrosion problems before and during desulphurization [3].

Microbial coal desulphurization, namely biodesulphurization, before combustion may have the following advantages over physical and chemical methods:

1) Biodesulphurization usually requires lower capital and operating costs compared to high cost of chemical processes and flue gas desulphurization.

2) Unlike physical desulphurization methods, finely distributed sulfur compounds (pyritic and organic sulfur) can be removed by microbial catalysis, without causing any significant energy loss or coal refuse.

3) The biodesulphurization operates at relatively low temperatures (25°C-75°C) and atmospheric pressures and, therefore, is less energy intensive than chemical processes.

4) Microbial desulphurization of coal is associated with microbial growth and organisms have a heating value of 14,000 kJ/kg. When coal particles and microbial cells are separated from the process water at the end of the desulphurization, the microorganisms remaining in the solid phase contribute to the heating value of coal.

5) Since the organisms oxidizing pyritic sulfur are autotrophs and utilize carbon dioxide as carbon source, bioprocess provides a useful means for the utilization of stack gas as source of carbon dioxide (e.g., CO_2 released from coal combustion can be recycled to the bioprocess.

The application of sulfur removal via microbial desulphurization might be suitable for small scale operations where the use of coal-water slurries is proposed or for cleaning fines which arise from modern mining operations and coal preparation. Although bioprocess introduces the possibility of organic sulfur removal, whereas most physical processes separate only the inorganic pyrite, too slow desulphurization rates limit the large scale commercial applications.

2. Sulfur Compounds in Coal

Sulfur occurs in coal in both organic and inorganic forms. The inorganic sulfur is usually present in iron sulfides and sulfate. The iron sulfide is in the form of either pyrite or marcasite, which have the same chemical composition of FeS_2 but different crystalline structures [3]. Sulfate is usually formed as a result of chemical and biological oxidation of pyrite in the presence of air. The pyritic sulfur content of coal varies between 0.5% and 6%, depending on the type of coal [7]. Bituminous coal usually has higher pyrite content than subbituminous coal and lignite. The pyrite particles are randomly and physically distributed within the coal and the extent of pyritic sulfur removal by microbial catalysis strongly depends on how fine this distribution is. In certain cases, particle size of coal has to be reduced down to a very low level (-360 mesh) in order to expose the pyrite surfaces to microbial attack for effective sulfur removal. The sulfate content of freshly mined coal is usually less than 0.1%, and it increases with time upon long term exposition of coal to air. The presence of sulfate in coal is not a problem, since it is readily washable. Although it is not frequent, the occurrence of sulfides other than pyrite (e.g., sphalerite, chalcopyrite, galena) in some coal samples is possible [4]. However, the concentrations of these mineral sulfides are usually negligible, i.e., less than 0.5%.

Organic sulfur compounds are present as part of molecular structure of coal [1]. Most of the organic sulfur is bond to the aromatic ring structure of coal and can be either aliphatic or aromatic; however, the exact forms of these compounds in coal structure are not known [3]. Major organic sulfur compounds present in coal are:

1) Sulfides or thio ethers: R_1-S-R_2, where R_1 and R_2 are either alkyl or aryl groups (e.g., diethyl sulfide, C_2H_5-S-C_2H_5).
2) Disulfides: R_1-S-S-R_2 or $ø_1$-S-S-$ø_2$, where $ø_1$ and $ø_2$ are aromatic groups (e.g., diethyl disulfide, C_2H_5-S-S-C_2H_5).
3) Mercaptans or thiols: R-SH or ø-SH (e.g., ethanethiol, C_2H_5SH, or benzenethiol, C_6H_5SH).
4) Thiophenes (e.g., dibenzothiophene).

However, removal of organic sulfur from coal matrix in coal desulphurization is a problem. Chemical methods, such as hydrogenation or hydrodesulphurization, are effective for organic sulfur removal. But these methods are very costly, since they operate at high temperatures and pressures. Another problem arises in the quantitative analysis of the organic sulfur in the coal. It is usually determined by subtracting pyritic sulfur from total sulfur, but this is not an entirely satisfactory procedure. The reliability of the basic determination of pyrite in coals using standard procedures was discussed by Suhr and Given [8]. The presence of some elemental sulfur can complicate the issue further, and quoted figures for organic sulfur removal should consequently be treated with a certain amount of reserve.

3. Microorganisms for Biodesulphurization

There are many bacteria, mold and mixed cultures held responsible for the desulphurization of coal and research is still carried on new generations of these to increase both the reaction rate and removal percent.

3.1. BACTERIA

The most commonly considered and used microorganisms are *Thiobacillus ferrooxidans* and *Thiobacillus thiooxidans*. These have been shown to be effective in the removal of pyritic sulfur from coal [9]. The reported removal rates, however, have been too low to make a bioprocess economically attractive so far. *Sulfolobus acidocaldarius* is a chemoautotrophic and thermophilic (50°C - 80°C) microorganism isolated from acidic hot springs. It can also act as a heterotroph and has potential value as it removes some of the organic sulfur as well as inorganic pyrite. It has been demonstrated that it is capable of oxidizing both reduced sulfur compounds and ferrous ions to obtain energy for its life processes. CB1 is another microorganism developed by the Atlantic Research Corporation in the USA. It is a *Pseudomonas* and primarily attacks the thiophenic sulfur in coal. A number of other microorganisms considered and tested are given in Table 1.

TABLE 1. Colorless sulfur oxidizing bacteria [10]

Obligate chemolithoautotrophs	Facultative chemolithoautotrophs	Chemolithoheterotrophs
aerobes	**aerobes**	*Thiobacillus perometabolis*
		Pseudomonas spp
Thiobacillus neapolitanus	*Thiobacillus intermedius*	*Beggiatoa spp*
Thiobacillus thiooxidans	*Thiobacillus novellus*	
Thiobacillus concretivorus	*Thiobacillus acidophilus*	**heterotrophs**
Thiobacillus ferrooxidans	*Thiobacillus organoparus*	
Thiobacillus kabobis	*Sulfolobus acidocaldarius*	*Beggiatoa spp*
Thiomicrospira pelophila	*Sulfolobus brierleyi*	*Pseudomonas spp*
	Sulfolobus solfataricus	
		unclassified
facultative anaerobes	**facultative anaerobes**	
		Achromatium, Macromonas,
Thiobacillus denitrificans	*Thiobacillus A 2*	*Thiobacterium, Thioploca,*
Thiobacillus thioparus	*Thermothrix thiopara*	*Thiospira, Thiothrix,*
Thiomicrospira denitrificans	*Paracoccus dentrificans*	*Thiovulum*
	Thiosphaera pantotropha	

Sulfobacillus brierleyi and *Sulfobacillus solfataricus* are also very active bacteria species especially for the removal of organic sulfur integrated in the part of coal [11]. Our preliminary work showed that *Sulfobacillus solfataricus* had comparatively faster desulphurization rates and higher conversion percent for organic sulfur.

3.2. MOLDS

Several molds are reported to decompose some organic sulfur compounds. For example, sulfate can be released from the sulfonated phenol compounds by the sulfatases produced by fungi [12]. The major use of molds in desulphurization of coal is the removal of organic sulfur. However, there are no reports in the literature indicating the efficient use of molds for the removal of organic sulfur from coal. The isolation and identification of molds capable of oxidizing model organic sulfur compounds present in coal and the utilization of these molds for the removal of organic sulfur from coal is a challenging and promising area in coal desulphurization by microbial catalysis.

3.3. MIXED CULTURES

Several mixed cultures can be used for the removal of pyritic sulfur from coal. Dugan and Apel used a mixed culture of *T. ferrooxidans* and *T. thiooxidans* and proved that such a mixed culture is more effective than pure culture of *T. ferrooxidans* for the removal of pyritic sulfur from coal [13].

A similar mixed culture was also used by Andrews *et al.* for the removal of pyritic sulfur from coal [14]. Other mixed cultures such as *T. thiooxidans* and *Leptospirillum ferrooxidans* can be used for coal desulphurization [3]. *T. thiooxidans* can oxidize elemental sulfur to sulfate but cannot oxidize ferrous iron. On the other hand, *L. ferrooxidans* can oxidize ferrous iron but cannot oxidize elemental sulfur or sulfide compounds. In a mixed culture, *L. ferrooxidans* can oxidize ferrous iron to ferric and then ferric iron oxidizes pyritic sulfide to elemental sulfur, and *T. thiooxidans* can oxidize elemental sulfur to sulfate; therefore, the mixed culture can effectively remove pyrite from coal.

4. Microbial Removal of Sulfur from Coal

4.1. DESULPHURIZATION BY OXIDATION

Thiobacillus ferrooxidans and *Thiobacillus thiooxidans* can be found in association with coal mine effluents where oxidation of pyritic material produces the sulfuric acid present. The microbial action was shown to accelerate the chemical oxidation by a factor of 500,000 [15]. The role of bacteria in the oxidation of pyrite and other metal sulfides is well established, although the mechanism of dissolution is still not completely understood [5]. The reaction mechanisms are discussed in two major papers [16, 17].

The reactions involved are represented by the equations;

$$2\ FeS_2 + 7\ O_2 + 2\ H_2O \longrightarrow 2\ FeSO_4 + 2\ H_2SO_4 \qquad (1)$$

(chemical reaction plus bio-reaction)

$$4\ FeSO_4 + O_2 + 2\ H_2SO_4 \longrightarrow 2\ Fe_2(SO_4)_3 + 2\ H_2O \qquad (2)$$

and take place in the presence of air and water. The bacteria derive energy from the oxidation of the ferrous iron to ferric iron.

Then ferric ions react with additional pyrite;

$$FeS_2 + Fe_2(SO_4)_3 \longrightarrow FeSO_4 + 2\ S \qquad (3)$$

and the elemental sulfur is oxidized by microbial action to sulfuric acid.

Other metal sulfides are attacked by the ferric sulfate to release ions of the appropriate metal together with more elemental sulfur. There is disagreement in the literature as to whether *Thiobacillus thiooxidans* can oxidize pyrite independently. It is generally considered that it simply enhances the effect of *Thiobacillus ferrooxidans* through its oxidation of the free sulfur formed.

The overall biological oxidation reaction can be given as;

$$FeS_2 + 15\ O_2 + 2\ H_2O \longrightarrow 2\ Fe_2(SO_4)_3 + 2\ H_2SO_4 \qquad (4)$$

A considerable amount of work has been undertaken to study this reaction covering a period of over twenty years [5, 18]. To enable the bacteria to make the sulfur soluble, the coal had to be milled so that most of the pyrite particles were exposed to attack. It is clear that up to 90% of pyritic sulfur can be removed by this method and this commonly accounts for about half the total sulfur present in the coal. Further results are presented in Table 2 where the reduction obtained using*Thiobacillus ferrooxidans* is compared with that of utilizing *Sulfolobus acidocaldarius* [5].

TABLE 2. The effect of some biological and physical parameters on the rate and extent of coal desulphurization [5]

Culture(s)	pH	T(°C)	Coal Source	size (μm)	Slurry %	Inorganic sulfur Initial percentage	Percentage removed	Time (days)
T. ferrooxidans ATCC 19859	2.5	25	Eastern Us	43-74	« 20	4.9	90-98	8-12
T. ferrooxidans from acid mine waters	2.0	35	Ohio	<37	25	4.1	77 -	isolated
	2.0	35	New Mexico	<37	25	1.95	83	-
S .acidocaldarius (strain 98-3)	2.5	70	Pennsylvania	100-149	5	2.1	30	10
S .acidocaldarius	1.5	70	Pennsylvania	<48	20	2.1	90	-
S .acidocaldarius and	1.5-2	28	Illinois No 2	<74	20	0.98	90	14
Ferrolobus spp	1.5-2	60	Illinois No 2	<74	20	1.89	90	6
T. ferrooxidans and	2.5	25±3	W. Virginia	<74	20	0.89	90	14
T. thiooxidans	2.0-2.5	ambient	W. Virginia	74-300	20	3.1	97	5

4.2. FACTORS AFFECTING MICROBIAL ACTIVITY

Investigations with both coal and with pure pyrite have revealed a number of critical variables which affect the reactions and hence any potential desulphurization process. These include the biological parameters such as nutrient supplies and supplements, control of the physical environment (and such factors as pH, temperature and redox potential together with consideration of the source size of the coal), and the particular strain of bacteria used.

Figure 1 lists the process variables involved in microbial pyrite removal, and they all require investigation in order to define the conditions for a commercial process [19].

Microbial coal desulphurization is a complex process involving gas, liquid and solid phases as well as biological components. Organic nutrients, mineral salts and dissolved carbon dioxide in the oxidation medium all affect the rate and extent of desulphurization. The process involves microbial growth on the surfaces of coal particles and the diffusion of dissolved nutrients including oxygen and carbon dioxide through the product layer surrounding the coal particles [9].

Oxygen is the primary terminal electron acceptor for iron and sulfur oxidation by species of *Thiobacillus* and *Sulfolobus*, and must be made available to the organisms for oxidation. *T. ferrooxidans* shows a broad optimum for dissolved oxygen, centered around 8 ppm (near saturation) in desulphurization of coal. Carbon dioxide serves as the major source of cell carbon for growing *T. ferrooxidans*. Kargı found that the addition of CO_2 to reactor sparging air increased microbial removal of sulfur from a Pennsylvania coal .

An another feature of the reactions taking place with *Thiobacillus ferrooxidans* is that they proceed only under acidic conditions, that is at low pH values.The optimum pH is found as 2.0- 2.5.

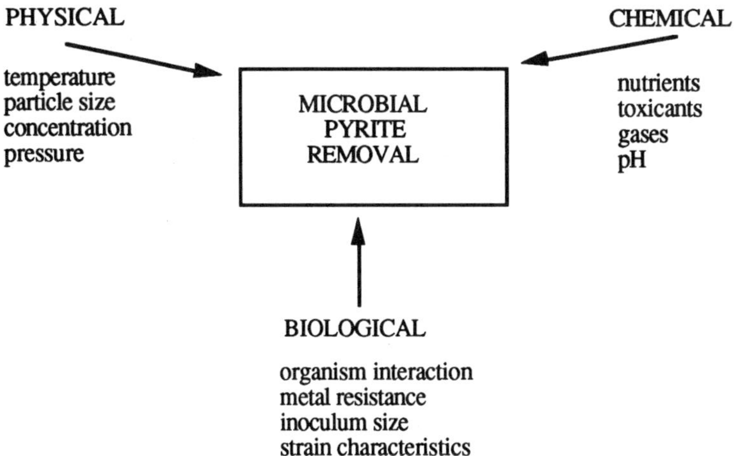

Figure 1. Bioprocess variables in pyrite removal [19]

The reactions also proceed satisfactorily within a narrow temperature range of 25-30°C. *Sulfolobus* also has an optimum growth pH in the range of 2 to 3, coinciding with its maximum coal desulphurization activity. Various results are reported, and the optimum conditions depend on the particular bacterial strain being used, and the nature of the coal.

High coal concentrations in the reaction medium together with high microbial cell concentrations improve the potential process economics [9]. Slurries with a high coal concentration are, however, difficult to handle. In addition, at high coal and microbial cell concentrations the removal of metabolic heat could be a potential problem, although this effect is likely to be more marked with *Sulfolobus* [9]. A concentration of about 20% seems to strike a balance between reducing the total volume and ease of handling [20] while Olsen and others [21] found that densities around 25% yielded the highest desulphurization rates.

Particle size is a major factor affecting the rate of pyrite solubilization since pyrite is present in coal as discrete inclusions and microbial oxidation occurs at particle surface. Torma and Murr found that the smallest particle size tested (-400 mesh, <37 μm) showed the greatest extent and most rapid rate of iron solubilization from pyrite in coal in shake flasks[22] . Kargı discussed the relationship between particle size and the number of microorganisms in the reaction medium as it affects microbial pyrite oxidation rates [23] .

Temperature is another important factor. *T. ferrooxidans* is mesophilic (optimum growth at moderate temperatures of 20-35°C), and optimum temperatures for desulphurization of coal by this microorganism are in the range of 28-35°C. However, moderately thermophilic *Thiobacillus* like bacteria and thermophilic bacteria such as *Sulfolobus* are also promising for accelerated leaching of sulfur from coal. These moderate to extreme thermophiles carry out biodesulphurization of coal at 50 to 75°C. Advantages to using thermophilic bacteria such as *Sulfolobus* include accelerated reaction rates and less aseptic conditions of operation. However, problems with corrosion of container and equipment materials will likely increase at elevated temperatures.

Electrochemical potentials of the reaction medium has an effect on the rate of pyritic sulfur leaching from coal. Pyrite oxidation is more favorable at high values of electrochemical potential since pyrite leaching requires oxidative environmental conditions. However, this value must be less than 400 mV in order to minimize the deposition of oxidative products (mainly $Fe(OH)_3$ or jarosite) on coal particle surface. In order to do that pH and DO levels are controlled during bioprocess as well as addition of ammonium salts is made.

Also high concentrations of Fe^{+3} (>100 ppm) were regarded to be inhibitory to microbial activity; the media containing 1-10 ppm Fe^{+3} was found to improve the rise of pyritic sulfuric acid[24].

An extensive study on reaction kinetics looked at different strains of *Thiobacillus ferrooxidans* and *Thiobacillus thiooxidans* [25]. The results showed high levels of desulphurization of 90-98% in coal samples with high pyrite content within eight to twelve days.

4.3. MODIFICATION OF THE SURFACE PROPERTIES OF COAL PARTICLES

Thiobacillus ferrooxidans can also be used to modify the surface behaviour of pyrite as an aid to its separation during conventional coal cleaning. It can improve the efficiency of the flotation process and the results indicate that roughly a 40% reduction in pyrite flotation has been achieved with less than 15 minutes of bacterial preconditioning [26].

Mixtures of rom coal and of waste coals have been tested to determine if the bacteria could enhance sulfur removal from rom coal during flotation associated with oil agglomeration [18]. A much better separation of pyrite was achieved in systems containing *Thiobacillus ferrooxidans* than using chemical depressants. With chemical depressants, the maximum pyrite rejection was around 50%. With systems containing *Thiobacillus ferrooxidans*, 80-90% pyrite rejection was achieved (80-90% sinks to the bottom and is removed in the tailings) and cleaned material retains only 10-20% of the original pyrite. The pH must be raised during agglomeration for the greatest pyrite rejection. The process apparently relies on surface oxidation which modifies the wetting properties of the materials. In these experiments, treatment times were between one and three days.

In the North Staffordshire work , the standard bacterial surface conditioning time for the pyrite was only 2.5 minutes, and it is claimed that this can be reduced even further, and perhaps to as little as 10 seconds. It is certainly a short enough residence time to be amenable to industrial process applications where the tests were carried out at a 2% pulp density [27].

The work suggests that the use of biological techniques for the selective modification of the surface properties of pyrite to enhance separation during flotation is quite promising. Since a change in surface characteristics can in principle be achieved with the oxidation of only a few molecular layers on the mineral surface, it requires short residence times compared with bioleaching of the whole bulk of mineral particles. The work however is not nearly as advanced as that on the oxidation route, and most is based on laboratory tests.

4.4. MAJOR CULTURES FOR ORGANIC SULFUR REMOVAL

4.4.1. *Sulfolobus acidocaldarius*. The organism was first reported in the early 1970s as oxidizing reduced sulfur compounds, and also ferrous iron, in order to obtain energy for its life processes [28, 29]. It is of particular interest because its optimum temperature is in the 50-80°C range, and operations at elevated temperatures could permit high cell and coal concentrations and speed the chemical oxidation of inorganic sulfur by ferric iron produced by the microbial oxidation of pyrite, FeS_2 [3]. High temperature operation also reduced the chance of contamination of the reaction medium. Up to 50% of sulfur removal is reported and its pH requirements are similar to those for *Thiobacillus ferrooxidans* , i.e., the optimum is between pH 2 and 3. The higher temperature of operation might pose problems of heat removal and/or of process control, since the optimum conditions for reaction are narrowly defined. Optimum slurry concentration was around 15% resulting in the maximum surface rate of sulfur removal for these particular conditions. Other variables examined were the effects of adding organic nutrients (mainly yeast extract and peptone) and of adding a chemical oxidizing agent ($FeCl_3$). In each case the addition had adverse effects on the removal of sulfur from the coal being used.

A comparison was made between coal refuse with 11% sulfur and an untreated coal with 4% sulfur. While the total sulfur removed is greater with 11% of sulfur in the feed, more was removed on the percentage basis from the untreated coal slurry containing about 4% sulfur. The maximum removal obtained in the tests was about 50% of the total sulfur present, including 96% of the pyritic sulfur [9].

Sulfolobus acidocaldarius is reported as removing about 75% of the sulfur from a finely ground Appalachian coal in three to six days [30]. The sulfur content of the coal was about 4% consisting of 0.2-0.3% sulfate, 1.6-1.8% pyrite and 1.9-2.2% organic. It was ground to 70-75% below 200 mesh. The laboratory work confirmed the need for fine grinding to increase the available surface area.

4.4.2. *Coal bug 1 (CB1)*. Atlantic Research Corporation of USA have reported the development of a microorganism named CB1 and CB2 capable of removing a proportion of the organic sulfur from coal [31]. The development is the result of mutagenic alterations of the microbial DNA in a mixed culture of naturally occurring soil organisms. It is a *Pseudomonas,* but it is a different strain from other *Pseudomonas* that have been tested against it [32]. CB1 is capable of oxidizing the sulfur in dibenzothiophene, DBT, although it does not break the carbon ring structure. The organic product has been identified as 2,2'dihydroxybiphenly using gas chromatography/mass spectra analysis.

A continuous bench unit has been operated employing CB1 on different coals and the results are presented in Table 3. In subsequent work it has been shown that maximum desulphurization occurs in about 9 hours [32]. A larger 1134 kg/d continuous pilot scale unit has been built and tested.

TABLE 3. Coal desulphurization using CB1 in bench scale plant [31]

Coal feed	Total sulfur, %	Pyrite, %	Sulfate, %	Organic sulfur, %
Illinois No 6	3.5	1.3	.1	2.1
Consolidation	3.8	1.6	.2	2.0
Illinois No 6	3.65	1.2	.3	2.15
Homer City underflow	1.4 (input)	0.7	-	0.7

Results	Retention time, h	Plant operation	Solids %	Total sulfur, % before	Total sulfur, % after	Organic sulfur reduction, %
Illinois No 6	48	5 weeks	10	3.5	2.9	24%
	24	"	10	3.5	2.8	29%
Consolidation	24	2 weeks	10	3.8	3.1	25%
	12	"	10	3.8	3.1	25%
Illinois No 6	12	1 weeks	10	3.65	2.95	19%
Homer City underflow	12	1 weeks	26	1.4	1.0	57%

Initial laboratory studies determined the effects of temperature on the oxidation of DBT. Tests showed that efficient microbial oxidation of DBT could take place between 25 and 40°C and the best result was obtained at 35°C.

Organic sulfur removal by CB1 appears to depend on how much thiophenic sulfur there is in a coal and on the ability of the microbes to get at it. Particle size is important since desulphurization increases with decreasing particle size and surface oxidation tends to inhibit desulphurization.

Organic sulfur removal varies from 19% to 57% depending on the type of coal, and the performance of an organism such as CB1 is likely to be coal specific. The fermentation with CB1 takes place at a pH of around 7.0 and combustion analysis of coal before and after treatment indicated that the heat content of the samples was virtually unaffected by the biotreatment.

The Atlantic Research Corporation work has been continued and a second microorganism developed, named coal bug 2 (CB2), which directly attacks diphenly sulphide. It has been tested on coals and can result in an organic sulfur reduction of up to 30%. Result will inevitably be coal specific, depending on the nature of the organic compounds in a particular coal .

4.4.3. *Sulfolobus solfataricus*. Our preliminary work showed that *Sulfobacillus solfataricus* (see Table 1), had comparatively faster desulphurization rates and higher conversion percent for organic sulfur in Turkish lignites. First objective of this study is to determine the effects of temperature, initial pH of the growth media, agitation rate, percent of inoculum and initial concentrations of glucose and ammonium sulfate on specific growth rate of *Sulfolobus solfataricus*. The second is to investigate the operational parameters of bioconversion which offer the highest sulfur removal percent and desulphurization rate. The ultimate goal through these studies is to obtain the optimum conditions for microbial removal of organic sulfur as well as others from coal.

Microorganism and culture medium: A pure culture of *Sulfolobus solfataricus* was obtained from DSM (Deutsche Sammlung von Mikroorganismen and Zellkulturen GmbH) with strain no: 1616. The inoculum culture used in experiments was grown in a medium which had the following composition (g/l): Yeast extract 1.0; KH_2PO_4 3.1; $(NH_4)_2SO_4$ 2.5; $MgSO_4.7H_2O$ 0.2; $CaCl_2.H_2O$ 0.25; $MnCl_2.4H_2O$ 1.8 .10^{-3}; $Na_2B_4O_7.10\ H_2O$ 4.5.10^{-3}; $ZnSO_4.7H_2O$ 0.22.10^{-3}; CuCl .$2H_2O$ 0.05 . 10^{-3}; $Na_2MoO_4.2\ H_2O$ 0.03.10^{-3}; $VOSO_4.2H_2O$ 0.03.10^{-3}; $CoSO_4.7H_2O$ 0.01.10^{-3}. The initial medium pH was adjusted to 4.0-4.2 with 10 N H_2SO_4, and the medium was autoclaved at 121°C for 15 minutes. A 24 hour grown culture was used for inoculation.

Growth conditions: Experiments were perfomed in 250 ml cotton plugged shake flasks with 100 ml of working volumes at different temperatures and stirring rates in temperature controlled water bath shakers.

Method of assay: In order to determine the amount of microorganism, 2 ml of samples were drawn at constant time intervals from the flasks, and the growth was followed by absorbance measurements at 500 nm. In coal desulphurization runs, total sulfur, inorganic and pyritic sulfur forms of Beypazarı lignite were determined by Eschka Method according to the ASTM D3177 and ASTM D2492 respectively, and organic sulfur content was calculated from the difference.

Box-Wilson Experimental Design: Box-Wilson experimental design was used for the determination of the growth kinetics of *Sulfolobus solfataricus* in this study [33]. Six bioreaction variables; temperature (T), pH, initial glucose (S_{G_0}) and initial ammonium sulfate (S_{N_0}) concentrations of growth media, inoculum percent (I%) and shaking rate (n) were chosen as independent variables for specific growth rate , μ, of the culture. Experimental design parameters used in this study and the specific growth rates obtained in each run are presented in Table 4.

TABLE 4. Fermentation parameters and specific growth rates for *Sulfolobus solfataricus*

Run No	T (^oC)	pH	S_{Go} (g/l)	S_{No} (g/l)	I% (v/v)	n (rpm)	μ (l/h)
01	55	4.0	15	5	6	55	0.0144
02	64	4.0	15	5	6	55	0.0172
03	70	4.0	15	5	6	55	0.0178
04	85	4.0	15	5	6	55	0.0120
05	70	1.5	15	5	6	55	0.0077
06	70	3.0	15	5	6	55	0.0184
07	70	5.0	15	5	6	55	0.0177
08	70	6.5	15	5	6	55	0.0164
09	70	4.0	9	5	6	55	0.0157
10	70	4.0	21	5	6	55	0.0188
11	70	4.0	30	5	6	55	0.0193
12	70	4.0	15	5	6	0	0.0148
13	70	4.0	15	5	6	33	0.0257
14	70	4.0	15	5	6	77	0.0126
15	70	4.0	15	5	4	55	0.0124
16	70	4.0	15	5	8	55	0.0139
17	70	4.0	15	5	11	55	0.0132
18	70	4.0	15	0	6	55	0.0122
19	70	4.0	15	3	6	55	0.0222
20	70	4.0	15	7	6	55	0.0155
21	70	4.0	15	10	6	55	0.0150
22	70	4.0	0	5	6	55	0.0060
23	70	4.0	15	5	0	55	0.0000
24	70	4.0	15	5	6	110	0.0137

Effect of temperature: Temperature effect on specific growth rate of *Sulfolobus solfataricus* was investigated in the range of 55-85°C. It was observed that specific growth rate was increased with temperature up to 70°C, whereas it dropped sharply beyond this value. Specific growth rate values were correlated to temperature employing least square regression techniques in the form;

$$\mu = -0.0786 + 0.0028\ T - 2.062\ .10^{-5}\ T \qquad (5)$$

Effect of pH: Initial medium pH effect was investigated in the range of 1.5-6.5 at 70°C and optimum initial pH value was found to be 3.0 for this thermophilic microorganism. Again least square regression offers the relation as;

$$\mu = 0.408\ pH/[53.940 + pH + (pH^2/0.489)(1 + pH/62.323)] \qquad (6)$$

Effect of initial glucose concentration: Glucose is the main carbohydrate source for *S. solfataricus* in growth experiments. The initial glucose concentration effect on microorganism growth was experimented in the range of 0-30 g/l. It was observed that the data fitted to Monod kinetic model with a high correlation coefficient [34]. The maximum specific growth rate and Monod constant were calculated as $\mu_{max} = 0.022\ h^{-1}$ and $K_s = 3.311$ g/l respectively from least squares regression analysis in the equation;

$$\mu = (0.022 \,.\, S_{G_o})/(3.311 + S_{G_o}) \tag{7}$$

Effect of initial ammonium sulfate concentration: The effect of ammonium sulfate (nitrogen source) concentration on specific microorganism growth was investigated in the range of 0-10 g/l. Optimum initial concentration was found as 3 g /l in the specified medium formulation for *S. solfataricus*. The least square regression analysis defines the relationship as;

$$\mu = 3.808\ S_{N_o}/[10.46+S_{N_o}+(S_{N_o}{}^2/5.652)(1+S_{N_o}/0.1021.10^{-4})] - 0.1747.10^{-5} \tag{8}$$

Effect of inoculum ratio: Effect of inoculum ratio was investigated in the range of 0 -11% (v/v) of the medium volume. Maximum specific growth rate was then observed by using 6% of inoculum. Microorganism concentration decreased at higher inoculum ratios. Following equation was obtained by correlating μ values to inoculum percentage using least square regression method;

$$\mu = 0.2849\ I\%/[0.7413+I\%+(I\%^2/0.4089)(1+I\%/2.01.10^{-3})] \tag{9}$$

Effect of agitation rate: Agitation rate effects were investigated in the range of 0-110 rpm at constant temperature of 70°C and initial pH of 4.0. The results showed that specific growth rate was naturally low for static culture, maximum specific growth rate was obtained around 33 rpm, and then growth of microorganism decreased at higher stirring rates mainly due to mechanical denaturation of cells. Specific growth rates calculated were correlated to the agitation rate by the least squares regression method in the form ;

$$\mu = 0.01493 + 0.0008477\ n - 0.00002025\ n^2 + 0.1132 \,.\, 10^{-6}\ n^3 \tag{10}$$

Box-Wilson [33] experimental design applied for the determination of the growth kinetics of *S. solfataricus* will formulate a general model comprising the six independent bioprocess parameters to calculate the optimum conditions for growth.

Biodesulphurization runs for a high organic sulfur lignite: In this study Beypazarı lignite was selected for its high organic sulfur content. The lignite sample was ground in a ball mill and sized to 53-88 μm, and was analyzed for ultimate, proximate and sulfur forms. The results are presented in Table 5.

These experiments were carried out at the temperature of 70°C, and shaking rate of 40 rpm, in a growth medium containing 10 g/l of glucose, and 3.2 g/l of ammonium sulfate and the other medium constituents at the fixed concentrations given above by using an 24-30 hour grown inoculum of 6.5% (v/v), and 4% (w/v) of lignite.

TABLE 5. Analyses of Beypazarı Lignite

Proximate Analysis (%)	
Moisture	20.5
Volatile Matter	22.8
Ash	31.6
Fixed Carbon	25.1
Ultimate Analysis (daf%)	
Carbon	64.8
Hydrogen	5.1
Sulfur (total)	9.3
Nitrogen	2.0
Oxygen (by difference)	18.8
Analysis of Sulfur Forms (mf%)	
Total	5.57
Inorganic	0.97
Pyritic	0.35
Organic	4.25

Desulphurization results: The changes in the pH and absorbance values of the medium (after filtering the coal); and total sulfur, inorganic sulfur, pyritic sulfur, and organic sulfur reduction values are shown as a function of time in Table 6. A rapid increase in inorganic, pyritic and organic sulfur reduction values with increasing reaction time were observed. It was also noticed that organic sulfur reduction rates remained almost constant up to 5 days. It is clear from Figure 2 that all sulfur forms are reduced significantly by the microbial treatment of coal compare to those of other strains, CB1 or *Sulfolobus acidocaldarius* (see Table 2 and 3). Maximum reduction was realized in inorganic sulfur form and pyritic sulfur removal is not as high as achieved by *Thiobacillus* microorganism group. Moreover, organic sulfur removal was more than 1% on formula basis and more upgraded results are anticipated after the completion of the optimization study.

TABLE 6. Results of a desulphurization experiment

			Sulfur Reduction, %			
Days	pH	Medium Absorbance	Total	Inorganic	Pyritic	Organic
0.0	5.1	0.015	0.0	0.0	0.0	0.0
2.5	4.3	0.111	9.7	4.1	4.3	11.6
5.0	4.7	0.148	20.5	13.3	14.0	22.9
7.5	4.3	0.143	27.2	39.3	30.0	23.7
10.0	4.0	0.207	39.9	85.6	39.4	25.5
12.5	3.8	0.215	45.7	91.9	51.7	33.0
15.0	3.4	0.284	48.8	96.0	66.3	35.0

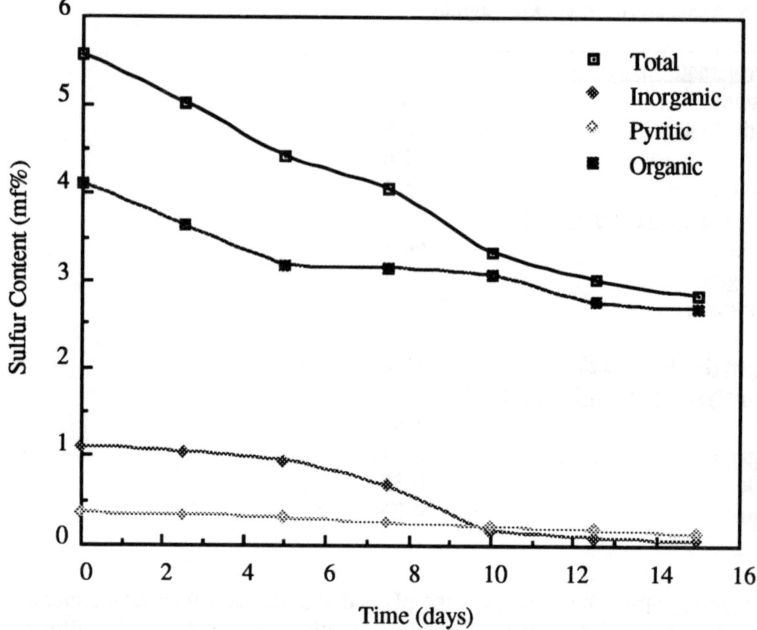

Figure 2. Decrease of various sulfur forms by time for Beypazarı lignite.

5. Process Schemes

There are several routes for incorporation of bioprocesses for the reduction of sulfur in coal. The first one which has been extensively referred and considered is based on the microbially enhanced oxidation of pyrite to ferric iron and to sulfate. The second, as mentioned earlier , involves the use of microorganisms to change the surface characteristic of the particles such that performance of conventional separation methods like flotation and oil agglomeration is enhanced.

Relatively little work has been done on detailed reactor design and on scale up for biodesulphurization, and one major problem encountered is the potential production of a relatively dilute but corrosive acidic effluent. The effluent will also be contaminated with superheats, biomass and leached trace elements, making a potentially difficult waste to treat. Another problem might be the aeration of large volumes of warm liquids with the associated heat loss and oxygen transfer. Requirement of fine particle sizes for enhanced reaction rate is a major energy consumption in crushing and grinding, nevertheless for bulk or heap treatment size reduction is not severe. The low pH values in possible reactors mean corrosive media, so any moving parts in the system should be avoided. For large amounts to be processed in industrial scale, the system should be a continuous one allowing recycle so that acclimated culture retains in the process and conversion percent increases. The design and scale up of possible reactors are studied by Huber *et al.* [35] and reported by Bos *et al*. [36].

Although much depends on the properties and requirements of particular coals, the design conditions for a possible reactor utilizing *Thiobacillus* and *Sulfolobus* species are roughly established as;

- A coal feed size of less than 100 µm
- Operating temperature around 30°C (70°C for *Sulfolobus* type)
- pH around 2
- Coal slurry density around 20%.
- An adequate air supply to satisfy the stoichiometric oxygen requirements
- Adequate agitation

A typical process is depicted in Figure 3.

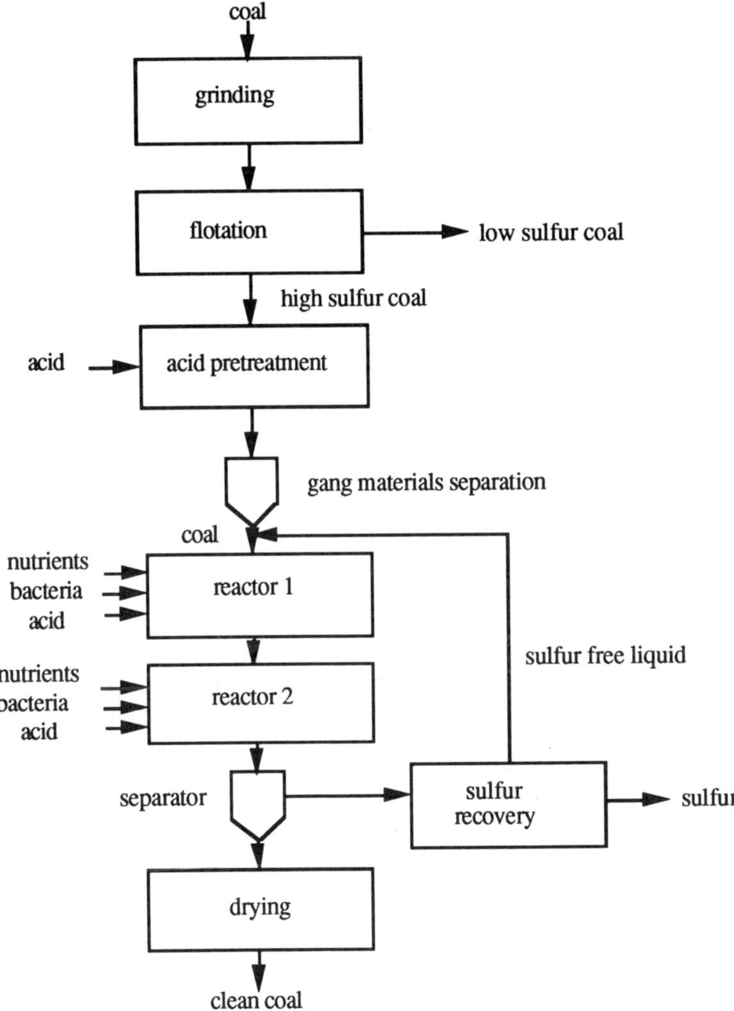

Figure 3. Possible process scheme for biodesulphurization [3]

The raw coal would be ground to increase the surface area for microbial attack. A preliminary separation by conventional flotation would provide a higher sulfur content feed to bioprocess. It might be necessary to remove alkaline materials such as calcium carbonate prior to reactor and this could be done by an acid treatment followed by separation of the gang materials either by sedimentation or filtration. Then pH is adjusted to around 2.5 and nutrients, mainly ammonium sulfate and potassium hypophosphate, and bacteria inoculum are supplied. Air agitated reactors are almost undoubtedly preferred to mechanically stirred vessels due to agitation by air is producing less shear rate that is less harmful to microorganisms. A consecutive reactor in series could involve different cultures and it might be necessary to adjust the optimum conditions of operations from one reactor to the next. For example a two stage scheme could use *Thiobacillus ferrooxidans* in the first reactor operating at around 30°C to remove pyritic sulfur and then *Sulfolobus acidocaldarius* in the second operating at 70°C to convert organic sulfur. Final stages may comprise separation and sulfur recovery.

Several different types of reactors such as airlift, horizontal rotating bioreactor and lagoons have also been considered in the integrity of biodesulphurization process.

Airlift reactors offer advantages over conventional stirred tanks in that lower shear stress causes less cell damage and reduces cell separation from coal particles. However, slurry density is crucially important in the stage of aeration and mixing [37].

A detailed design of possible horizontal reactor was studied by Vaseen [38] and conical types of reactors aerated at the bottom were also offered by Huber [39].

An alternative approach based on the lagoons as reactors has been proposed by Sproul *et al.* [40]. It is based on treating 8000 t/day of pulverized coal and reducing the sulfur content from 4% to 2% while pyritic sulfur removal is assumed 90%. The major drawbacks in this application are the vast area required and heat losses from the lagoon especially for organic sulfur removal process.

The contribution of microbial processes to conventional physical sulfur separation techniques is also available. In conventional flotation, pyrite tends to concentrate in the cleaned coal since it is hydrofobic like the coal where dirt content is hydrophilic. The pyrite can be made to be hydrophilic by introducing certain bacteria to the flotation cells, which changes the surface properties of the coal. The required residence time is very short and is measured in terms of minutes [27]. Because of its simplicity and the fact that flotation is already widely used, this method could find commercial application before the oxidation process is utilized.

Recently a combined process route has been proposed in Figure 4 to enhance pyrite separation using *Thiobacillus ferrooxidans* with the subsequent removal of the organic sulfur using the culture CB1 [31]. In this process, organic sulfur removal is carried out by the bacteria CB1 after the separation of pyrite and ash using a Flotaire column in which the surface property differences are microbially enhanced using *Thiobacillus ferrooxidans*.*Thiobacillus ferrooxidans* is grown in the fermenter on a continuous basis along with *Thiobacillus thiooxidans*. The main feed constituents are tailings from the processing stream which are high in pyrite content. Ammonium sulfate is added as a nitrogen source and carbondioxide is obtained by sparging air through. After a specified growth period, the stirring and aeration systems are turned off and particles are allowed to settle, recovered liquor containing microorganisms is fed back to the contact tank and mixed with coal. The contact time was 30 to 120 minutes and then slurry is fed to a conditioning tank where pH is adjusted and some agents added and then on to the flotation columns. This enables both froth flotation and gravity separation.

Another practical possible scheme is to use the residence time during pipeline transport to achieve the necessary conditions. Reduced capital costs would be an obvious benefit, however process control down the length of the pipeline and deposition of pyrite oxidation products would be major problems. Moreover, the effect of corrosion associated with low pH values might be a further problem. Microbial desulphurization at the coal source or at the point of use is probably better option from the technological and economic point of view.

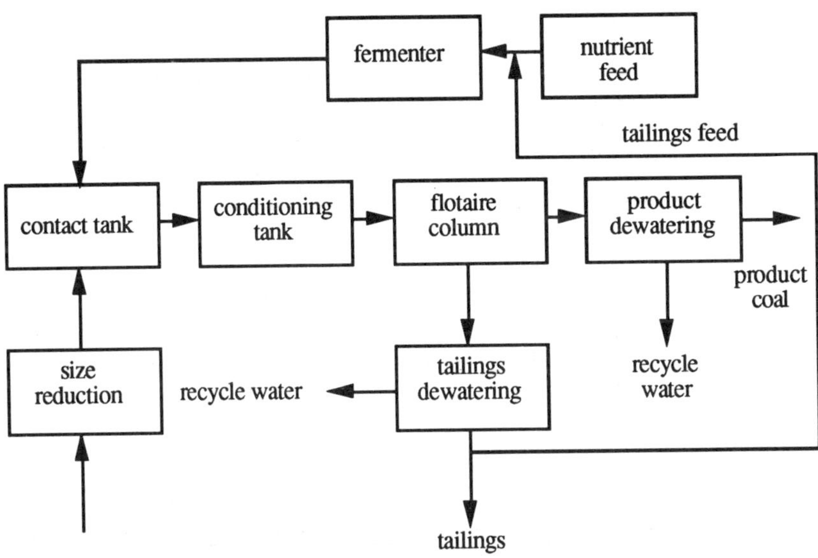

Figure 4. A combined process scheme for microbial desulphurization [31]

There have been a few published studies on the potential economics of microbial desulphurization and the data available are mainly from small scale laboratory experiments based on batch testing. Pilot scale continuous data are scarce and mostly kept confidential.

As early as 1979, estimation of a microbial desulphurization plant operating on a coal containing 2% pyritic sulfur at 8000 t/day feed rate was made as 10-14 $/t depending on pulverizing the feed and pelletizing the product by Detz and Barvinchak [20]. In Table 7, with the prices of 1979, microbial desulphurization is compared with the cost of other precombustion processes.

TABLE 7. Comparative desulphurization costs [20].

Process	Cost ($/t)
Microbial desulphurization process	14-20
TRW ferric leaching (Meyers) process	20
Battelle hydrothermal process	20
Kennecot oxygen leaching process	22
Solvent refined coal	30

Another study treating coal fines is shown in Figure 5 and it indicates costs in the range of 21.1-25.5 $/t, depending on the moisture content of the coal product [41]. The breakdown of these costs is given in the figure and the calculations is based on treating 100 t/hour coal fines and producing 0.32 Mt/year of cleaned coal. As seen clearly, organic sulfur removal accounts for almost half the costs.

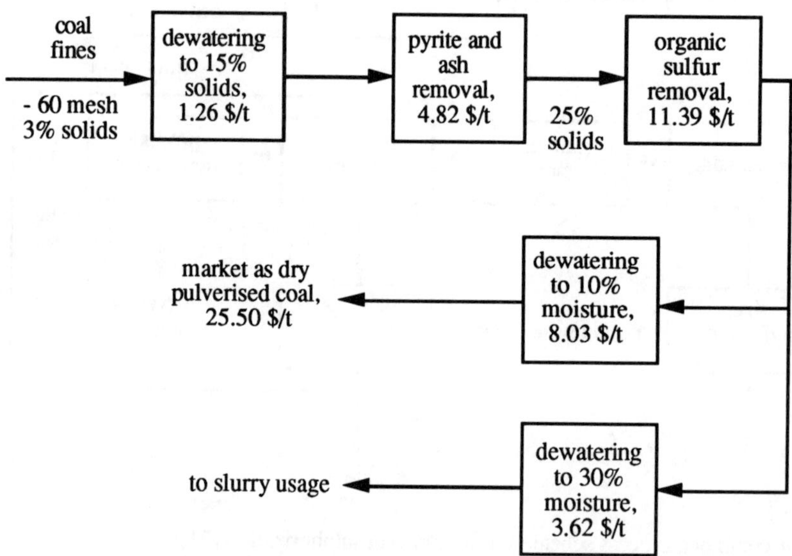

Figure 5. Process costs for a plant to produce 0.32 Mt/y of clean coal [31]

At the present state, flue gas desulphurization via chemical methods will continue to be the most economic option for large scale applications. However on small scale, in the treatment of high sulphur streams from various coals or where existing units are required to meet more stringent environmental requirements, biodesulphurization of the feed coal may be a viable alternative. Nevertheless, the costings are not sufficiently well founded nor are they attractive enough to justify large capital expenditures but sufficiently encouraging to support further process development work.

6. Conclusion

A great deal of work that has been done has looked at the biodesulphurization of coal. Microbial action can reduce pyritic sulfur and organic sulfur. It is possible to remove around 90 % of the pyritic sulfur and 50 % of the organic one from finely milled coal by microbial oxidation. Residence times measured in days for the bioreactions are necessary.

Alternatively, microorganisms can improve the separation of pyrite in a conventional flotation process in a matter of minutes since surface characteristics of coal do not take too much time to alter. Beneficial side effects of microbial desulphurization can be the simultaneous removal of some metal ions and of ash, and increased ash fusion temperature.

No commercial plant has yet been founded but an economically attractive microbial desulphurization process for certain coals could well emerge during next few years, and the state of the art justifies further development work. Some preliminary costings have indicated that biodesulphurization might cost a sum in the region of 10-15 $/t and if these figures can be substantiated keeping the low end of the range, the bioprocess is likely to find practical application. The cost of effluent treatment may emerge as a significant disincentive and as yet, little work has been done on this.

7. References

1. Eliot, R. C. (ed) "Coal desulphurization prior to combustion", New Jersey, Noyes Data Corporation, 1978, pp: V-VI, 33-42, 141-153.
2. Meyers, R. A. "Coal desulphurization", New York, Marcel Dekker Inc., 1977, pp: 26-40, 61-83, 187-191.
3. Kargı, F. (1982) "Microbiological coal desulphurization", Enzyme Microb. Tech., 4, 13-19.
4. Wheelock, T.D. (ed) "Coal desulphurization: Chemical and physical Methods", ACS Symposium Series 64, Washington, DC, 1977, pp: IX-XI, 101-120.
5. Monticello, D. J.and Finnerty, W. R. (1985) "Microbial desulphurization of fossil fuels", Annual Review of Microbiology, 39: 371-389.
6. Markuszewski, R., Miller, L. J., Straszheim, W. E., Fan, C. W., Whellock, T. D. and Greer, R. T. (1981) "New approaches to coal chemistry", ACS Symposium Series 169, pp: 401-412.
7. Roffman, H. K. (1979) "Proceedings of the institute of environmental Science", pp: 266-270.
8. Suhr, N. and Given, P. H. (1981) "Reliability of determination of pyritic iron in coals using standard procedures", Fuel, 60, 541-542.
9. Kargı, F. and Robinson, J. M. (1982) "Removal of sulfur compounds from coal by the thermophilic organism *Sulfolobus acidocaldarius*", Applied and Environmental Microbiology, 44 (4), 878-883.
10. Bos, P. and Kuenen, J. G. (1983) "Microbiology of sulfur oxidizing bacteria", in Microbial Corrosion (eds. Mercer; A. D., Tiller, A. K., Wilson, R. W., The Metal Society , London, UK., pp: 18-27.
11. Ollson, G., Larsson, L., Holst, O. and Karlsson, H. (1989) "Microorganisms for desulphurization of coal: The influence of leaching compounds on their growth", Fuel, 68, 1270-1274.
12. Starkey, R. L. (1956) "Transformation of sulfur by microorganisms", Indus. Eng. Chem., 48, 1429-1956.
13. Dugan, P. R. and Apel, W. A. "Microbiological desulphurization of coal", in Murr, L. E., Torma, A. E., Brierly J. A., (eds), "Metallurgical applications of bacterial leaching and related microbiological phenomena", New York, Academic Press, 1978, pp: 223-250.
14. Andrews, G. F. and Maczuga, J. (1982) "Bacterial coal desulphurization", Biotechnology and Bioengineering Symposium, 12, 337.
15. Granger, Q. P. (1984) "Bacterial leaching of minerals", Colliery Guardian, 232 (6), 212-214.

16. Ingledew, W. J. (1986) "Ferrous iron oxidation by *Thiobacillus ferrooxidans*", in Biotech. and Bioeng. Symposium No: 16, John Wiley and Sons, New York, USA, pp: 23-24.
17. Lundgren, D. G., Valkova-Valchanova, M.and Reed, R. (1986) "Chemical reactions important in bioleaching and bioaccumulation", Biotech. and Bioeng. Symposium No: 16, John Wiley and Sons, New York, USA., pp: 7-22.
18. Capes, C. E., Mcilhinney, A. E., Sirianni, A. F. and Puddington, I. E. (1973) "Bacterial oxidation in upgrading pyritic coals", Canadian Mining and Metallurgical Bulletin, 66, 88-91.
19. Olson, G. J. and Brinckman, F. E. (1986) "Bioprocessing of coal", Fuel, 65, 1638-1646.
20. Detz, C. M. and Barvinchak, G. (1979) "Microbial desulphurization of coal", Mining Congress Journal, 65 (6), pp: 75-82.
21. Olsen, T. M., Ashman, P. K., Torma, A. E. and Murr, L. E. (1980) "Biochemistry of ancient and modern environments" , (ed. Trudinger, P.), Springer-Verlag, New York, USA, 723 pp.
22. Torma, A. E. and Murr, L. E. Report EMP 2-67- 3319, New Mexico Energy and Minerals Department, Santa Fe, 1981, pp: 58.
23. Kargı, F. and Weissman, J. G. (1984) " A dynamic mathematical model for microbial removal of pyritic sulfur from coal", Biotechnology and Bioengineering, 26, 604-612.
24. Torma, A. E. "The Role of *T. ferrooxidans* in hydrometallurgical process", in (ed. Fiechter, A.), Advances in Biochemical Eng., 6:1, Zurich, Springer-Verlag, 1977.
25. Hoffmann, M. R., Faust, B. C., Panda, F. A., Koo, H. H. and Tsuchiya, H. M. (1981) "Kinetics of the removal of iron pyrite from coal by microbial catalysis", Applied and Environmental Microbiology, 42 (2), 259-271.
26. Attia, Y. A. and Elzeky, M. A. (1985) "Biosurface modification in the separation of pyrite from coal by froth flotation", in First International Conference on Processing and Utilization of High Sulfur Coals, Colombus, Ohio, USA, Oct 13-17, Elsevier Science Publishers.
27. Atkins, A. S., Davis, A. J., Townsley, C. C., Bridgwood, E. W. and Pooley, F. D. (1985) "Production of sulfur concentrates from the bioflotation of high pyritic coals", International Conference, Sulfur 85, London, UK, November pp. 83-104.
28. Brierley, C. L.and Brierley, J. A.(1973) "A chemoautotrophic and a thermophilic microorganism isolated from an acid hot spring", Canadian Journal of Microbiology, 19 (2), 183-188.
29. Brock, T. D., Brock, K. M., Belly, R. T. and Weiss, R. C. (1972) "*Sulfolobus*, a new genus of sulfur oxidation bacteria living at low pH and high temperature", Archives of Microbiology, 84, 54-68.
30. Murphy, J., Riestenberg, E., Mohler, R., Marek, D., Beck, B. and Skidmore, D. (1985) "Coal desulphurization by microbial processing" in First International conference on processing and utilization of high sulfur coals, Columbus, Ohio, October 13-17, Elsevier Science Publishers, Amsterdam, The Netherlands, pp. 643-652.
31. Isbister, J. D. and Kobylinski, E. A. (1985) "Microbial desulphurization of coal" in International Conference on Process Utilization of High Sulfur Coals, Columbus, Ohio, USA, October 13-17, Elsevier Science Publishers, Amsterdam, The Netherlands, pp. 627-642.
32. Isbister, J. D. (1986) "Biological removal of organic sulfur from coal" in Workshop on Biological Treatment of Coals, Washington, USA, July 23-25.
33. Perry, R. H. and Chilton, C. H. (1973) "Chemical Engineerings' Handbook", 5th edition, McGraw-Hill Kogakusha, Ltd., pp.2-74, Tokyo.
34. Monod, J. (1949) "The growth of bacterial cultures", Ann. Review of Microbiol., 3, 371.

35. Huber, T. F., Kossen, N. W. F., Bos, P. and Kuenan, J. G. (1983) "Modelling , design and scale up of a reactor for microbial desulphurization of coal" in (eds. Rossi, G. and Torma, A. E.) Recent progress in biohydrometallurgy, Cagliari, Italy, pp. 279-289.
36. Bos, P., Huber, T. F., Kos, C. H. and Doddema, H. J. (1986) "Microbial desulphurization of coal", Delft University of Technology, The Netherlands, p. 187.
37. Kargı, F. and Cervoni, T. D. (1983) "An airlift recycle fermenter for microbial desulphurization of coal", Biotechnology Letters, 5(1), 33-38.
38. Vasseen, V. A. (1985) "Commercial microbial desulphurization of coal" in First International Conference on Processing and Utilization of High Sulfur Coals, Columbus, Ohio, USA October, Elsevier Science Publishers, Amsterdam, The Netherlands, pp. 699-715.
39. Huber, T. F., Ras, C. and Kossen, N. W. F. (1984) "Design and scale-up of a reactor for the microbial desulphurization of coal: A kinetic model for bacterial growth and pyrite oxidation" in Third European Congress on Biotechnology, Munich, September 10-14, Verlag Chemie, Weinheim, Germany, pp. 151-159.
40. Sproull, R. D., Francis, H. J., Krishna, C. R. and Dodge, D. J. (1986) "Enhancement of coal quality by microbial deminerilasition and desulphurization" in Workshop on Biological Treatment of Coals, Washington, USA, July 23-25.
41. Isbister, J. D., Doyle, R. E., Kobylinkski, E. A., Ayres, D. M. and Kitchens, J. F. (1985) "Companion process for removal of sulfur and ash from coal" in Second Annual Pittsburg Coal Conference, September 16-20, MEMS, POB 2270, Greensburg, Pa, USA, pp. 103-110.

UPGRADING OF COAL-DERIVED LIQUIDS

L. L. ANDERSON
Department of Fuels Engineering
University of Utah
Salt Lake City, U.T. 84112
U.S.A.

ABSTRACT. Upgrading of coal-derived liquids is an important aspect of direct coal liquefaction and may determine when such liquefaction will be economically feasible. Coal derived liquids (CDL) are largely unacceptable as refinery feedstocks because of their high heteroatom and metals content and their aromatic nature. Heteroatomic removal and metals reduction are part of some petroleum refinery processing, but the presence of condensed ring polyaromatics is unique to liquids from coal. Successful upgrading processes will have to achieve significant reductions in the content of these aromatic components. Methods will also have to be found to reduce the buildup of coke on the catalyst surface, which has been the main reason for decreasing catalyst-bed activity during upgrading processing.

1. Introduction

Liquids derived from coal are typically lower in hydrogen content and contain more nonhydrocarbon materials than petroleum products. The nonhydrocarbon impurities consist mostly of nitrogen, sulfur, oxygen and inorganic materials. If coal-derived liquids (CDL) are to be used to make specification-grade fuels, upgrading by both addition of hydrogen and heteroatom removal will be necessary. Many processes to obtain suitable refinery feed liquids from CDL involve reactions and methods presently used in the treatment of heavy fractions from the distillation of petroleum. These include thermal cracking, hydrotreatment and hydrocracking. The composition of petroleum usually does not require extensive hydrodesulfurization (HDS), hydrodenitrogenation (HDN) or hydrodeoxygenation (HDO) as may be necessary for CDL. Finally, CDL differ from petroleum liquids in one other aspect, their higher content of polynuclear aromatics. It is evident that appropriate upgrading of CDL to materials that can be handled in a conventional petroleum refinery will require detailed analysis and characterization of such CDL as well as development of technologies for the upgrading processes.

2. Coal-Derived Liquid Composition

Literature on the properties and composition of CDL suggest that these may be at least partially dependent upon the composition and rank of the coal used to produce such liquids. However, since CDL can be produced by a variety of methods at different temperatures, pressures and with different sources of hydrogen present, specific data on this generalization are not reliable. In this discussion some general properties

Y. Yürüm (ed.), Clean Utilization of Coal, 207–212.

and some specific examples will be used to indicate the types and extent of upgrading reactions necessary for obtaining acceptable liquids which could be utilized as feedstocks for conventional refining.

Sturm et al. (1) performed a systematic study on the relationship of coal rank to coal liquid composition utilizing eight CDL prepared from six different coals. Properties of the coals and their liquids are given in Table 1. Sturm and coworkers also upgraded the liquids under conditions designed to decrease the nitrogen content to 0.2-0.3 weight percent using presulfided Aero HDS-3A catalyst. They observed that such upgrading yielded liquids quite similar in composition from all of the coals. Differences did not correlate with coal rank systematically. Table 2 shows the properties of upgraded liquids including their distillation fractions. Acid and base contents are not shown but were small and correlated well with oxygen and nitrogen contents, respectively. These workers found that liquids from lower rank coal were more easily upgraded and that liquids of comparable suitability as feedstocks for the production of refined distillate fuels could be produced from all of the coals of different rank that they studied.

TABLE 1. Analyses and properties of coal and their crude liquid products. Liquids were prepared by hydrogenation from coals in a batch autoclave at 2500 psi hydrogen, maximum temperature of 400°C using a Cyanamid Aero HDS-3 catalyst, [after Sturm et al. (1)].

Coal	Source	Analysis, wt %		Crude Liquid Composition
Pittsburg Seam	PA-WVA	C : 76.9		N : 0.44
HvbA		H : 5.1		O : 0.59
		N : 1.5		S < 0.01
		S : 1.6		
		O : 7.6		
		ash : 7.4		
Illiniois No. 6	Illinois	C : 75.2	Run 1	N : 1.45
Hvb B / C		H : 4.9		O : -
		N : 1.6		S : 0.17
		S : 1.5	Run 2	N : 1.10
		O : 9.7		O : 1.34
		ash : 7.1		S < 0.01
Kentucky.Bitum.	Western	C : 70.2		N : 1.28
Hvb B/C	Kentucky	H : 4.9		O : 2.09
		N : 1.4		S : 0.02
		S : 4.3		
		O : 8.8		
		ash : 13.5		
Colstrip Rosebud	Montana	C : 65.2		N : 0.64
Seam		H : 4.3		O : 1.61
Subb.A		N : 0.6		S < 0.01
		S : 1.8		
		O : 14.8		
		ash : 13.5		
Lower Wyodak	Wyoming	C : 66.8		N : 0.48
Subb.C		H : 4.8		O : 1.08
		N : 1.0		S < 0.01
		S : 0.5		
		O : 19.2		
		ash : 7.8		
Beulah Std. II	North Dakota	C : 63.5		N : 0.43
		H : 4.5		O : 1.55
		N : 0.9		S < 0.01
		S : 1.3		
		O : 19.1		
		ash : 10.7		

TABLE 2. Composition of upgraded liquids from eight coals, [after Sturm et.al. (1)].

Sp. gr., 16°C/16°C	0.993	1.006	0.992	0.989	0.987	0.955	0.922	0.983
SUS vis.@ 38°C	441	129	126	89	263	96	56	181
SUS vis.@ 54°C	189	65	86	-	123	-	-	123
Pourpoint, °C	-15	<-15	<-15	>-15	0	21	18	7
Carbon, wt %	89.2	90.1	88.6	88.2	88.1	88.9	88.0	89.0
Hydrogen, wt %	10.8	9.9	10.3	10.7	10.6	11.0	11.9	10.8
Sulfur, wt %	<0.01	0.03	0.2	<0.01	<0.01	<0.01	<0.01	<0.01
Nitrogen, wt %	0.20	0.444	0.250	0.287	0.192	0.095	0.008	0.250
Oxygen, wt %	0.28	0.50	0.19	0.32	0.34	0.17	0.04	0.61
Distillation, wt %								
<200°C	10.0	12.3	11.4	16.6	11.5	13.9	18.3	12.3
200°-325°C	21.7	27.3	27.9	26.1	21.5	26.7	30.3	24.0
325°-425°C	20.3	20.7	22.5	22.8	21.1	21.3	19.1	20.7
425°-540°C	26.6	19.1	23.7	19.7	20.9	25.2	25.3	21.0
>540°C	16.2	6.4	7.5	9.3	18.3	11.2	6.3	15.0
Asphaltenes, wt %	4.7	14.3	6.8	5.2	6.6	1.2	0.1	6.7
H / C (atomic)	1.45	1.32	1.40	1.46	1.44	1.48	1.62	1.46

3. Upgrading Reactions

As indicated earlier, coal-derived liquids pose particular problems for upgrading because they are typically high in nitrogen, oxygen, sulfur, aromatic compounds and metals. To upgrade such materials, multistage processes or multifunctional catalysts are necessary. This is a direct result of the compositional characterizations of CDL and the low probability that one set of conditions or a single function catalyst will be optimum for the various upgrading reactions necessary. The necessary reactions may be different depending on the desired final liquid product. This diversity makes generalizations about upgrading somewhat difficult.

In this communication rather than be comprehensive, examples of upgrading will be discussed and summarized. Upgrading to synthetic crude oil or to diesel fuel will be discussed as well as specific upgrading reactions (i.e., hydrocracking, HDN, HDO or HDS). In order to put these steps and reactions in proper perspective, some specific properties and analyses of CDL will be given.

Dooley et al. (2) examined five coal-derived liquids for elemental composition, acid and base content, boiling range, ring number distribution (mono-, di- and poly-aromatics). Data on these liquids are shown in Table 3. Ring number distributions for saturate, monoaromatic-, diaromatic- and polyaromatic-concentrates were determined and are summarized in Tables 4 and 5.

Although the analyses and ring-number distributions are helpful in showing the presence of various aromatics, the liquids cannot be compared in detail based on these data, since there were differences in cut points and other aspects (for example, the Synthoil residuum was not analyzed). The coal-derived liquids are typical of conventionally produced CDL; high in aromatics (saturates comprised 11-33 % of the liquids analyzed), contained significant heteroaromatic species such as acids (3.2-14.7 %) and bases (1.1-2.8 %). The properties of these liquids help one to understand the limiting factor in upgrading CDL is their highly aromatic character. In particular the presence of major quantities of condensed-ring polyaromatics is significant because each condensed aromatic ring must be saturated by hydrogenation before it can be cracked to produce light distillates. Nitrogen is the most difficult heteroatom to remove since it more often occurs in the aromatic ring structures.

TABLE 3. Physical and chemical properties of five CDL.

Liquid	Elemental Composition (wt %)				Distillate[a]		
	Sulfur	Nitrogen	Acids	Bases	<200°C	200-370°C	370-540°C
COED[b] (Utah Coal)	0.05	0.48	8.17	1.30	1.33	45.4	40.3
COED[c] (Western Kentucky Coal)	0.08	0.23	3.11	1.05	21.0	54.2	24.2
Synthoil[d] (West Virginia Coal)	0.42	0.79	10.40	2.75	4.4	42.6	27.3
H-Coal[e] Fuel Oil (Ill.No.6 Coal)	0.21	0.44	14.69	1.29	3.55	63.8	63.8
H-Coal Syncrude[f] (Ill.No.6 Coal)	0.27	0.63	13.81	2.73	34.6	65.4	65.4

[a] Approximate boiling ranges (the difference between these fraction totals and 100 was resid).
[b] About 10% of this material was not analyzed. 1.50% heteroatomic compounds determined in addition to acids and bases.
[c] About 8% of this material was not analyzed.
[d] About 4% (plus resid) of this material was not analyzed. 2.75% heteroatomic compounds determined in addition to acids and bases.
[e] Only overhead products analyzed. +200°C material boiled up to 460°C.
[f] Only overhead products analyzed. +200°C material boiled up to 510°C.

TABLE 4. Ring number distributions for coal-derived liquids [Saturates Concentrates (from Distillates Table 3)].

Total Rings	COED (UT)	COED (W.KY.) Illinois No.6	Synthoil	H-Coal Fuel Oil Illinois No.6	H-Coal Syncrude
0	8.70	4.30	1.90	1.70	1.40
1	2.50	4.30	2.60	2.00	1.10
2	2.20	3.90	1.80	1.90	1.00
3	2.20	3.20	1.80	1.20	0.62
4	2.20	2.10	1.20	0.60	0.45
5	4.00	0.93	0.31	-	-
6	1.30	0.44	-	-	-
<200°C (sat.frac.)	7.00	14.30	1.20	14.20	16.00
Totals	30.10	33.40	10.60	21.60	20.70

TABLE 5. Ring-number distributions for aromatic concentrates derived from coal liquids (wt.).

Total Rings	COED UT	COED W KY	Synthoil W VA	H-Coal Fuel Oil Illinois No.6	H-Coal Syncrude Illinois No.6
1	1.75	2.04	0.30	1.15	0.96
2	9.23	10.35	4.21	14.34	8.22
3	11.05	14.54	10.49	15.96	9.28
4	9.68	10.93	10.97	5.92	6.24
5	7.18	5.98	5.34	1.51	4.58
6	4.91	3.17	3.73	0.43	4.19
7	2.17	2.04	3.37	0.10	1.05
8	0.19	0.45	2.44	0.07	1.47
<8	-	0.07	2.17	0.03	1.31
Monoaromatics	2.79	4.75	1.21	12.18	10.85
Diaromatics	-	-	0.14	trace	0.06
Totals	48.95	54.33	44.37	51.69	48.21

TABLE 6. Upgrading studies for coal-derived liquids

Authors, Ref.	Catalyst (s) Used	Reaction Mode	Liquid Product Objective	Major Reaction(s)
Shmada, et al.(3), (1990) (JAPAN)	γ-zeolite supported Ni/Mo	Single-stage and two-stage	Gasoline	Hydrocracking
A.A.Krichko et al.(4), (1990) (U.S.S.R)	Ni/Mo/alumina	Single-stage	Distillates (BP: 55-425°C)	Hydrogenation
Perrot et al. (5), (1988) (FRANCE)	Cr/Mo/ activated alumina	Single-stage	Monocyclic (BTX)	Hydrocracking of bicyclic compounds
Martinez et al.(6), (1988) (SPAIN)	Ni/Mo and Ni/Ti	Single-stage	Upgraded liquids	Hydrogenation heteroatom removal
van der Watt (7), (HOLLAND)	Ni/Mo	Single-stage	Diesel fuel	Hydrotreatment
Mochida et al. (8), (1986) (JAPAN)	Ni/Mo/alumina	Single-stage and two-stage	Distillate fuels	Hydrocracking, HDN, HDO
Edwards et al. (9), (1986) (AUSTRALIA)	Ni/Mo	Single-stage (2nd of two)	Synthetic Crude oil	Heteroatom removal hydrocracking and ring hydrogenation

4. Examples of Upgrading Studies on CDL

Upgrading of CDL may involve reactions to accomplish one or more changes in the CDL. Table 6 gives some examples of recent work where different feedstocks and products are involved. In a total upgrading procedure where gasoline or other refinery products are produced, both upgrading and refining reactions are carried out. Usually upgrading would involve only those treatment which would result in a crude oil which would be acceptable in a conventional oil refinery. From Table 6, one can observe that molybdenum and nickel catalysts are the main upgrading catalysts. While these metals are expensive their use may be justified by appropriate catalyst life performance. The current approximate relative costs of various metals used or considered as catalytic materials is shown in Table 7. This shows the incentive for research directed toward iron and iron-based catalysts as well as catalyst recovery methods for many of the catalysts used for upgrading processes.

TABLE 7. Relative costs of various metals, used in catalytic upgrading of coal-derived liquids, (Iron: $70-80/ton, December 1991).

Iron	1.0	Molybdenum	65
Manganese	7.0	Nickel	100
Aluminum	14.0	Cobalt	185
Zinc	17.0	Tin	5,194
Copper	23.0	Palladium	37,000
Magnesium	30.0	Platinum	144,000
Tungsten	46.0	Rhodium	785,000

5. References

1. Sturm, G.P., Jr., Dooley, J.E., Thompson, J.S. Woodward, P.W., and Vogh, J.W. (1981) 'The Composition of Liquids from Coals of Different Rank', ACS Symp. Series, 156, R.F. Sullivan (ed.), 1-37.
2. Dooley, J.E., Lanning, W.C., and Thompson, C.J. (1979) 'Characterization Data for Syncrudes and Their Implication for Refining', Adv. in Chem. Series, 179, 1-24.
3. Shimada, K., Sato, T., Yoshimura, Y., Hinata, A., Yoshitomi, S., Mares, A.C. and Nishifima, A. (1990) 'Application of Zeolite-Based Catalyst to Hydrocracking of Coal-Derived Liquids', National Chem. Lab. for Industry, Tsukuba, Ibaraki 305, (JAPAN) ,Fuel Proc. Technol., 25,153-165.
4. Krichko, A.A., Malotetnev, A.S., Mau, O.A., Slvinskaya, J.J., Jacobsen, A.C. and Christensen, H.(1990) 'Hydrogenation of Distillate Products from Coal Liquefaction', Instit. for Fossil Fuels, Leninsky Prospect, Moscow, USSR, Fuel, 69, 344-348.
5. Perrot, J.M., Ayadai, A., Bastick, M. and Bastick, J. (1988) 'Upgrading of Coal Pyrolysis Products: Catalytic Hydrocracking of Bicyclic Compounds', Laboratoire de Physico-Chimie Industrielle-ENSIC, Fuel. Proc. Technol., 20, 223-232.
6. Martinez, M.T., Miranda, J.L., Juan, R., 'Catalytic Hydrotreating of Coal Liquids', Instit. de Caraboquimica, CSIC, Zaragoza, (SPAIN), Fuel, V.67, September.
7. van der Watt, J.G. (1985) 'Upgrading Coal-Derived Liquids to Diesel Fuel', Fuel Proc. Technol., 11, 101-112.
8. Mochida, I., Sakanishi, K., Korai, Y., and Fujitsu, H. (1986) 'Two-stage Catalytic Upgrading of Vacuum Residue of a Wandoan Coal Liquid', Fuel, 65, 1090-1093.
9. Mochida, I., Sakanishi, K., Korai, Y., and Fujitsu, H. (1986) 'Two-stage Hydrodenitrogenation of Heavy Distillate in a Coal Liquid', Fuel, 65, 633-635.
10. Yoshida, R., Yoshida, T., Narita, H. and Maekawa, Y. (1986) 'Upgrading of Coal-Derived Liquids: Characterization of Upgraded Liquids by Field-Ionization Mass Spectrometry', Fuel, 65, 425-428.
11. Yoshida, R., Miyazawa, M., Yoshida, T., Ishizaki, K., Shinrik, N. and Maekawa, Y. (1986) 'Upgrading of Coal-Derived Liquids: Characterization of Upgraded Liquids by Thin-Layer Chromatography Combined with Flame-Ionization Detection', Fuel, 65, 412-424.
12. Edwards, J.H., Schluter, K., and Tyler, R.J. (1986) 'Upgrading of Flash Pyrolysis Tars to Synthetic Crude Oil: Second-stage Hydrotreatment Using Nickel/Molybdenum Catalysts', Fuel, 65, 202-207.

GAS UPGRADING, CLEANING AND PURIFICATION

E. EKİNCİ
Department of Chemical Engineering
İstanbul Technical University
Ayazağa
İstanbul 80626
Türkiye

ABSTRACT. Gas upgrading, cleaning and purification play an important role in synfuel proecesses. At present, there are abundant data available on the low temperature and low pressure applications of scrubber, electrostatic precipitators and cyclones. There is an urgent need to transform these techniques to high temperature and pressure and special applications. The research and development work in particulate and gas treating methods comply with strict statutory controls and economics.

1. Introduction

There are severe pressures on synthetic fuel industry to develop substitute sources for the replacement of depleting crude oil reserves. There is also an urgency to provide sources for globally increasing demand for energy. Even after the superficial relaxation is felt, after the Gulf war, inherent energy crisis based on the dependence on petroleum still keeps its iceberg character. Synthetic fuel processes also feel the pressure of enviromental concern which is centered around issues such as acid rain, green house effect, ozone layer, carcinogenic emissions and occupational hazards. The synfuels processes are to obey strict enviromental controls. Urgency, pollution control and economy of synthetic fuel processes compose a difficult challenge for chemists, chemical engineers, fuel and energy specialists, mechanical engineers and professionals of related fields.

2. Definitions

Gas upgrading is the process where by the concentration of a gas or gas mixture is raised above a specified level. This may be due to a limiting calorific value or preparation of the gas for a subsequent use. For example, for Clauss process at least 20% H_2S concentration is necessary in the feed gas for acceptable sulphur recovery economics (1). It is reported that if all the acid gas in the gas of Kopper-Totzek (K-T) process is absorbed and it is fed to Clauss process, the feed contains only about 18% H_2S. On the other hand, the gases existing Lurgi process contains only 3.5% H_2S and refinery sour gas contains 93% H_2S. Therefore, for Clauss sulphur recovery the acid gases needed to be upgraded to above 40% H_2S concentration for K-T and Lurgi processes.

The gaseous products of synthetic fuel processes may contain solid, liquid or gaseous components which may not be desired to be above a certain concentration. This limit may be

Y. Yürüm (ed.), Clean Utilization of Coal, 213–220.

imposed to meet statutory limits or to meet conditions for a subsequent use as a product or mediatory chemical. In this case the usual procedure is cooling to remove condensables, scrubbing particulates and removal of undesired impurities. Condensation of tars, scrubbing of particulates and removal of acid gases from a coal gasifier is a series of gas cleaning operations.

Gas purification is an overlapping terminolgy with gas cleaning in certain cases. It generally implies attaining relatively high concentration in a specified component. For certain cases, it is necessary to use a combination of techniques to achieve the desired purification level.

General purification and cleaning costs are relatively high. For the case of coal gasification plants, the cost for cleaning and purifying the hot dusty and broad mixture of gases exceed the cost of any other section of the plant, amounting to 15-20% of the total plant cost (2).

The urgent need to develop gas cleaning and purification processes in an enviromentally acceptable and economic way prompted developers to use the experience gained in natural gas sweetening, petroleum refining, hydrogen and ammonia technologies. To adapt these technologies to synfuels from solid fuels posed serious problems, since for example coal and refinery gases differ considerably in composition and other aspects. To solve the extra purification and cleaning problems on individual basis may cause severe economical penalties (2).

By terminology gas upgrading, cleaning and purification may all be named in a single process depending on the component and aims of the operation. It would be helpful to classify different tools and techniques that are used in gas upgrading, cleaning and purification processes.

3. Particulate Cleanup

Particulates are objectionable for a number of reasons and their removal are often necessary for statutory or other subsequent usage restrictions. Turkish Air Quality Assurance Legislation restricts the particulate emission for coal gasification plants (3). Particles larger than about 5 μm are generally easy to remove from gases since they are susceptible to physical methods. However, generally the particle sizes below 5 μm are of health concern due to their long stay in atmosphere. The smaller size particles such as resulting from coal combustion and gasification processes contain high pollution risks. Major methods that are used for removal of particulates from gases are as follows (4): a) Wet scrubbing, b) Electrostatic precipitation (ESP), c) Fabric filtration, d) Inertial separation.

Among the group of the listed particulate removal technologies, the first three are capable of removing particulates less than 3 μm diameter. The inertial separators which are effective above 10μm are sometimes used as precleaners. Electrostatic precipitation on the other hand are utilized for dust, mist and fume control. Sometimes they are used to recover tars that are target products from fluidized bed pyrolysis of oil shales (5).

A convenient parameter for describing the capacity of a particle scrubber performance is the cut diameter. Cut diameter is the diameter of particles which are collected at 50% efficiency by the given scrubber.

3.1. WET SCRUBBERS

The type of scrubbers may be classified according to their contacting system as: plate, fixed bed, fibrous, preformed spray, gas atomized spray, centrifugal baffle impingement and entrainment, mechanical aided, moving bed and combination. The current running limits of some of these systems are given in Table 1.

Table 1. Efficiency of some of the scrubber systems (4).

Type of System	Particle Cut Diameter	Contact Device and Conditions
Packed tower	1.5 μm	1in Berl saddles or Raschig rings
	0.7 μm	1/2in Berl saddles or Raschig rings
Spray scrubbers	2.0 μm	Moderate liquid/gas ratio
	0.7 μm	High velocity
Centrifugal	4.0-5.0 μm	No spray
	0.7 μm	With spray
Baffle type	5.0-10.0 μm	Louvers, zigzag baffles, dish or dougnat baffles

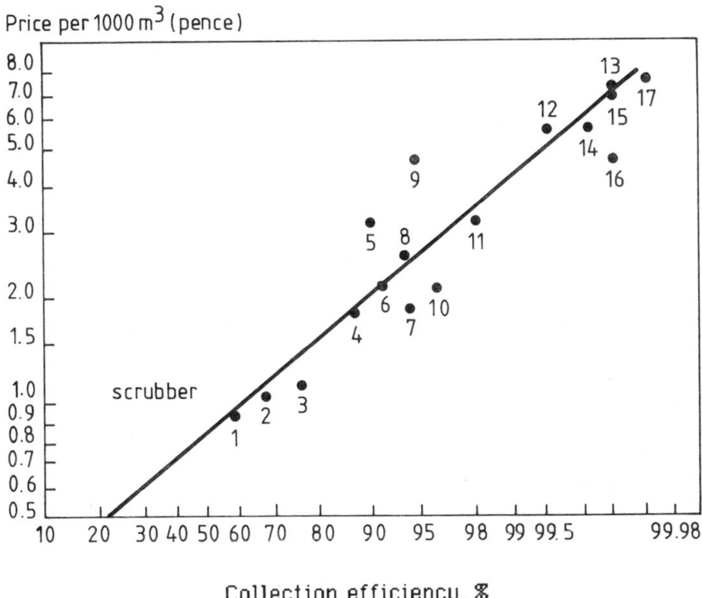

Figure 1. Efficiency and cost of gas cleaning methods. 1. Inertial collector, 2. Medium efficiency cyclone, 3. Low resistance cellular cyclones, 4. High efficiency cyclone, 5. Impingement scrubber (Doyle type), 6. Irrigated cyclones, 7. Tubular cyclones, 8. Self-induced spray deduster, 9. Void spray tower, 10. Fluidized bed, 11. Irrigated target scrubber (Peabody type), 12. Electrostatic precipitator, 13. Irrigated electrostatic precipitator, 14. Venturi scrubber-medium energy, 15. Venturi scrubber-high energy, 16. Low velocity fabric filter, 17. Reverse jet fabric filter.

The relative scrubbing efficiency and economics of various cleaning processes are shown in Figure 1. The data in Figure 1 was prepared for a gas volume of 100,000 m^3/h at 20°C, for fine industrial dust containing only 12% particles less than 2.5 μm in size. Assumptions also include no allowance for the treatment of liquid effluent (7). For high gas volumes greater than 50,000 Nm^3/h and for high dust sizing and high gas cleanliness greater than 20 mg/Nm^3 ESP becomes cost competitive compared to fabric filter. Wet and sticky or pyrophobic particulates require scrubbers or an irrigated ESP.

The choice of any one or combination of these cleaning methods depends on the required cleanliness, gas characteristics such as temperature, dew point, toxic hazards and particulate characteristics such as concentration, sizing and composition, and economics. The breakdown of UK gas cleaning practices show that scrubbers, electrostatic precipitators, fabric filtration and inertial separation have 11, 63, 23 and 3 percent of the national share, respectively (6).

As was pointed out previuosly the infant synthetic fuels technology is transferring air pollution control methods from earlier developed industries such as mining and oil industry. Some of the applications of scrubber technologies on synthetic fuel plants are the Cold Lake project on Athabasca bitumen, General Electric Corporation Research and Development Center for the integrated gasification and combined heat and power (CHP) generation system and U.S. Department of Energy (DOE)'s refined coal process and all pilot plant systems utilizing wet scrubbing systems (8).

3.2. ELECTROSTATIC PRECIPITATORS (ESP)

Removal of particulates by electrostatic precipitation passes the gas through a series of discharges and collecting electrodes. The corona discharge created by DC provide the electric field for the charged particles. The particles collected at the electrodes are transferred to the hopper which is situated below the electrodes. ESP is an efficient gas cleaning method for submicron sized fume to levels of 30 g/Nm^3 and less. This high performance cleaning is usually achieved at a low operating cost. However the capital investment for ESP is very high therefore its application becomes economical only at gas flow rates of 50,000 Nm^3/h (6).

ESPs are widely used in steel, cement and combustion systems. The collection efficiencies are 99% as reported by Gooch et al. (10) for several types of coals and ESP installations. It is now necessary to extend the performance of ESP to high temperature and pressure for gasification. (CHP) integrated systems, pyrolysis and atmospheric and pressurized fluidized bed combustion. A research program on high temperature and pressure ESP includes the following topics: discharge elctrode type, cleaning method and intensity, electric field strength, polarity of applied voltage, operating pressure and temperature and gas velocity (9). The problem areas realized from the current research are: back corona, retainment during cleaning, optimization of operating cycles and dust resistivity effects. Apart from electrostatic precipitators, electrostatic entrancement of particulate removal from gases is employed in conjunction with many other well known separators such as electrostatic augmented cyclone and electrostatic augmented fabric filter (9). The fall-off of collection efficiency with decreasing particle size and increasing equipment size has led to performance augmentation by electrostatic applications. Through these applications General Electric developed an air shielded electrocyclone with 4.3 m diameter that has a particulate removal efficiency equivalent to that of conventional 15 cm i.d. cyclone with a capacity of 113 kg/s of dirty gas compared to 0.45 kg/s.

3.3. FABRIC FILTRATION

Fabric filtration removes a good proportion of particulates within the pores of the fabric. Due to accumulation of the particles the cake formation cause an appreciable increase in pressure. Therefore it is necessary to clean the fabric periodically. Filtration technique is much the same for many different filters such as polyesters, acrylic-copolymer, acrylic homopolymer, aromatic-polyamide, glass and PTFE. There are mainly three different methods for the bag cleaning operation. First of these methods is the pulse jet filter which is cleaned by a short pulse (0.06 s) through a venturi pipe. Cleaning to particulate concentration less than 10 mg/Nm^3 can be guaranteed using needle felted fabric (6). The second method is the reverse air filter which uses a reverse flow of low pressure air for about 10 s. Similar to pulse jet method needle felted bags can be used. The performance and costs are comparable. The last cleaning method is the mechanically shaken filter. Cleaning is accomplished usually by a mechanism which imparts a short stroke. Gas normally travels from the inside of the filter bags outwards. The dust collected inside the bag is removed with the bag filter off-line or the filter is compartmented and cleaned in sections (6).

At present the research interest is concentrated on the production of fabrics with improved properties such as resistance to acid attack, operability at high temperatures and pressures. Use of bag filters are restricted to gases which have dew point problems. Fabric filters are generally economical for small to medium gas flow rates up to 50,000 Nm^3/h.

There has been extensive research on the electrostatic augmentation of fabric filters in literature starting as early as 1930's (9). However Apitron (American Precision Industries) produces the only commercial electrostatically augmented fabric filter at present. This is essentially a pulse induced reverse air flow fabric filter with a wire-pipe type electrostatic precipitator as a precollector and particle charger.

3.4. INERTIAL SEPARATORS

Cyclones and settling chambers have been used in industry for more than 100 years. They are popular due to their simplicity, ease of construction and economy. However they are only capable of removing particulates above 10 μm. Therefore they mey be used as prefilters to ease the removal duty on the subsequent cleaning operations. There has been extensive research to improve the removal efficiency of particulates. Some of these research are in the form of novel designs cyclones in series or using it coupled with other systems.

Several stages of cyclones in series with several parallel units are required for pressurized fluidized bed combustors. For turbine safety, at least three stages were needed. For U.S. Enviromental Protection Agency (EPA) - National Standards for Power Sources (NSPS) restrictions four or more stages are required in which case economical advantage may be lost. Historical cyclones are designed to recover particulates larger tahn 10 μm; however in pressurized fluidized bed systems 90% of particles as small as 2 μm were removed.

Due to economical advantages cyclones receive a lot of research interest at present which include electrocyclones, tanjet cyclones, cyclone centrifuges and centrifugal wedge separators.

Gas cleanup for atmospheric fluidized bed combustion (AFBC) is provided by a modest extension of existing pulverized coal-fired boiler technology cyclone-bag house system. If it is desired to transform this technology to oil shales or other pyrolysis processes for the production of synthetic crude oil, then precleaning of particulates is essential. The failure in removal of the micron size particulates from the condensates causes aggravated filtration problems in the condensate. Therefore high efficiency scrubbing systems working at relatively high temperatures are needed to be developed.

3.5. OTHER SYSTEMS

Besides the particulate removal systems discussed above, fixed bed, fluidized bed, fixed/fluidized bed and moving bed electrostatically augmented particulate removal systems are being extensively studied and developed for different applications (9).

Particulate cleanup from gas streams at low temperature and pressure is in a well handled position at processes that are related to synfuel production from solid fuels. In synfuel products usually particulate cleanup at high temperatures and pressures may be essential. This is a difficult application, therefore research and development work on cleanup systems at high temperature and pressure should be the ultimate goal.

4. Gas Phase Upgrading, Cleaning and Purification

In gasification, pyrolysis, cracking, hydrocracking and hydrorefining, H_2S is a usual component that is desired to be removed. Also COS, CS_2, alkyl sulphides, alkylmercaptanes which are undesirable may be present. Synfuel gases may also contain contaminants which are highly toxic such as benzene, toluene, xylene, phenols, thiocyanates, mercury, hydrogen cyanide and carbonyl sulfide (2). Production of nitrogen, oxygen and hydrogen are essential parts of synfuel processes; two of the latter are the subject of other chapters. Among the so many resultant gases any one or combination of some may be chosen as a target product (5). Production of methane by pressure swing adsorption may be given as an example. One of the most well known H_2S removal processes is the Clauss process which has the advantage of producing a valuable end product.

The problem with removal of one or more components from a gas stream forms hundreds of possible combinations. The product gas characteristics, composition and level of removal all need to be considered. Therefore, the task of assigning processes may be much more difficult compared to gas scrubbing in a single straight forward process.

4.1. ACID GASES REMOVAL

Processes available for removal of CO_2 and H_2 are given by Massey and Brown (2) as: (a) processes involving chemical reaction, (b) physical adsorption, (c) condensation and (d) combination of these.

Chemical reactions are mainly Hot Carbonate Process, Alkazid Wash, Adip Wash, Sulfinol Wash and in the general non-selective, low pressure amine based processes found in the petroleum and natural gas industries. There are advances in the application of membrane technology.

In physical adsorption, absorption solvent is chosen as a liquid of high density, low molecular weight and low viscosity. Water is used in pressurized coal gasification, natural gas and refinery gas applications with limited success. Gum formation and failure in the removal of hydrogen cyanide were main problems (2). The trace components found in coal gas and similar streams accumulate in the recycling solvent which depress the absorption process. In Table 2 some of the commercially successful absorption processes are shown.

Condensation is defined as cooling a gas to its dew point to separate it from other components which have a difference in their dew point temperatures. Co-condensation and condensate freezing temperature should be noted for thermodynamic efficiency. In order to produce pure CO_2 and an acid gas stream rich in H_2S for sulphur recovery triple point crystallisation is being reached (2).

Combination of different acid gas removal systems is necessary when the concentration of acid gas is low. Exxon also developed absorption processes to remove acid gases from petroleum refineries and gas plants (15). In one the processes Flexsorb SE selective amin treating agent, rather than the previously used non-selective agents, such as MEA or DEA, is used. As a result in the process H_2S, not CO_2, preferentially absorbed. The enriched H_2S stream may be converted to elemental sulphur using Clauss process. It is the state of the art technology for H_2S and provides advantages such as high acid gas loadings, good cleanup, high solubility in water, low volatility and excellent stability.

The Flexsorb PS process was developed to replace methyldiethanol amine (MDEA) which is biodegradable therefore can be handled in waste treatment facilities. This process was commercialized in 1983. Flexsorb PS absorbs H_2S, CO_2 and COS, and has potential to be used in an enhanced oil recovery operation to desulphurize CO_2 used in well injection and in natural gas sweetening. This process has 40% circulation rate compared to MDEA and considerable energy savings.

TABLE 2. Features of Physical Absorption Acid Gas Removal Processes

Process	Solvent	T/P	Selectivity		Solvent Loss	
			H_2S/CO_2	CO_2/HC		
Rectisol	Methanol	Low/High	Good	Poor	Mod	Low
Selexol	DMEPG*	Low/High	Good	Mod	Low	Low
Purisol	NMP**	Mod/High	Good	Mod	Low	Low
Flour	PC***	Mod/High	Mod	Mod	Low	Low

* dimethyl ether of polyethylene glycohol, ** n-methyl-2-2-pyrrolidone, *** propylene carbonate

The Flexsorb PS process can handle 40% higher acid capacity compared to conventional amine technology. It is also claimed to have smaller equipment size, stable, high resistance to chemical degradation, low leak and retrofiable to existing plants. Applications are made to the removal of CO_2 and H_2S from synthetic and natural gases.

Furimsky and Yamura (16) dealt with the problem of H_2S removal from hot gases using alkaline earth metals and transition metal adsorbents. Alkaline metal oxides contained calcium oxide, limestone, dolomite and calcium silicate. Transition metal adsorbents included iron oxide alone or in combination with some supports such as zinc oxide, zinc ferrite and manganese oxide. Also a group of adsorbent contained a combination of alkaline earth and transition metals such as manganese nodules. Fly ash and the reject from aluminum industry have been tested. At present due to energy conservation considerations, purification from H_2S is carried out near gasification temperatures.

4.2. OTHER GAS SEPARATION APPLICATIONS

Gas separation other than CO_2 and H_2S concentrate mainly on non-cryogenic systems after 1970's, some of which are vacuum, pressure and temperature swing adsorption systems on separation of O_2, H_2, N_2 and CH_4 from various gas mixtures. CH_4 from land fill gases, H_2 from coke oven gas, O_2 and H_2 from water splitting applications, N_2 from air and inert gases production from various sources are some other general examples.

5. Conclusion

There is abundant information on certain applications of gas upbringing, cleaning and purification such as petroleum, steel, gasification and industrial gas production processes. These techniques are important in setting enviromentally acceptable levels of treatment of gases. However, differing character of specific applications may cause technical and economical problems.

Among particulate cleanup techniques wet scrubbers, electrostatic precipitators and fabric filtration are good for particulate sizes that extend to less than 2 μm. The choice among these three techniques depends on the capacities, operational expenses and specific problems. Generally for capacities higher than 200,000 m^3/h electrostatic precipitators are economical. Below this capacity fabric filters and inertial separators are generally preferred.

Cyclones are cheap and easy to control. They are generally used for scrubbing of particles larger than 10 μm but lately novel designs made it possible to scrubbe particle diameter around 2 μm.

There has been extensive research to extend performance of the known particulate scrubbing technologies to high pressure and temperature applications in which considerable success has been reached with the pressurized fluidized bed combustion system with and without combined cycle. Combination of the different cleaning techniques are also finding extending application.

In gaseous phase cleaning there has been extensive research efforts on scrubbing of acid gases. Selective and nonselective processes developed according to the needs of the operation.

There has been growing interest in adsorbent characterization, manufacture and regeneration.

6. References

1. Ghassemi, M., Strehler, D., Crawford, K. and Quinlivan S. (178), Applicability of Petroleum Refinery Control Technologies to Coal Conversion NTIS, No.PB-288 630.
2. Massey, L.G. and Brown, W.R. (1983), 'Overview of coal gas purification technology', Energy Technology 10, 1142-1151.
3. Air Quality Assurance Legislation, Turkish Official Gazette, No: 19269, 2-145 (Tuv).
4. Colvert, S. and Parker, R.D. (1983), 'Dust control' in J.J. Mcketta (ed.), Encyclopedia of Chemical Processing and Design, Dekker, New York.
5. Carter, S.D., Rubel, A.M., Robl, T.N. and Taulbee, D.N. (1990), 'The Development of the Kentort II Process for Eastern U.S. DOE', Final Report.
6. Lee, R.W. (1984), 'Selection of equipment for removing particulates from gases', Chemistry and Industry 13, 476-480.
7. Stairmand, C.V. (1970), 'Selection of gas cleaning equipment: A study of basic concepts', Filtration and Separation Jan./Feb. 53-63.
8. Mcilvaine, R.W. (1983), 'New developments in scrubber particulate technology', 76th Annual Meeting of the Air Pollution Control Association 83.53.3.
9. Saxena, S.C., Henry, R.F. and Podalski, W.F. (1986), 'Electrostatically augmented particulate removal devices for high temperature and pressure applications', Adv. in Transp. Processes 14, 465-505.
10. Gooch, D.S., Gullary, J.L. and Plater, F.M. (1980), 'Electrostatic enhancement of particulate control technology' Particulate Control Devices Vol.3, EPA-600/9-80-039C, pp. 289-308.
11. Ariman, T., Ojalvo, M.S. and Drehmel, D.C. (1979), 'Novel concepts, methods and advanced technology in particulate/gas separation: Report on a Workshop', J.Air Pollution Control association 22, 818-822.
12. Daniel, J.D. (1984), 'Removal of particulate matter from air by use of wet scrubbers', Chemistry and Industry 126-129.
13. Henry, R.F., Podolski, W.F. and Saxena, S.C. (1982), 'Review of cleaning systems for fluid bed coal combustor', World Filtration Congress 3, 98-105.
14. Ciuksza, A., Theobald, a., Kingston, M. and Duggan, J. (1987), Steel Times, June, 296-302.
15. Goldstein, A.M., Brown, E.C. and say, G.R. (1986), 'New flexsorb gas realing technology for acid gas removal', Energy Progress 6, 67.
16. Furimsky, E. and Yamura, M. (1986), 'Solid adsorbents for removal of hydrogen sulphide from hot gases', Erdol und Kohle-Erdgas-Petrochemie Verenigt mit Brennstoff-Chemie 39, 163-172.

DESULFURIZATION OF COAL GAS AT HIGH TEMPERATURES WITH REGENERABLE SORBENTS

A. T. ATIMTAY
Middle East Technical University
Environmental Engineering Department
06531 Ankara
Turkiye

ABSTRACT. Hydrogen sulfide, which is the most abundant sulfur-containing compound in coal gas, can react readily with the oxides of alkali earth and transition metals. Economic and environmental requirements for advanced power generating systems mandate the high-temperature removal of corrosive and abrasive compounds from coal gases such as sulfur and nitrogen containing compounds. Regenerable sorbents for sulfur removal from coal gas developed for high temperature applications have been reviewed in this paper. Also, the results of sulfidation and regeneration experiments conducted using a simulated coal gas mixture at temperatures 1250°-1600°F (675° - 870°C) are reported for a novel sorbent having different loadings of copper and manganese oxides on a high silica-zeolite.

1. Introduction

In recent years production of clean synthetic fuels has gained a tremendous importance because of environmental regulations getting more stringent every year. The main purpose of synthetic fuel production is to produce clean energy. Among the different kinds of clean energy, one of the most desirable one is the electrical energy.

Electrical energy is mostly being produced today with conventional systems (pulverized coal-fired or stoker-fired boilers + steam turbine/generator system). The efficiency of conversion in the conventional systems is around 30-35 %, which is not great.

Considerable amount of work has been carried out to increase the efficiency of generating electrical energy in power plants and new technologies are being developed in the United States of America and other industrialized countries. In the mean time, it is important to keep the emission levels below emission standards.

The Integrated Gasification Combined Cycle (IGCC) system is one of the most promising energy systems for producing electrical energy from coal gas because it can provide noticeable improvement in thermal efficiency (about 50 %). Additionally, the environmental pollution created by thermal power plants is prevented to a great extent with the IGCC system. The technology for this system is currently being developed at the Morgantown Energy Technology Center (METC) of the U.S. Department of Energy. A large number of research projects on this subject funded by DOE is also being conducted at various U.S. Universities [1].

Y. Yürüm (ed.), Clean Utilization of Coal, 221–237.

Coal is primarily made of carbon and mixed with varying amounts of mineral matter and chemical compounds. When coal is burned or converted to combustible gases, chemical compounds are released as gaseous contaminants, and mineral matter is released as particulates. These contaminants must be removed to protect the equipment in advanced-power generating systems from corrosion and erosion, and to simultaneously prevent environmental pollution [2].

In the IGCC system the main process is the gasification of coal. When coal is gasified, reducing conditions are created. Under reducing conditions most of the sulfur in the coal is converted to H_2S. The H_2S concentration from a typical gasifier is around 5000 ppm. H_2S is a toxic gas. It also contributes to the formation of acid rain when it is oxidized to SO_2 and/or SO_3. It has to be removed from the coal gas as much as possible. Additionally, turbines and related equipment need to be protected from corrosive action of sulfurous compounds in the coal gas. A gas turbine in an IGCC system can tolerate a H_2S concentration around 150 ppm [3]. Therefore, H_2S concentration in the coal gas should be decreased from 5000 ppm to 150 ppm with a suitable system [2].

A Gas Stream Cleanup (GSC) program at the U.S. Department of Energy's (DOE) Morgantown Energy Technology Center (METC) has been developed to economically remove contaminants from hot gas stream. The project is aimed at gas streams generated at pressures greater than 6 atmospheres and temperatures greater than 550°C. It is important to conduct GSC under these conditions because of the following advantages [1]:

1. Expensive heat-recovery equipment is eliminated.
2. Efficiency losses are avoided due to fuel-gas quenching.
3. Tars are not condensed since the fuel gas is kept hot. Tars can be burned downstream in a combustor and adds to the heat of the system.
4. Problems related with the disposal and treatment of wastewater is eliminated.
5. The volume of gas that must be processed for cleanup is reduced since the contaminants are removed before combustion.
6. The removal efficiency of contaminants is increased since the contaminant concentration is more in the gas before the gas cleanup system.

The advanced energy conversion systems for which hot contaminant control technologies are being developed have been identified as having significant efficiency and economic advantages over conventional pulverized coal-fired and stoker-fired boilers to produce electricity. The GSC Program is presently addressing contaminant control for the following energy conversion systems(1):

* Pressurized fluidized-bed combustion combined cycle
* Integrated gasification combined cycle
* Gasification molten-carbonate fuel cell
* Direct coal-fueled turbine.

A brief description of each system with its schematic diagram is given below in the next section.

2. Energy Conversion Systems

2.1. PRESSURIZED FLUIDIZED BED COMBUSTION COMBINED-CYCLE SYSTEM (PFBC)

A schematic diagram of a PFBC system is shown in Figure 1. The bed consists of inert granules fluidized by air. Sand or ash particles are used as inert granules. Air enters the bed from the bottom of the combustor through a perforated plate. The system operates at a temperature of 1500° - 1700°F (815° - 920°C) and at a pressure of 7 - 17 Atmg. Coal is fed into the bed as fuel. Limestone or dolomite particles are also fed into the bed as sorbent. As the coal burns in the fluidized bed, the sulfur in the coal forms the sulfur oxides in the gas phase. These sulfur oxides react in-situ with the sorbent in the bed and forms $CaSO_4$, a solid end product. This material is removed at the bottom of the combustor together with the coal ash.

The heat released during the combustion is removed with heat exchanger tubes placed in the fluidized bed. High temperature - high pressure steam is produced in heat exchangers and it is used to drive a steam turbine which generates 65 to 75 % of the total power output of the system [1].

In a combined cycle system gas turbines are used in addition to steam turbines. HTHP combustion gases exit the combustor and are expanded in a gas turbine to produce additional power. The power output in the gas turbine is 25 to 35 % of the total plant output. Altogether, the estimated PFBC combined cycle conversion efficiency reaches to 40 %, while this efficiency is only 30 % for conventional coal-fired boilers [1].

The gas turbine system should be protected from particulates and corrosive gases like sulfur oxides. Therefore, combustion gases are passed through a "hot gas cleanup" system which consists of particulate removal and sulfur oxide polishing units.

2.2. INTEGRATED GASIFICATION COMBINED CYCLE SYSTEM (IGCC)

A schematic diagram of the IGCC system is shown in Figure 2. This system gasifies the solid coal as opposed to combusting it in the PFBC system. IGCC system consists of a gasifier, hot particulate and H2S removal systems, heat exchangers for steam generation, and a gas and a steam turbine with their generators. In a fixed bed gasifier coal is reacted with oxygen and steam at high temperatures. Reducing conditions prevail in the gasifier as opposed to oxidizing conditions in the combustor. Under reducing conditions, sulfur in the coal is mainly converted to H_2S and particulates are formed due to ash. Therefore, coal gas should be cleaned from particulates and H_2S before it is combusted and used in the gas turbine/generator unit [1].

Hot gases coming out of the gasifier directly enters the "hot particulate removal" and "hot H_2S removal" units. This way , gas turbine blades are protected from erosion and corrosion. Heat is recovered from the gas turbine's exhaust and is used to generate steam in a "heat exchanger" system. The steam generated is used to run a steam turbine and produce additional power [1].

The total conversion efficiency of an IGCC system can approach to 50 %, out of this total output 25 to 35 % is produced by the steam cycle and 65 to 75 % is produced by the gas turbine generator.

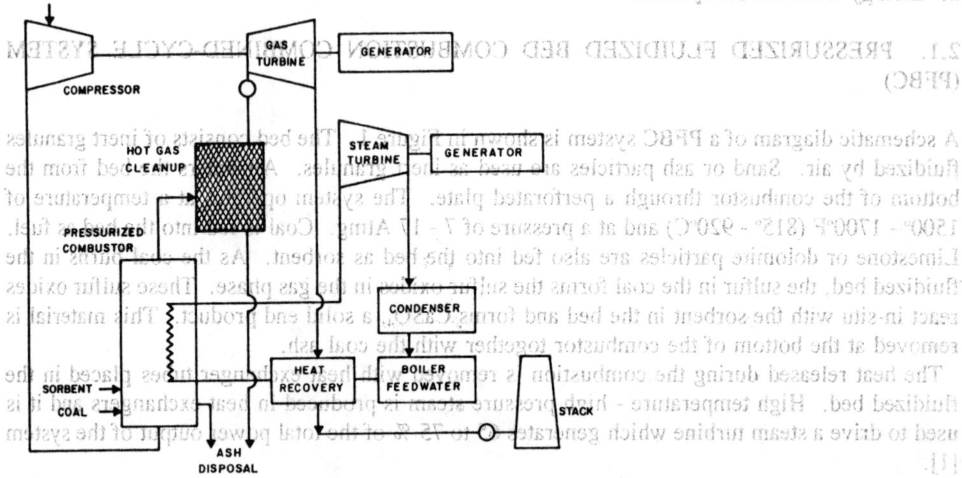

FIGURE 1. PFBC STEAM-COOLED SYSTEM (Ref:1)

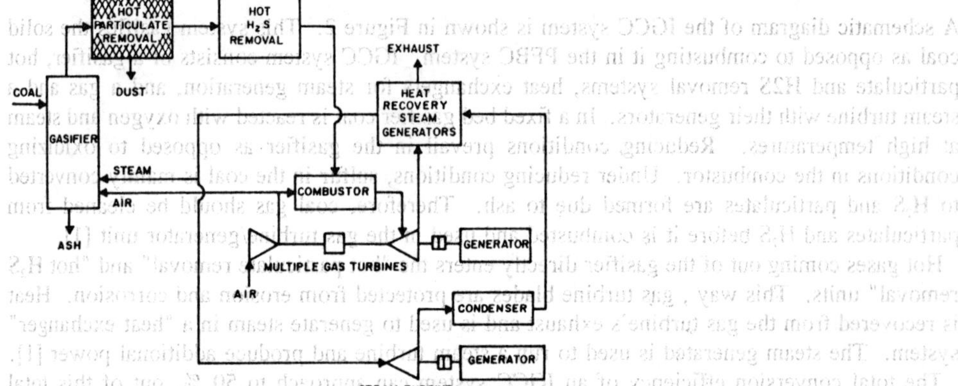

FIGURE 2. INTEGRATED GASIFICATION COMBINED CYCLE (Ref:1)

2.3. GASIFICATION MOLTEN CARBONATE FUEL CELL SYSTEM (MCFC)

A simplified diagram of a MCFC system is shown in Figure 3. This system is quite different from a PFBC or an IGCC system. Hydrogen and carbon monoxide generated in a gasification system are used to produce electricity directly. The system consists of a gasifier, a hot gas cleanup system (high temperature-high pressure), a molten carbonate fuel cell, a steam and a gas turbine, and a heat recovery system. In a molten carbonate fuel cell there are two electrodes separated by an electrolyte. Gases cleaned from their particulates and sulfurous compounds are fed to the anode, where a catalytical oxidation takes place. An oxidant (air) is fed to the cathode, where it is catalytically reduced. The electron flow through the outer circuit due to the red-ox reaction in the fuel cell produces direct-current(dc) electricity (1).

The electrodes in the fuel cell are very sensitive to contaminants. A small amount of H_2S will cause poisoning in the electrodes. Therefore, the H_2S concentration in the fuel gas should be removed below 10 ppm. Particulates will also cause clogging in the anode if they are not removed to very low levels.

A steam turbine is run with the steam generated in the heat recovery unit. It is estimated that the total energy conversion efficiency for a MCFC system can exceed 50 % [1].

2.4. DIRECT COAL-FIRED TURBINE SYSTEM (DCFT) [1]

A schematic diagram of a DCFT system is shown in Figure 4. A DCFT system is in the initial development stage. Research activities at DOE are continuing to develop an optimum configuration for the system. The system simply consists of a combustor, HTHP desulfurization and particulate removal units, a heat recovery and steam turbine cycle, and an expander. The system uses dry pulverized coal or a coal slurry as the fuel. A major advantage of the system is its potential for using a wide range of fuels.

It is estimated that a total energy conversion efficiency exceeding 50 % is possible for a DCFT system.

3. Coal Gas Desulfurization

Desulfurization of coal gas is very important from both economical as well as environmental point of view. All coal gas applications should comply with environmental regulations. The degree of sulfur removal requirement depends on the system used for energy production. For example, MCFC requires a fuel gas which contains less than 10 ppm H_2S. However, a gas turbine in an IGCC system can tolerate an H_2S concentration of around 150 ppm. Nevertheless, sulfur removal to below 10 ppm could also be useful in turbine systems in IGCC, because sulfur and alkali compounds interact with each other to promote corrosion, deposition, or both on turbine blades.

3.1. BACKGROUND

Hydrogen sulfide, which is the most abundant sulfur-containing compound in coal gas, can react readily with the oxides of alkali earth and transition metals [4]. There are different processes available which operate within the hot-gas desulfurization range. This range is from 840° -

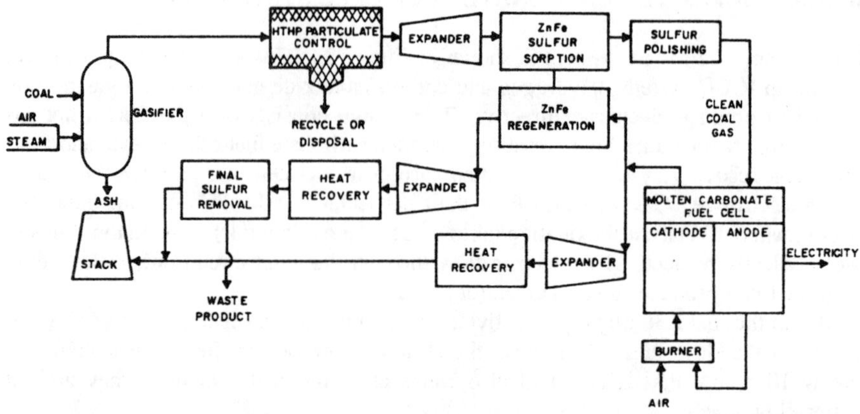

FIGURE 3. GASIFIER/MOLTEN-CARBONATE FUEL-CELL SYSTEM (Ref: 1)

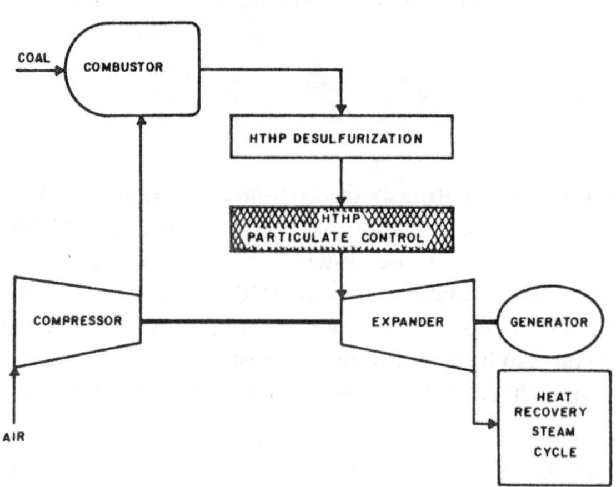

FIGURE 4. SIMPLIFIED DIRECT COAL-FUELED TURBINE SYSTEM (Ref: 1)

2000°F (450°C to 1100°C). The sulfur removal efficiencies of these processes are shown in Figure 5.

3.1.1. Iron Oxide

Iron oxide was one of the first metal oxides tested for removal of H_2S from the coal gas. DOE/METC tested iron oxide in a fixed bed gasifier in 1970's. About 90 % sulfur removal was achieved. However, in some systems better than 90 % removal was required. Therefore, other metal oxides were sought for more efficient removal of H_2S.

3.1.2. Zinc Ferrite

Westmoreland, et al.[5] reported the results of a thermodynamic screening of the high temperature desulfurization potential of 28 single metal oxides by using the free energy minimization method. Also, kinetic limits for ZnO, MnO, CaO, and V_2O_3 were reported by Westmoreland, et al.[6] for desulfurization reactions. ZnO showed a very favorable thermodynamic equilibrium with H_2S [7] and iron oxide showed easy regenerability with air. Results regarding structural changes in the pure ZnO sorbent at high temperatures were reported by Ranade and Harrison [8]. They showed that sintering actually causes the particle to shrink radially, thus for the same number of particles the surface area available for reaction is smaller. Additionally, zinc metal produced by reduction of ZnO in coal-gas volatilizes above 1290°F (699°C).

Favorable properties of zinc oxide and iron oxide towards absorbing H_2S were combined at the in-house research program of Department of Energy's Morgantown Energy and Research Center. The combined metal oxide sorbent, zinc ferrite, was found to be a superior candidate for hot gas desulfurization than iron oxide. It was found to be capable of removing sulfur compounds in coal gas to below 10 ppm [9] and also easily regenerable with air. Zinc ferrite was originally developed for MCFC. Later on it was understood that application to IGCC would be very attractive from the efficiency point of view in energy production. Therefore, the research efforts were concentrated in understanding the characteristics of zinc ferrite in IGCC system.

The hopes were very high for zinc ferrite at the beginning. However, research results have shown that zinc ferrite has a temperature limitation. The maximum operating temperature for zinc ferrite is about 1200°F (649° C). Focht, et al. [10] showed that the sorbent lost its pore volume during reduction, and subsequent loss of reactivity was observed during sulfidation. The sorbent also showed a substantial amount of decrepitation above 1200°F (649°C).

Improvements in the formulation of zinc ferrite have been attempted in an effort to overcome its limitations. An inorganic binder, bentonite, was added to zinc ferrite. Thus, a sorbent with comparable reactivity and capacity to the original sorbent, but with a higher temperature limit than the original zinc ferrite was obtained. This sorbent can withstand temperatures up to 1275°F (691°C), which is not very satisfactory for high temperature fluidized and entrained bed gasifiers.

3.1.3. Metal Oxides

The need for developing a sorbent having better properties (durability and reactivity) than the original or modified zinc ferrite led researchers to try other metal oxides. Various combination of metal oxides, such as Zn, Cu, Al, Ti, Fe, Co, Mo, and V were tried at Massachusetts Institute of Technology (MIT) and Jet Propulsion Laboratory (JPL) in an effort to develop a durable and

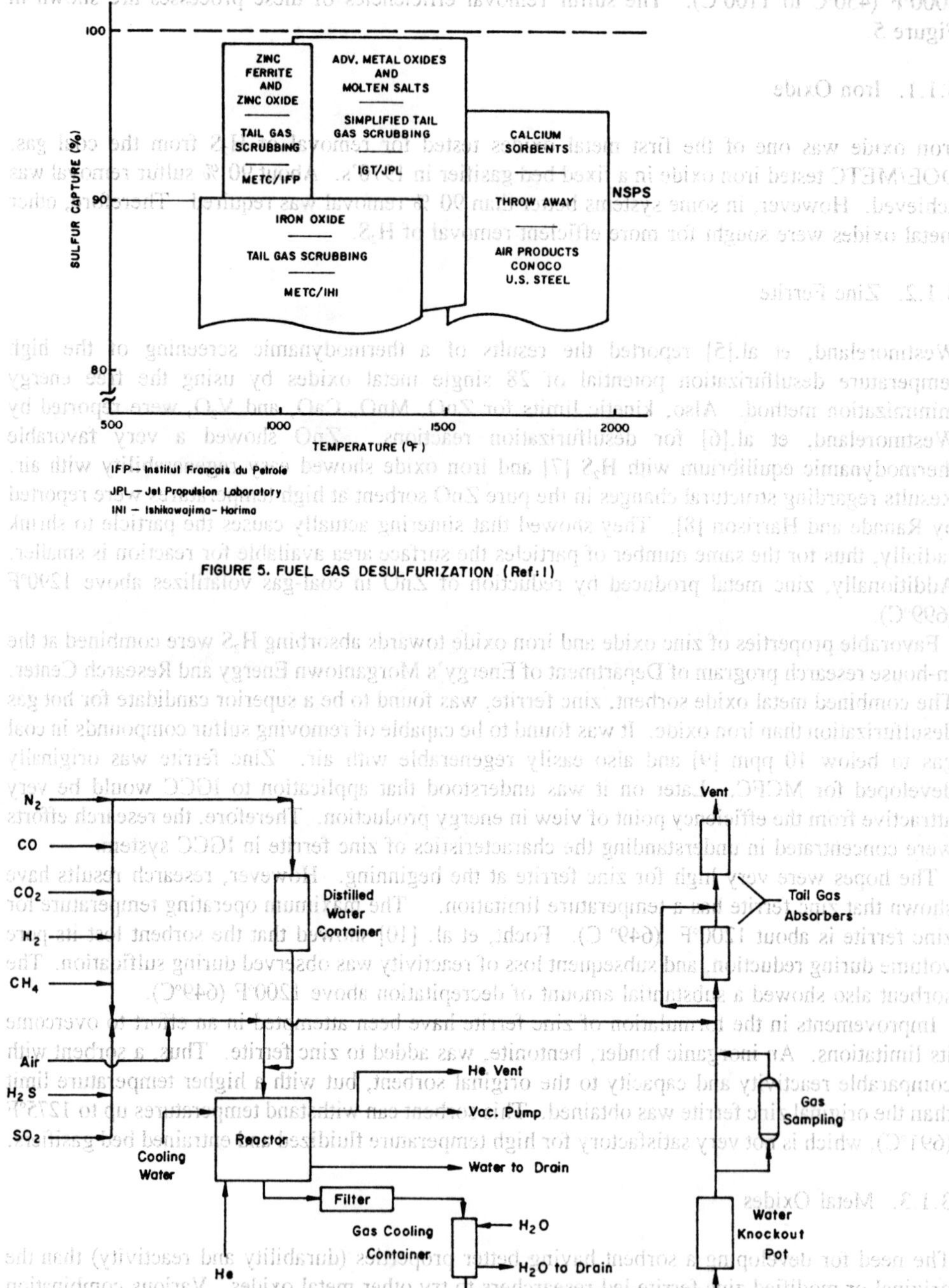

FIGURE 5. FUEL GAS DESULFURIZATION (Ref: 1)

FIGURE 6. FLOW DIAGRAM OF FIXED BED REACTOR SYSTEM (Ref: 2)

high capacity sorbent for removal of H_2S from hot coal-derived gases [11]. Basically, the sorbents tested were divided into two groups: 1. Zinc-based sorbents, and 2. Copper-based sorbents. Zinc-based sorbents included zinc ferrite (ZF), zinc-copper ferrite (ZCF), and zinc titanate (ZT). Copper-based sorbents included copper aluminate (CA) and copper-iron aluminate (CFA). Research results [11] showed that zinc-based sorbents were more promising than copper-based sorbents in the temperature range of 1000° - 1350° F (538° - 732°C). One of the most promising sorbents among the zinc-based sorbents was zinc titanate (maximum operating temperature of 1340°F (727°C)). It was found that the TiO_2 phase helps to stabilize the ZnO phase, and volatilization of Zn at temperatures higher than 1290°F (699°C) has been reduced (about 1 wt % Zn is lost in 1000 hours of operation). However, TiO_2 does not absorb H_2S. Therefore, the capacity of zinc titanate is lower than that of zinc ferrite due to dilution effect.

Extrudates of regenerable mixed metal oxide sorbents (copper-modified zinc ferrite, zinc titanate, copper aluminate and copper-iron aluminate were tested in a high temperature-high pressure (HTHP) fixed bed reactor by Research Triangle Institute (RTI) [12]. Reductions in H_2S concentration from > 10,000 ppmv to < 1 to 50 ppmv were achieved. The copper-modified zinc ferrite sorbent reduced the H_2S concentration to less than 1 ppmv at up to 1100°F (593°C) with 20 % steam in the gas. The zinc ferrite sorbent was also tested in the same reactor. Zinc titanate outperformed the other mixed metal oxide sorbents in structural strength and capacity at steam levels as low as 5 % and temperatures as high as 1350°F (732°C). However, entrained and fluidized bed gasifiers which are presently being developed have outlet temperatures as high as 1500°F (816°C) to 1800°F (982°C). Furthermore, for continuous desulfurization of a full scale gasification stream, a fluidized bed sorbent system instead of a fixed bed would improve the process efficiency considerably by providing a means for continuous solids regeneration. Zinc ferrite sinters at temperatures higher than 1200°F (649°C). The present zinc ferrite formulations do not have sufficient attrition resistance to withstand a fluidized bed environment. Research Triangle Institute is working on the development of reactive and durable zinc ferrite sorbent for fluidized bed application. Zinc titanate can withstand temperatures up to 1350°F (732°C), and maintains good strength, but the sorbent sulfur capacity is low, i.e. 17 - 24 % compared to 40 % for zinc ferrite. Therefore, novel sorbents are still being sought which will show improved resistance to higher temperatures and attrition. Recently, copper manganate sorbents (bulk) are under study by ElectroChem Inc.[13].

The objective of this research was to develop a novel regenerable sorbent which will show improved resistance to higher temperatures and attrition. The potential advantages of supported sorbents, as compared to bulk sorbents, are higher surface area, less resistance for diffusional transport, and higher temperature and attrition resistance. The sorbent investigated in this study is copper and manganese oxides supported on SP-115 zeolite. Results of sulfidation and regeneration experiments conducted using a simulated coal gas mixture at temperatures of 1000° - 1600°F (538° - 871°C) are reported.

4. Experimental Section

4.1. SORBENT PREPARATION [2]

The sorbents investigated in this study were prepared by a coprecipitation process. SP-115 zeolite (1/8" extrudates, Union Carbide) was used as the support. SP-115 had the following characteristics : SiO_2 > 99 % by weight, surface area = 482 m^2/g , crush strength = 6.6 kg/cm, pore diameter = 5.4 A°. The coprecipitation process included : 1) Impregnation of the zeolite with a solution of cupric acetate and manganese acetate in a rotary vacuum evaporator, 2) Dehydration of the sorbent in a vacuum oven at 158°F (70°C) and 0.1 inch Hg, 3) Calcination of the sorbent in a muffle furnace at 1382°F (750°C) for 7 hours. Three samples were prepared with this process (SP1C5M5, SP2C5M5, and SP2C9M1). The atomic absorption spectrophotometer analysis showed that Cu and Mn loadings on the sorbent SP1C5M5 were 1.82 % and 1.64 % by weight, respectively. SP2C5M5 had 4.39 % by wt. Cu and 6.11 % by wt. Mn. The corresponding values for SP2C9M1 were 5.15 % and 0.573 % by weight, respectively. The uncertainty in the atomic absorption analysis is $\pm 1\%$.

4.2. EXPERIMENTAL [2]

A schematic diagram of the experimental system is shown in Figure 6. The system consisted of a gas mixing system, a fixed bed reactor, a condenser and a water knock-out pot, a gas sampling system, and a tail-gas cleanup system. Gases were supplied from gas cylinders and the flow rates were monitored through mass flow controllers. After the gases were mixed in a manifold, they were fed into the fixed bed reactor. This gas mixture simulated the partially quenched exit gas of an air-blown KRW fluidized bed coal gasifier which had the following molar composition: 42.5 % N_2, 11 % CO_2, 12.5 % CO, 13.8 % H_2, 1 % CH_4, and 0.2 % H_2S. 19 % H_2O in the gas mixture was provided by adding a predetermined amount of water into the gas stream before it entered the reactor via a high pressure pump.

The reactor which consisted of a zirconia tube , with 0.875-inch inside diameter and 12-inch length, was placed vertically in an electric furnace equipped with graphite heating elements. The furnace was water cooled. The graphite heating elements were located in a helium atmosphere to protect them from oxidation. The temperatures of the gas at the inlet and the center of the reactor were measured by K-type thermocouples located along the central axis. The reactor was designed to withstand a pressure of 15 psig and a temperature of 3000°F (1649°C). The sorbent sample was placed in the center of the reactor and the bed height was 2 inches. The gas lines between the reactor outlet and the condenser inlet were heated via heating tapes and heavily insulated to prevent the condensation of steam in the gas lines. The exit gas from the reactor was filtered, cooled down in the condenser, and sampled for gas analysis. The tail gas was sent through the absorbers to remove the sulfurous gases before it was discharged to the atmosphere.

The experiments consisted of a sequence of sulfidation and regeneration runs. The reactor temperature was held constant during the sulfidation and regeneration experiments. Reactor pressure was maintained at 5 $\pm$ 3 psig. Three temperatures were tested: 1000, 1250, and 1600° F (538°, 677°, and 871°C). The space velocities for the sulfidation and the regeneration runs were 2000 and 600 hr^{-1}, respectively. Regeneration of the sulfided sorbent was conducted with air only and with a 50 % air/50 % steam mixture. Breakthrough during sulfidation runs was defined as 200 ppmv H_2S in the effluent gas. After the sulfidation run was stopped, the reactor was purged with nitrogen for at least 15 minutes. Regeneration was carried out at the reaction temperature. The regeneration cycle was stopped when the SO_2 concentration of the effluent gas was below 500 ppmv. The sorbents were subjected to five sulfidation/regeneration cycles to establish sulfur sorption capacity and regenerability.

Selected sorbent samples were analyzed for total weight, metal and sulfur content, crush strength, surface area,average pore diameter, pore volume, and pore size distribution both before and after reaction. Gas grab samples obtained during experimentation were analyzed by gas chromatography. Also Gas-tech precision gas detection tubes were used to determine the H_2S and SO_2 concentrations in the inlet and outlet gas streams.

5. Results [2]

The effect of temperature on sorbent capacity is shown in Figure 7. The H_2S breakthrough curves indicate that the breakthrough times were 37, 45, and 45 minutes at 1000°F (538°C), 1250°F (677°C), and 1600°F (871°C), respectively. The sorbent has greater capacity at 1250°F (677°C) and 1600°F (871°C) than at 1000°F (538°C). However, the prebreakthrough H_2S levels are lower at 1000°F (538°C) and 1250°F (677°C) than at 1600°F (871°C). Although the prebreakthrough H_2S level is lower at 1000°F (538°C), the sorbent has more total capacity at the higher temperatures.

Two runs were conducted at 1250°F (677°C) with different samples from the same batch of SP1C5M5 sorbent, to establish the extent of reproducibility of the experimental method. Twelve detector tube readings of H_2S were taken, 5 min apart, for each run. The average difference between readings for the two runs was 30 ppmv. Thus, the experimental method appears to be reasonably reproducible.

Sorbent regenerability was tested at 1250°F (677°C). Two and one-half cycles were conducted with only air as the regenerant gas, and five cycles were studied using a 50 mole % steam/50 mole % air mixture for regeneration. Two criterion were used to assess the regenerability of the sorbent: (1) a comparison of the H_2S breakthrough curves, and (2) a comparison of the SO_2 outlet curves during the sulfidations.

When the sorbent was regenerated with air, the second and third sulfidation breakthrough curves were somewhat similar. However, the prebreakthrough average H_2S outlet concentration was approximately 50 ppmv less for the third sulfidation than for the second sulfidation. Figure 8 shows the H_2S breakthrough curves after the sorbent was regenerated with steam/air. The curve for the first sulfidation was not shown, as the reactor bypass was accidentally left open during the run. The third through sixth sulfidations, conducted with steam/air regeneration, gave similar breakthrough curves. Thus, Figure 3 shows that the 50 % steam/50 % air (by volume) regeneration appears to give good regenerability.

Outlet SO_2 concentrations were compared for the second and third sulfidations (set I) using air regeneration, with the second, third, and fourth sulfidations (set II) using steam/air regeneration. After 15 min on-stream, set I showed outlet SO_2 concentrations between 500 and 600 ppmv. After the same period, set II yielded SO_2 concentrations ranging between 30 and 450 ppmv. These results indicate that more sulfate was formed during air regeneration than during steam/air regeneration. This is not unexpected, since the presence of steam reduces the overall oxidative properties of the regenerant stream. In general, if SO_2 is released in large quantities when reducing gases are introduced after oxidative regeneration, this indicates that sulfates have been formed during the regeneration. The release of such sulfurous gases must be avoided during sulfidation. It has been suggested in studies on zinc ferrite that a clean reducing gas could be used for reductive regeneration of the sorbent after oxidative regeneration, and prior to sulfidation. However, the economics for carrying out a reductive regeneration step are poor, due to the capital cost of additional vessels and the introduction of complications into the overall operation of the system [14].

Although no carbonyl sulfide was fed to the reactor in the simulated gas mixture, appreciable concentrations were detected in the outlet gas during sulfidation. The COS concentration was

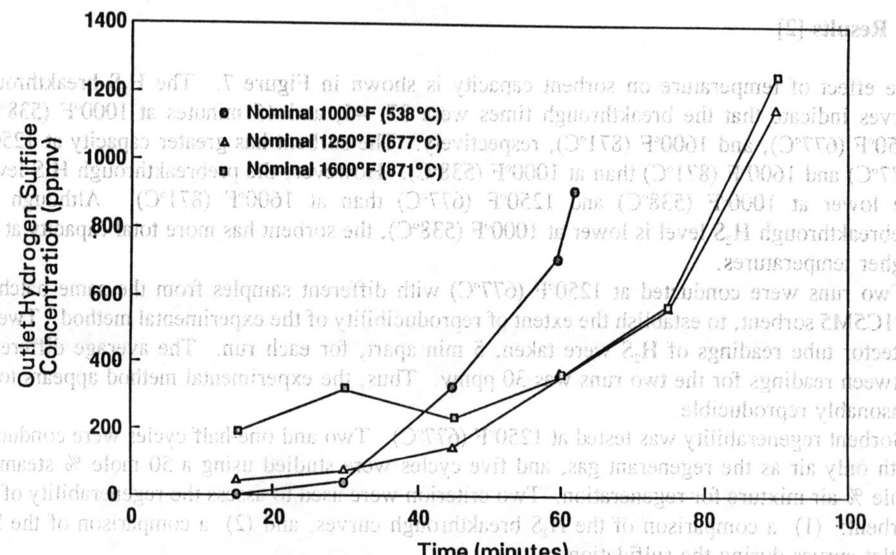

Figure 7. The effect of temperature on hydrogen sulfide breakthrough curve. (Ref: 2)

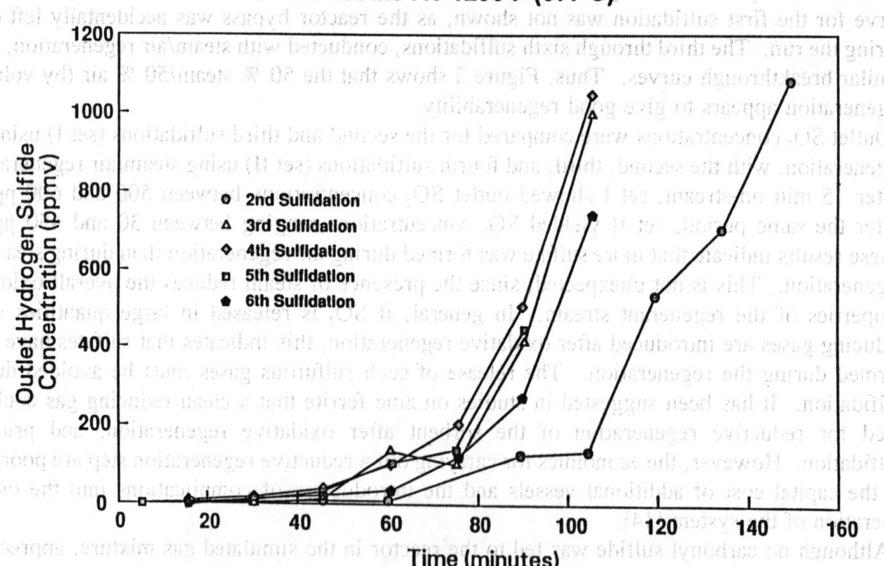

Figure 8. Regenerability of the sorbent at 1250°F (Ref: 2)

Regeneration Gas : 50% Air / 50% Steam

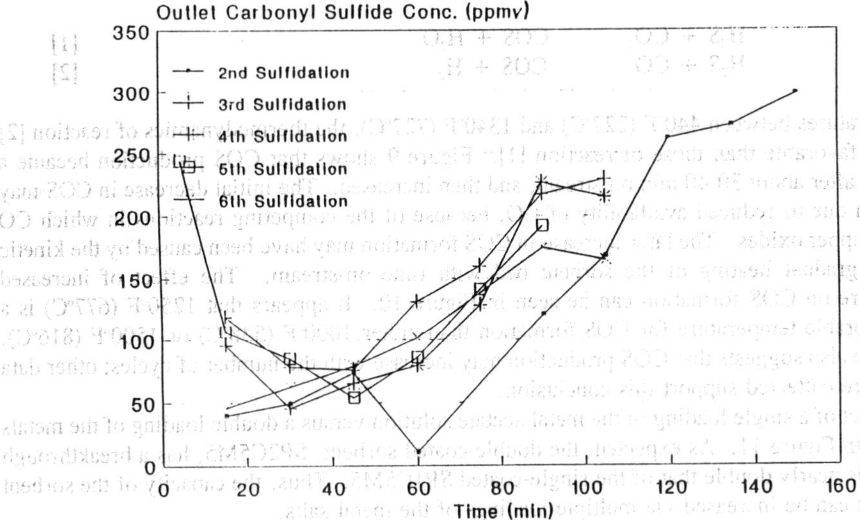

Figure 9. The effect of cycle number on carbonyl sulfide formation (Ref: 2)

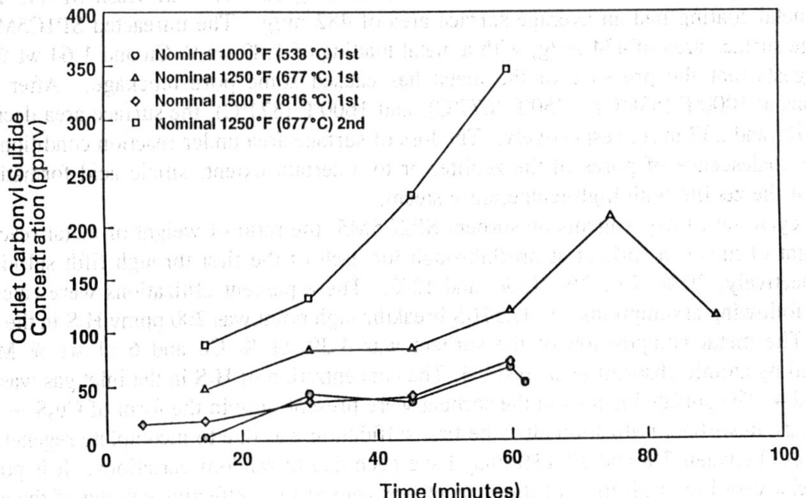

Figure 10. The effect of temperature and cycle number on COS formation (Ref: 2)

EFFECT OF TEMPERATURE AND CYCLE NUMBER

usually greater than 50 ppmv, and often as high as 100-250 ppmv in the outlet gas. It is well known that certain zeolites can catalyze the COS formation reactions. CO and CO_2 can react with H_2S according to the following reactions [15]:

$$H_2S + CO_2 \quad COS + H_2O \qquad [1]$$
$$H_2S + CO \quad COS + H_2 \qquad [2]$$

At temperatures between 440°F (227°C) and 1340°F (727°C), the thermodynamics of reaction [2] are more favorable than those of reaction [1]. Figure 9 shows that COS production became a minimum after about 30-40 min on-stream, and then increased. The initial decrease in COS may have been due to reduced availability of CO, because of the competing reactions in which CO reduces copper oxides. The later increase in COS formation may have been caused by the kinetic effect of gradual heating of the sorbent bed with time on-stream. The effect of increased temperature on COS formation can be seen in Figure 10. It appears that 1250°F (677°C) is a more favorable temperature for COS formation than either 1000°F (538°C) or 1500°F (816°C). The figure also suggests that COS production may increase with the number of cycles; other data which were collected support this conclusion.

The effect of a single loading of the metal acetate solution versus a double loading of the metals is shown in Figure 11. As expected, the double-coated sorbent, SP2C5M5, has a breakthrough time that is nearly double that of the single-coated SP1C5M5. Thus, the capacity of the sorbent extrudates can be increased via multiple loadings of the metal salts.

Sorbents SP2C5M5 and SP2C9M1 were also compared for their ability to remove H_2S from the simulated coal gas mixture. The second and third sulfidations of each sorbent are shown in Figure 12. The SP2C9M1 sorbent appears to have greater capacity than SP2C5M5 for H_2S removal.

The surface area of the sorbent was determined before and after the reaction by using the Langmuir adsorption method, with a maximum error of ± 1.5 %. The fresh SP-115 zeolite without metal coating had an average surface area of 482 m^2/g. The unreacted SP1C5M5 had an average surface area of 434 m^2/g, with a metal loading of 1.82 wt % Cu and 1.64 wt % Mn. This suggests that the presence of the metal has caused some pore blockage. After single sulfidations at 1000°F (538°C), 1250°F (677°C), and 1600°F (871°C), the surface area decreased to 379, 370, and 237 m^2/g, respectively. The loss of surface area under reaction conditions may be due to coalescence of pores of the zeolite, or to a certain extent, silicic acid formation by reaction of the zeolite with high-temperature steam.

For a 5-cycle set of experiments on sorbent SP2C5M5, the ratio of weight of metal reacted to total weight of metal in sorbent at breakthrough for each of the first through fifth sulfidations was, respectively, 20%, 7%, 7%, 13%, and 12%. These percent utilizations were calculated using the following assumptions: 1. The H_2S breakthrough point was 200 ppmv H_2S in the outlet gas, 2. The metal composition of the sorbent was 4.39 wt % Cu and 6.11 wt % Mn, as determined by atomic absorption analysis, 3. The concentration of H_2S in the inlet gas was 2000 ppmv, and 4. The sulfided metals in the sorbent were present only in the form of Cu_2S + MnS. The decrease in sorbent utilization after the first sulfidation was due to incomplete regeneration. The increase between 7% and 12-13% may have been due to random variations. It is possible that the relatively low utilization of the sorbent was caused by ineffective coating of the zeolite pores, i.e., the metals were deposited in several atomic layers on the outer surface of the extrudate, rather than inside the pores. Or, even if the pores were effectively coated, there may have been significant resistances to diffusion of H_2S into the pores.

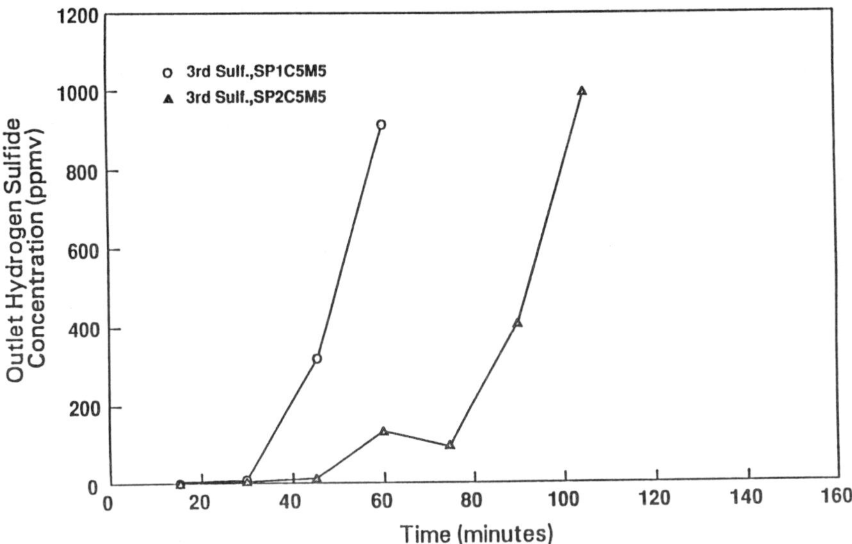

Figure 11. The effect of metal loading on hydrogen sulfide removal (Ref: 2)

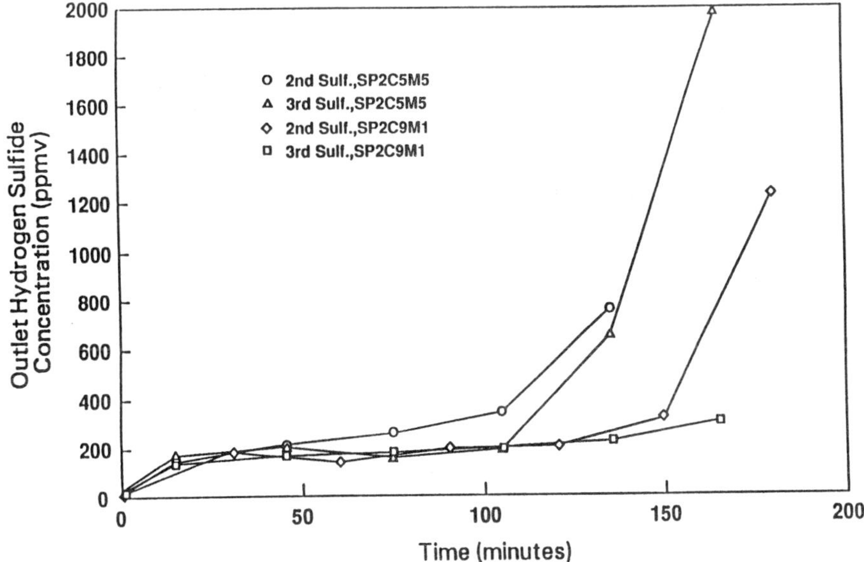

Figure 12. The effect of nominal molar ratios of Cu: Mn of 1:1 versus 9:1 (Ref: 2)

Six different samples of the bed material from the 5-cycle set performed with SP2C5M5 at 1250°F (677°C) and the 3-cycle set run with SP2C9M1 at 1250°F were analyzed using atomic absorption. There was no conclusive evidence for loss of Cu or Mn. Thus, these metals do not appear to undergo appreciable volatilization under these conditions of sulfidation and regeneration.

The crush strength of the sorbent extrudates is an important consideration for practical operation. Crush strength was determined by taking the average of the force per unit length of extrudate required to crush 15 extrudates; the range of error for these tests is $\pm$20-35%, within the first standard deviation. The fresh SP-115 zeolite without metal coating had an average crush strength of 6.58 kg/cm. The crush strength for fresh SP2C5M5 and SP2C9M1 were 29.6 and 11.1 kg/cm, respectively. The metal coating makes the extrudates much stronger; particularly the Mn appears to promote increased resistance to crushing. The average crush strength of 6 samples from the SP2C5M5 bed, after 5 cycles, decreased by only 6%. The average crush strength of 5 samples from the SP2C9M1 bed, after 3 cycles, increased by 3%. Therefore, the crush strength of the zeolite was increased by the metal loading, and essentially unchanged by reaction.

6. Conclusions

For the temperature range of 1000°-1600°F (538°-871°C), the SP-115 zeolite-supported Cu/Mn oxides sorbent had more total capacity at 1250°F (677°C) and 1600°F (871°C) than at 1000°F (538°C). Regeneration with a 50 mole % steam/50 mole % air mixture was preferred over air regeneration, due to decreased sulfate formation in the presence of steam. Further studies would be required to optimize regeneration conditions, so that minimal sulfate is formed during oxidative regeneration. Appreciable amounts of COS were formed at 1250°F (677°C), probably due to reactions catalyzed by the zeolite; however, at 1500°F (816°C) the production of COS greatly decreased. The sorbent containing Cu:Mn molar ratios of 9:1 had greater capacity for H_2S than those containing equimolar ratios of the metals. Sorbent utilization with repeated sulfidation/regeneration cycles was low ($\leq$13%) for SP2C5M5, but higher (58 - 68 %) for SP2C9M1. Surface area of the sorbent decreased with metal loading and/or removal of powder from processing, and with reaction temperatures between 1250°F (677°C) and 1600°F (871°C). It is uncertain whether the zeolite pores had been effectively coated or plugged by the metal acetate solutions during preparation. There was little or no volatilization of the metals at 1250°F (677°C).

7. References

[1] Gas Stream Cleanup (Technology Status Report) (1987) Report DOE/METC-87/0255, pp. 3-7.

[2] Atimtay, A.T., Gasper-Galvin, L.D., and Poston, J.A. (1990) 'A Supported Sorbent for Hot Gas Desulfurization', Preprints, Fuel Chemistry Division, American Chemical Society, Vol. 35, No.1, Boston, pp. 104.

[3] Woods, M.C., Gangwal, S.K., Jothimurugesan, K. and Harrison, D.P. (1990) 'Reaction Between Hydrogen Sulfide and Zinc Oxide-Titanium Oxide Sorbents. I. Single Pellet Kinetic Studies ', Ind. Eng. Chem. Res., 29, pp.1160.

[4] T. Kyotani, H. Kawashima, A. Tomita, A. Palmer, and E. Furimsky (1989) 'Removal of H2S From Hot Gas in the Presence of Cu-Containing Sorbents,' Fuel, 68, pp.74.

[5] P. R. Westmoreland and D. P. Harrison (1976) ' Evaluation of Candidate Solids for High-Temperature Desulfurization of Low-Btu Gases,' Env.Sci.Tech., 10, pp. 659.

[6] P. R. Westmoreland, J. B. Gibson, and D. P. Harrison (1977) ' Comparative Kinetics of High Temperature Reaction Between H2S and Selected Metal Oxides,' Env.Sci.Tech., 11, pp. 488.

[7] T. R. Rao and R. Kumar (1982) ' An Experimental Study of Oxidation of Zinc Sulfide Pellets,' Chem.Eng.Sci., 37, pp.987.

[8] P. V. Ranade and D. P. Harrison (1981) ' The Variable Property Grain Model Applied to the Zinc Oxide-Hydrogen Sulfide Reaction', Chem.Eng.Sci., 36, pp. 1079.

[9] T. Grindley and G. Steinfeld (1981) ' Development and Testing of Regenerable Hot Coal-Gas Desulfurization Sorbent', Report DOE/METC/16545-1125.

[10] G. D. Focht, et al. (1986) ' Structural Property Changes in Metal Oxide Hot Coal Gas Desulfurization Sorbents', Report DOE/MC/21166-2163 (DOE 86016041).

[11] M. Flytzani-Stephanopoulos, et al. (1987) ' Detailed Studies of Novel Regenerable Sorbents for High-Temperature Coal-Gas Desulfurization', Proceedings of the Seventh Annual Gasification and Gas Stream Cleanup Systems Contractors Review Meeting, DOE/METC-87/6079, Vol. 2, pp. 726.

[12] S. K. Gangwal, et al. (1988) ' Bench-Scale Testing of Novel High-Temperature Desulfurization Sorbents ', Proceedings of the Eighth Annual Gasification and Gas Stream Cleanup Systems Contractors Review Meeting, DOE/METC-88/6092, Vol. 1, pp. 103.

[13] V. Jalan and M. Desai (1988) ' Copper-Based Sorbents For Hot Gas Cleanup', Proceedings of the Eighth Annual Gasification and Gas Stream Cleanup Systems Contractors Review Meeting, DOE/METC-88/6092, Vol. 1, pp. 45.

[14] L. Bissett (1989) 'Aspects of Fixed Bed Gasification/Fixed Bed Zinc Ferrite Integration', Ninth Annual Gasification and Gas Stream Cleanup Systems Contractors Review Meeting, DOE/METC/PETC, June 27-29, Morgantown, WV.

[15] Chemistry of Hot Gas Cleanup in Coal Gasification and Combustion, Morgantown Energy Research Center (1978) MERC/SP-78/2.

OXYGEN PRODUCTION

E. EKİNCİ
Department of Chemical Engineering
İstanbul Technical University
Ayazağa
İstanbul 80626
Türkiye

ABSTRACT. Oxygen at present is one of the major chemicals. Cryogenic technology is most favorable for high purity and large scale oxygen production, with high capital cost and low flexibility. Chemical oxygen production which is in development stage at pilot plant scale has potential to overtake cryogenic method. Adsorption and membrane technologies have scope for lower concentration products for intense natural gas combustion applications and clinical use.

1. Introduction

Oxygen is an essential input in synfuels processes for various uses. One of its important utilization is in the upgrading of coal gasification products by elimination of nitrogen in air (1). Injection of oxygen enriched air to upgrade the gaseous products of underground coal gasification has been contemplated during 1970's (2). Oxygen enriched combustion at high temperatures is found to be a high efficiency combustion application for natural gas (3).

It is also a very important industrial and health product, therefore its production methods have been well developed and there is at present a strong oxygen market. In fact in 1985, with 14.2 billion kilograms of production oxygen was the fifth largest product, majority of which was produced cryogenically (3). Due to its main production method oxygen is known as a cryogenic gas together with carbon dioxide, argon and hydrogen.

Over the years, the range of industrial uses and consumption patterns have expanded from large continuous systems to include smaller and cyclical applications. The changes were also noticed at the purity level. These changes manifested itself on the research and development of oxygen production techniques other than the classical cryogenics.

2. Oxygen Proiuction Technologies

The current oxygen production technologies may be presented in four main groups:
-Cryogenic
-Adsorption
-Membranes
-Chemical.

In addition, splitting of water seems to be a future option. The first three of these technologies are currently used in industrial oxygen production. Among the three processes cryogenic is used by far the most for large scale production. Chemical oxygen technology is

Y. Yürüm (ed.), Clean Utilization of Coal, 239–246.

at present at pilot scale and oxygen production from water thermally or by solar application is a process for the coming years.

Even though cryogenics overwhlems the other technologies in the current production, there is distinct separation among the current technologies with reference to relative power consumption, oxygen purity limitation and oxygen recovery percentage. The relationship between power consumption, purity and recovery is shown in Figure 1 for four oxygen production technologies.

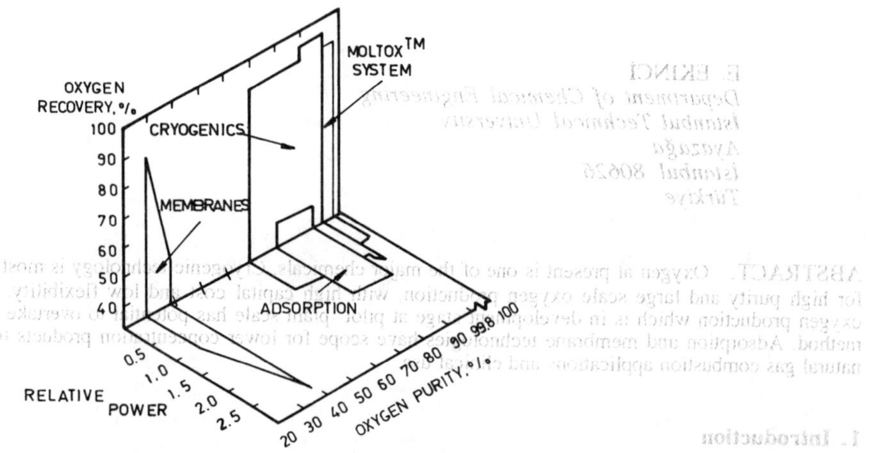

Figure 1. Technology comparison for oxygen production (3).

Power consumption is a major portion of the cost in oxygen production in larger plants, but decreases in importance as plant size decreases. Due to the trade off between power and capital costs, adsorption and membrane systems compete with liquid oxygen (lox) production for smaller production rates. As production rate increases and power costs begin to control, the cost of oxygen on-site cryogenic plants become economic. If high purity oxygen (>95%0) is required, cryogenic plants are the available technology (1). Chemical oxygen recovery system which is expected to be industriallized during 1990's may provide even higher purity than cryogenics.

TABLE 1. Key properties of some cryogenic gases (4).

	Air	Oxygen	Nitrogen
Molecular weight	28.96	32.00	28.01
Color/odor	none	none	none
Specific gravity (air=1)	1.00	1.11	0.97
Density (kg/m^3, 1 atm, 20°C)	1.197	1.327	0.646
Boiling point (^{o}C, normal, 1 atm)	-194.39	-182.9	-195.78
Heat of vaporization (kcal/kg)	49.09	50.99	47.59
Critical pressure (atm)	-	50.1	73.5
Critical temperature (^{o}C)	-	-181.1	-146.8
Heat capacity (kcal/K kg)	0.202	0.225	0.195
Viscosity (Cp)	184.7	206.4	178.0
Thermal conductivity (kcal/h m^2 ^{o}C-0°C)	0.112	0.339	0.112

2.1. CRYOGENICS

Oxygen has favorable characteristics for cryogenic processes as shown in Table 1. Due to the favorable characteristics and maturity of the cryogenic technique nearly all of the oxygen is produced by this method. Oxygen is separated from air by cryogenic distillation in various combinations. A diagram of a low pressure cryogenic cycle is shown in Figure 2.

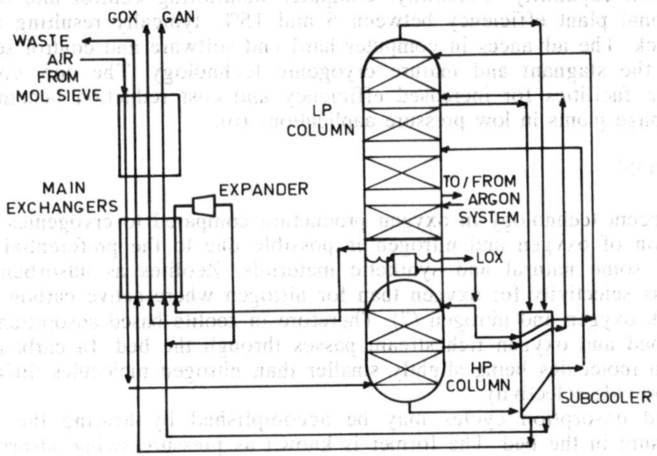

Figure 2. Low pressure cryogenic cycle (30).

The process shown in Figure 2 with certain variations has been used extensively to produce oxygen in 95-98% purity. Reversing heat exchangers which combine impurity removal with primary heat exchange or molecular sieve adsorption systems can be used to pretreat the air before it enters the distillation section of the plant. Below 95% purity, the specific power (Kwh/ton of contained oxygen) remains nearly constant for the low pressure cycle. Therefore, other process cycles have been developed to optimize the cost of producing oxygen at purities from 70-95% (5).

An advantage of cryogenic systems is their ability to produce nitrogen and argon by-products as well as liquefy a portion of the total output. The liquid is stored in vacuum insulated tanks where it is readily available for use during peak flow requirements, to serve as backup during outages, or for merchant sales. Operation of a cryogenic plant is not as flexible as an adsorption unit for oxygen-users with wide fluctuations in flow requirements. Depending on the length of time a cryogenic plant has been shutdown it can take several hours to restart, cool down and reach purity. If production schedules are well known, restart time may not adversely effect operations and LOX is available as backup; however, the unproductive power cost to restart the unit must also be considered in determining the actual cost of oxygen production (3).

Because cryogenics is a mature technology, drastic improvements in the cycle efficiencies are not likely in the future. The major improvements over the past few years and in the near future will come from advances in the operation of the plants such as the application of computer control. Air Products has pioneered this techmology and today over 100 plants that Air Products owns and operates have computer systems. Additionally, this technology has been applied to 15 plants that are owned and operated by others (3).

Cryogenic air separation plants are good candidates for process control computers, since the major production cost is electric power for driving compression equipment. A small savings in power consumption will offset the capital investment necessary to add a computer system, even when performed on a retrofit basis where pneumatic instrumentation must be replaced with computer-compatible instrumentation. Most air separation plants are operating to supply an online customer who has a variable product demand. A computer system can monitor this demand and quickly adjust production accordingly (3).

Computer control application to cryogenic oxygen production processes resulted in saving up to 14%. Savings obtained from computer application were higher at turndown due to greater turndown capability. Generally, computer monitoring control and optimization improves operational plant efficiency between 5 and 15%, typically resulting unless than a two year payback. The advances in computer hard and software and control sensors gave a great blow to the stagnant and mature cryogenic technology. The new computerized oxygen production facilities for increased efficiency and cost reduction become a reality especially for tonnage plants in low pressure aaplications (6).

2.2. ADSORPTION

Adsorption is a recent technology in oxygen production compared to cryogenics method. In principal separation of oxygen and nitrogen is possible due to the preferential adsorption characteristics of some natural and synthetic materials. Zeolites as adsorbent materials generally have less selectivity for oxygen than for nitrogen where active carbon has similar selectivity for both oxygen and nitrogen (7). Therefore in zeolite based adsoprtion processes, nitrogen is adsorbed and oxygen rich stream passes through the bed. In carbon molecular sieve beds oxygen molecules being slightly smaller than nitrogen molecules diffuse quicker than nitrogen to provide selectivity.

Adsorption and desorption cycles may be accomplished by heating the bed or by reducing the pressure in the bed. The former is known as pressure swing adsorption (PSA) and the latter as temperature swing adsorption (TSA). PSA is caused by adding or subtracting matter to the system while TSA is caused by direct heat transfer independent of material balance. Therefore, the two processes have no direct equivalence. However, it may be possible to combine pressure and temperature changes in a combined cycle gas system to obtain advantages of isotherm shift due to temperature, pressure and inert component to allow a reflux ratio because of expansion of the reflux gas (7). In some processes in the cycle a vacuum purge is used in which case the process is called vacuum swing adsorption (VSA). In VSA the reflux ratio is supressed and a larger recovery is attained. However, this benefit is achieved at the expence of investment for the pump and time needed to reach low pressure. The basic of PSA operating cycle is shown in Figure 3 for a single column operation (5).

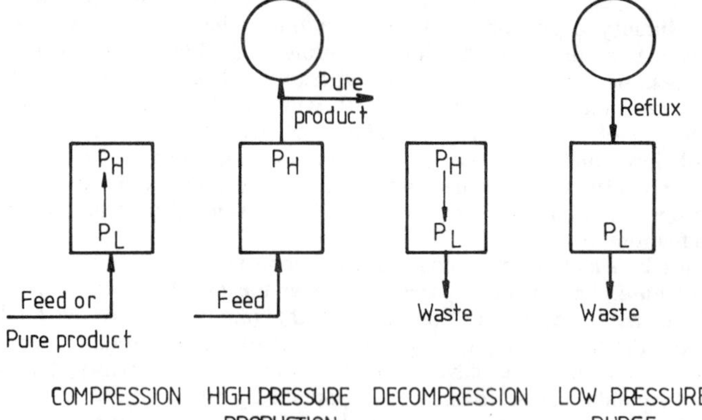

Figure 3. A standard pressure-swing operation cycle (5).

For oxygen production the adsorption bed may be carbon molecular sieve type. The first step is compression in which column is pressurized and mixture is started to be fed. In the second step, the feed flows out as an oxygen rich product while nitrogen front starts to move up until it breaks at the product end. Decompression is the third step in which column is emptied down to pressure P_C through the other end. Nitrogen is desorped as waste or secondary product. At the fourth and final stage remaining nitrogen is purged out with the previously stored oxygen. Of course vacuum application may also be employed. If VSA is used the operation would be essentially between atmospheric pressure and vacuum.

There are many reported literature and patents on adsorptive production of oxygen by various researchers and companies (7). Oxygen purity is limited to a maximum of 93-95% for these adsorption processes. Up to capacities of 100 ton/day the capital cost is relatively low for adsorption systems therefore compensation for the higher energy usage compared to cryogenics is an advantage (8).

2.3. MEMBRANES

Membranes are used to enrich oxygen in air for specific aplications. Membrane separation of oxygen from air is based on the difference in rates of diffusion of oxygen and nitrogen. Flux and selectivity are the two properties that determine the economics of membrane systems. The bottleneck of the system is the low selectivity of available membranes to produce medium or high purity oxygen. The oxygen concentration that may be produced by membrane technology is limited at 50% oxygen. Research on active or facilitated transport membranes which incorporates an oxygen-complexing agent may increase oxygen selectivity which will increase the oxygen concentration to higher levels.

At present due to the relatively low oxygen concentrations membrane generated oxygen-enriched air does not have a significant market. The main user of membrane product oxygen-enriched air is home medical consumption. In Japan, research on oxygen-enriched air combustion of natural gas was successful in attaining high combustion efficiencies (9). Provided that there is substantial development in high temperature burner design, the membrane oxygen process will have an expanded market for natural gas combustion, which is termed as sleeping giant (9).

A technology assesment of the relative economics of available oxygen shows that 35% oxygen enriched air costs $28/ton price, membrane produced oxygen is claimed to be attractive in high efficiency natural gas burners since it is technically feasible to save about 40% of natural gas fuel cost by operating with 30% oxygen feed (10). Membrane processes may also have an important market growth from the biochemical industry for sterile oxygen enriched air (7).

2.4. OXYGEN PRODUCTION BY CHEMICAL METHOD (5)

A new continuous liquid phase absorption/desorption chemical oxygen production process has been developed. This new chemical method significantly reduces electric consumption compared to cryogenic cycles. This process namely MOLTOX was sponsored partly by U.S. Department of Energy and 1/4 ton/day pilot plant experimental set up was started.

The two basic MOLTOX oxygen systems can be based on PSA and TSA. The actual design and operation can be a combination of these processes. Dry, carbondioxide free air enters the absorber at a temperature of 900°C to 1200°C and a pressure of 1.6 to 12.6 atm where it contacts molten liquid salt. Oxygen in the air reacts chemically with the salt and is removed with the liquid salt leaving the bottom of the absorber. The oxygen bearing salt flows to the desorber vessel where it is heated and/or reduced in pressure, generating gaseous oxygen. Lean salt is cooled before returning to the absorber to complete the loop (3).

The MOLTOX oxygen system's major advantage is that the chemical reaction does not consume the compression energy of nitrogen which is the major portion of the air stream. In the PSA mode, nitrogen leaving the absorber is pressurized high enough to overcome the pressure drop through a heat exchanger and absorption equipment. The TSA process is especially applicable for oxygen-users who have waste heat sources, low value fuel streams or boiler that can be integrated with the MOLTOX system.

Splitting of water is a potential process for the production of hydrogen as one of the future global energy source. Oxygen is a valuable by-product which will support the economics of this pollution free and virtually unlimited energy resource. The thermal cracking is very expensive due to high heating cost. Therefore it will be necessary to see development of solar hydrogen processes.

2.5. COMPARISON OF THE CURRENT OXYGEN PRODUCTION TECHNOLOGIES

So far, cryogenic process is the major oxygen production method which generally has a high operational cost and is a high purity process. The newly developed molten salt technology will be able to compete with cryogenic process in capacity and purity together with much lower energy consumption. In view of the unstable energy prices the importance of chemical oxygen production method may grow in coming years. Membrane technology does not claim great share from oxygen market with respect to capacity and purity, however, its new application in high efficiency natural gas combustion and the development of new membranes may increase its contribution substantially. The comparison of membrane, chemical, cryogenic and adsorption methods on relative power (to produce a contained ton of oxygen) normalized for an oxygen supply pressure of 0 psig is shown in Figure 4. The characteristics of current commercial processes with regard to some important parameters are given in Table 2.

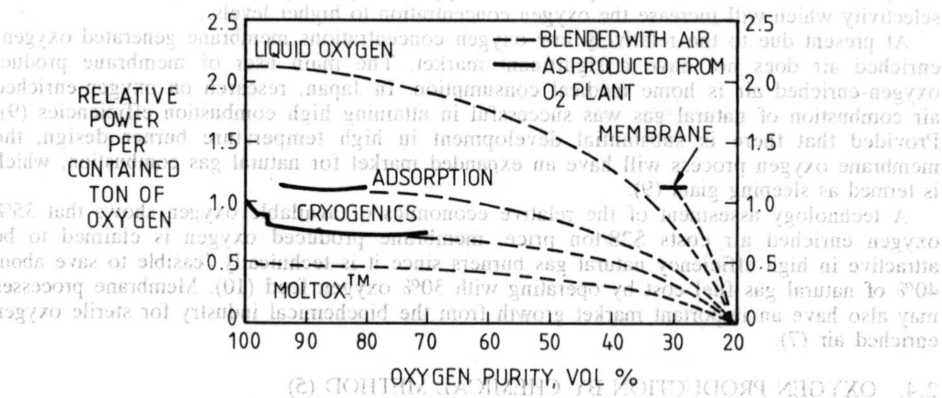

Figure 4. Oxygen technology comparison at 0 psig (3).

3. Conclusion

Oxygen is an important industrial and health product which will grow in importance with increasing utilization in synfuel processes such as oxygen enriched combustion and gasification. At present the overwhelmong proportion of oxygen is produced by cryogenic method. However, the present trend is to develop adsorption, membrane and chemical methods of production for lower capacities. Cryogenic system is amature technology and used for the production of oxygen from 70 to 95%. The well established technique has the benefit of computer control reaching up to 14% saving in operational costs.

TABLE 2. Characteristics of air separation technologies.

	Oxygen Production Technology				
	Adsorption	Chemical	Cryogenic	LOX	Membrane
Status	semi-mature	pilot plant	mature	mature	developing
Economic range	<100 T/D	>500 T/D	>20 T/D	<50 T/D	<20 T/D
Specific power ranking (1=lowest)	3	1	2	5	4
Purity limit	95%	99.9%+	99.8%	99.5%+	ca. 40%
Storage/Backup	add-on or haul in	add-on or haul in	STD	STD	add-on or haul in
Startup time	minutes	hours	hours	minutes	minutes
Flexible production ranking (1=highest)	3	5	4	1	2

Due to capital cost advantages up to 100 tons/day capacity and purity limited to a maximum of 93-95%, adsorption processes are finding industrial applications. In adsorptive processes, zeolites and carbon molecular sieves are used in pressure, temperature and/or vacuum swing adsorption cycles.

Thre tremendous development in membrane science and technology reflected as an alternative oxygen production technique. At present the low selectivity of the membranes limits the oxygen production purity to 50% oxygen. Improvements in active and facilitated transport membranes is expected to raise the impurity levels to higher levels.

The most recent oxygen production technology being developed is a liquid phase absorption/desorption chemical process. This process of at present is at pilot plant capacity of 1/4 ton/day. This process is capable of producing highest purity at lower relative power consumption.

6. References

1. Mcilvaine, R.W. (1983), 'New developments in scrubber particulate technology', 76th Annual Meeting of the Air Pollution Control Association 83.53.3.
2. Krantz, W.B. and Grunn, R.D. (1983), 'Underground coal gasification, the state of the art', A.I.Ch.E., Symp. Ser. No. 226, Vol. 79, pp. 1-3.
3. Murphy, K.J., Odorski, A.P., Smith, A.R. and Ward, T.J. (1988). 'Oxygen production techniques for non-ferrous smelting applications', in, Proc. Int. Symp. Impact Oxygen Prod. Non-Ferrous Metal Processes, G. Kachaniwsky (ed.), Allentown, USA.
4. Blakey, P.G. (1988), Liquefied (Cryogenic) Gases, in Encyclopedia of Chemical Processing and Design, J.J.McKetta (ed.), pp. 166-187.
5. Bernstein, J.T. (1984), Low Purity Oxygen Production, Study Report AP-3499.
6. Mookerjee, I. and Ganguli, S.N. (1987), 'Energy conversation in oxygen production technology', CEW 12, 63-65.
7. Tondeur, D. and Wankat, P.C. (1985), 'Gas purification by pressure swing adsorption separation and purification methods', 14, 157-212.

8. Puffenburger, J. (1986), 'Review of Oxygen Production Technologies for IGCC Power Plants', EPRI Report, RP 8000-5.
9. Baker, R.W., Cussler, E.L., Eykamp, W., Koros, W.J., Riley, R.L. and Strathmann, H. (1988), 'Membrane Separation Systems, A Research and Development Needs Assessment', US DOE Contract No. DEACO1-88ER30133.
10. Matson, S.L., ward, W.I., Kimuna, S.G. and Browall, W.R. (1986), 'Membrane oxygen enrichment, II. Economics Assessment', J.Membr.Sci. 29, 79.

FLUIDISED BED COMBUSTION OF COALS AND ENVIRONMENTAL PROTECTION

E.EKINCI
Istanbul Technical University
Department of Chemical Engineering
Ayazaga Istanbul 80626
TURKIYE

ABSTRACT. Fluidised bed technology immerged as a clean coal technology which reached to power generation capacity within the last two decades. It has brought distinct advantages on SO_2 control, NO_x and CO emission. The present trend is to achieve higher combustion efficiencies, and higher capacities. Stricter environmental pressures dictate combined heat and power systems for fluidised bed combustors like all other fuel combustion systems.

1. INTRODUCTION

Fluidised bed (FB) combustion is a process in which coal particles are suspended and mixed in a stream of air. This physical state is achieved above minimum fluidisation velocity in which the drag force of the fluid on the particles counterweights the particle weight and particles become naturally bouyoant. Above the minimum fluidisation velocity the excess gas passes the bed in the form of bubbles, which are reponsible for the mixing. If the temperature of the bed material is held above the coal ignition temperature then the system becomes a self sustained coal burner. If the inert bed material is choosen to be limestone or dolomite, at the fluid bed combustion temperatures the calcined dolomite or limestone capture SO_2 efficiently at the region of its formation with the help of increased particle/gas contact. In order to keep the SO_2 emission below acceptable limits depending on the sulphur content of the coal, the characteristics of the absorbent, and process conditions, limestone or dolomite Ca/S mole ratios of 1-4 is necessary. Emission of NO_x is also controlled by the low combustion temperatures of 800-900^oC which eliminates the nitrogen fixation reaction of

Y. Yürüm (ed.), Clean Utilization of Coal, 247–259.

combustion air. Therefore there is a 50 % reduction in NO_X emission compared to conventional coal combustion systems.

2. DEVELOPMENTS IN FLUIDISED BED COMBUSTION SYSTEMS

The first application of fluidised bed combustion dates back to 1928 [1]. The idea of fluidised bed combustion was revisited mainly as academic interest and patent work on and off till 1970's accumulating a considerable amount of scientific and technical knowledge. After the outbreak of first petroleum crisis fluidised bed combustion research and development work concentrated on system design and development in the areas of: (1) atmospheric fluidised bed combustion systems (AFBC) (2) pressurised fluidised bed combustion systems (PFBC) for combined cycle power generation (120 MW and 200 MW) (3) industrial shell boiler (<23000 kg steam/hr) and (4) packed water tube boiler (23000-90000 kg steam/hr and 120-MW) were aimed as conceptional design studies [2]. The fluidised bed combustion research and development work uptil 1975 was excellently reviewed by Nach et al [2]. At the time of the review the fluidised bed combustion was about to establish itself as an industrial scale system. However it took about another decade before fluidised bed combustion was listed as an available technology of 200 MW for atmospheric and 100 MW for the pressurised systems in years 1991 and 1992 respectively at 400 MW claimed to be available by 1997 for both systems. The IV. International Fluidised Bed Combustion Symposium organised by the Institute of Energy was on the theme of "Fluidised Combustion in Practice: Clean, Versalite and Economic?" In this conference the fludised bed combustion systems at capacities as high as 160 MW were reported. The system was claimed to be an industrial success [4].From the proceedings it is understood that research and development efforts were directed towards higher capacity systems using higher space velocities therefore utilising circulating fluidised bed. Also fluidised bed which are specificly designed and developed such as multi-fuel firing are referred [5]. The capacity of fluidised beds in the control of SO_2, NO_X and CO is well endorsed. However, there are specific issues which may be raised as fundemental problems such as acid rain, ozone depletion, global warming and carcinogenic emissions.

The low temperature combustion in fluidised bed provides advantages as well as some disadvantages. Two of these disadvantages are the polyaromatic hydrocarbon (PAH) and N_2O emissions. The former was raised by Kita and Puff [6]. It was found that PAH concentrations are higher than that obtained with other types of coal combustion processes (<30 ppb) though their levels remained relatively low (<200 ppb). In view of the fact that compounds that are active carcinogens are detected in the bottom and cyclone ash [7] and PAH concentration of the fly ash is many times higher than that of bottom ash [8, 9], questions will be raised in the future. Second concern is the relatively higher N_2O emissions in FBC systems compared to pulverized firing which was addressed by a recent EC report prepared by a select committee [10]. The concern over N_2O is of course due to its important contribution to the ozone depletion reaction in the atmosphere. Both PAH and N_2O levels can be reduced considerably in the light of available information and techniques developed so far.

Based on research and development work in the University of Nova Scotia Canadian government comissioned a 165 MW circulating fluidised bed this year. It has been announced by the government that this will be the last of fluidised bed projects. Due to the problem of CO_2 increase all the future supports will be available to fluidised bed systems involving combined heat and power generation [11].

3. FUNDAMENTAL STUDIES

3.1. Segregation

Segregation is a measure of deterioration in mixing in fluidised beds. A good mixing in a fluidised bed leads to a good combustion, heat transfer and environmental control. Segregation is not a problem for a narrow size distribution of particles in which there is not a great deal of density difference between the components. If there is a difference in the particle sizes of the components and/or densities there is a strong tendency for the bed to segregate and consequently loose fluidisation and combustion. Segregation is a relevant phenomena in lignite fired FBC systems, because of the high ash content and readily changing characteristics of lignites. Also during the combustion of volatile matter particles are subject to spend most of their time at the upper sections of the bed, thus have a segregation tendency [12, 13, 14]. In order to control fluidised bed combustion process it is necessary to measure the extent of segregation in fluidised bed combustors. There are various methods of detecting segregation. These include: (i) slumping the bed and then removing and sieving the components section by section, (ii) measuring the pressure drop in the bed, (iii) x-ray cinephotography, (iv) differential temperature readings and (v) radioactive tracer techniques [15]. Differential temperature readings seem to be the most appropriate method among these techniques. It is possible to relate the segregation to the point where coal concentration and thus release of heat is intense.

It has been proved that differential temperature and pressure drop measurements give parallel indications in the dedection of segregation tendencies [16] as shown in Figure 1.

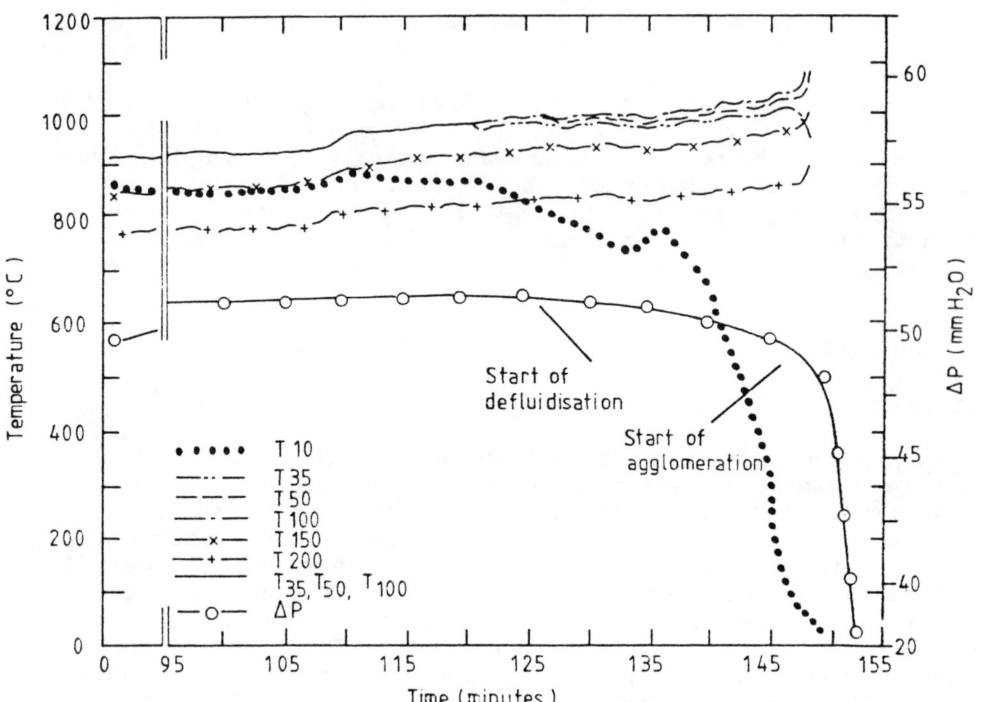

Figure 1. Determination of Defluidisation Using Temperature Profiles and Pressure Drop

This method was developed as an efficient and indirect tool in detecting the mixing efficiency and segregation tendency in fluidised beds. The differential temperature data obtained for lignite fired fluidised bed systems showed that the segregation behavior for the hot bed is consistent with the well established segregation models for the cold fluidised beds [15, 16, 17, 18, 19]. The differential temperature method was based on temperature measurements in the vicinity of the distributor plate, in the bed and the freeboard region. The differential between the bed and the distributor plate temperatures were related to ash flotsam jetsam rich systems and the differential between the bed and the freeboard temperatures were more relevant to ash jetsam rich type systems. Both differential temperatures could be used to detect the two types of segregation models. The differential temperature concept is believed to be an effective tool to control the segregation and agglomeration in solid fuel fired fluidised bed combustors.

3.2. Agglomeration

Fluidised bed combustors operating under controlled conditions can burn solids in a similar manner to fuel oil burners. In fact coal and air make up a power-low type nonnewtonian fluid under fluidised conditions [19]. However, if the bed agglomerates the contents of the bed solidify and need to be broken and removed. Agglomeration control is therefore of paramount importance in fluidised bed combustors.

Agglomeration may occur mainly for two reasons: (1) due to segregation hot pockets may occur (2) bed temperature at any part may be higher than the sintering temperature. If for any reason temperature in the bed reaches the sintering temperature of coal and necessary precautions are not taken then the coal particles stick to each other and/or to the inert particles so that whole bed agglomerates. Previously the ash characteristic temperatures were taken as the basis for agglomeration control and temperatures 200-250 °C, lower than the ash fusion temperature, were accepted as safe operational temperatures [20]. Later Basu [21] proposed that the sintering temperature would be a better and direct indicator of agglomeration temperature.

Basu [21] measured sintering temperatures of a type of coal using a vertical dilatometer which is shown in Figure 2. In his study, he also determined the sintering temperature in a fluidised bed by using defluidisation method. The sintering temperature is extrapolated to the minimum fluidisation velocity. The measured and extrapolated results are found to be in good agreement. Sintering temperature of some Turkish lignites are also determined using the defluidisation method [22].

It is found that the differential temperature method which was used to detect segregation tendencies is also used as a tool to control and analyze the agglomeration of the fluidised bed [23, 24]. A strongly lignite flotsam-rich system gives the largest temperature difference between the bed and the freeboard due to the relatively high heat release in the freeboard. Conversely, for strongly lignite jetsam rich system, the temperature difference between the bed and the vicinity of the distributor plate were the largest because the lignite concentrates and agglomerates in the lower parts of the bed.

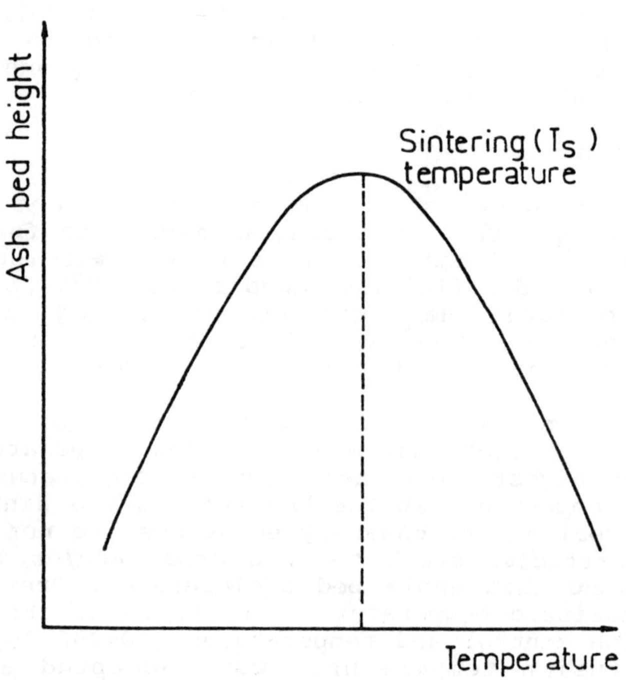

Figure 2.Determination of Sintering Temperature Using A Dilatomer

3.3. Volatile Matter Combustion

When coal is heated to temperatures higher than 350 °C, it undergoes a rapid thermal decomposition during which about 20-50 % of coal can be released in the form of volatiles. Devolatilization takes place at temperatures ranging from 200 to 400°C. The amount and composition of volatile matters (VM) varry with type of coal, particle size and heating rate. The volatiles mainly consist of carbon oxides (CO and CO_2), chemical water, hydrogen, methane, ethane and higher hydrocarbons. The composition of VM may change during decomposition; the chemical water and carbon oxides are released mainly in the early stages of the process while the hydrocarbons are released later [25].

The VM can account to as much as 40% of the heating value of coal [26]. Laboratory and pilot plant studies showed that VM also contribute to a major fraction of CO emission from fluidised bed combustors. Most of NOx is formed by oxidation of the nitrogenous groups present in the volatile matter [25]. On the other hand, coal behaves hydrodynamically in a different manner than the residual char during the release of VM [12, 13, 14]. Hence it is essential to understand the mechanism of devolatilization of coal and VM combustion in order to establish criterion for a proper fluidised bed combustor design and operation. For example, the size and position of heat transfer surfaces are strongly influenced by the combustion of VM and the behavior of the coal particle during combustion. Excess of volatiles near feed point could result in corrosion due to the reducing atmosphere. It can occur at the region near the inlet point for small particles which release their volatiles very rapidly. This region would be extended for larger particles. There has been a lot of studies in the last decade devoted to this subject. It was understood from these investigations that, devolatilization is a strong function of coal type, structure, particle size and size distribution and operational conditions.

Devolatilization of small particles (100 microns) is generally isothermal and kinetically controlled [27, 28]. Since there is a tendency to burn larger coal particles in commercial fluidised bed combustors, data on VM combustion of large particles are necessary. Most of the studies have been directed to understand the devolatilization of smaller coal particles, typically smaller than 1 mm [29]. The devolatilization of large coal particles is significantly difficult than that of pulverized coal particles due to the diffusion resistance and relative importance of heat and mass transfer. Many scientists claim that the coal particles tend to remain near the surface during the devolatilization and combustion of volatile matters and then they sink deep into the bed when VM combustion is completed [12, 13]. Pillai [30], depending on results obtained from experimental studies of combustion of coal particles whose circulation was limited, suggests that VM from coal particles burn in a plume-like cloud surrounding them. Stubington [12] pointed out that coal particles are dispersed rapidly in axial direction, but they are very slow in radial direction which confirm plume-like model combustion of VM. The radial oxygen concentration have significant effects on the combustion of volatile matter as well as char combustion. The combustion and distribution of VM across the bed may be shaped by the devolatilization time. For shorter devolatilization times, volatiles are expected to concentrate in a restricted area along the vertical axis from coal injection point. As the coal particles move upwards, the diameter of the volatile cloud region increases as a result of radial displacement of particles. This results in a relatively more uniform distribution of volatiles across the bed [12].

Atimtay [26] who conducted one of the most elaborated studies on the volatile combustion in FB combustors indicated that the burnout time of volatiles increases with mass charge, particle size, volatile matter content and decreases with air velocity. It is reported that devolatilization and burnout time is a function of temperature. The investigation of devolatilization of large coal particles (4.7 mesh) showed that for a devolatilization degree up to 90%, the extent of devolatilization is nearly linear with respect to the final combustion temperature [31]. Devolatilization of large coal particles is generally accepted to be controlled by internal heat transfer, hence devolatilization time increases with initial coal particle size [14, 31, 32]. The initial particle size of coal may decrease during the combustion due to the fragmentation which increases the devolatilization time and burning rate of the particles. Fragmentation is defined as the breaking up of a single particle into two or more pieces. This is not attrition in which a number are produced but the original particle size does not decrease significantly. Fragmentation may occur during devolatilization and combustion of char. The former is called primary fragmentation and the latter is called secondary fragmentation. Stubington and Linjewile [32] found that rate of devolatilization increase for fragmented particles. This was attributed to the rapid heat transfer to the particle due to decreased particle size which enhances the devolatilization. The same authors claimed that the effect of coal type on devolatilization may be related to coal plasticity, in particular to the swelling number. The lower the swelling number, the higher the fragmentation and hence the shorter the devolatilization time. Depending upon the structure and composition of coal and its physical behavior while burning, resistances due to the presence of ash or due to the size and distribution of pores may be a controlling factor for devolatilization [13].

The intensity of volatile combustion and flame is enhanced by inlet oxygen concentration due to increasing heat transfer rate. The devolatilization period of large coal particles increases in fluidised bed combustion performed with very low inlet oxygen concentration [33]. This is confirmed by the devolatilization carried out in air and nitrogen. The evolution rate of volatiles for coal combustion in air is reported to be 2-3 times higher than that for coal pyrolysis in nitrogen gas [34].

Devolatilization time measurements conducted by following the gas evolution show that gas evolution may still occur even after flame extinction [26]. However, the difference may disapppear with a decrease in coal particle size due to the short diffusion path and rapid release of volatiles.

It is well known that devolatilization is affected significantly by coal particle size. A power law relation between devolatilization time and particle diameter was proposed by several authors [12, 13, 14, 35]. This is formulated as $t_v = a\,d^n$, where t_v is the volatile burnout time (s), d is the particle size (mm) and a and n are the appropriate regression constants. The exponent n is equal to unity for kinetic control and value 2 for diffusion control. However, the n value may differ from these values and lies between 0 and 2 depending on the way in which combustion takes place [13]. For coal particle size typical for fluidised bed combustion, i.e. bigger than 1 mm, n values are reported to be bigger than 1. In a fluidised bed combustion study six coal of particle size ranging from 1 to 6.3 mm were experimented [14]. It has been found that the burnout time and n vary between 5 and 60 s, and 0.9 and 1.5 respectively. As the particle size increases, both burnout time and n increase [14]. However a distinct difference was observed between values of different coal types. The values for n were reported to be 1.4-1.9 for three types of coal of different swelling numbers [32].

4. OTHER STUDIES

In order to provide the fundamental data for the fast growing FB industry, intensive research work is being carried out on appropriate materials for bed internals, scrubbing of particulates, corrosion and errosion, heating up system and automatic control.

Higher sulphur coals like Turkish lignites consume large quantities of limestone and dolomite which may be a land pollution concern, therefore regenerable synthetic absorbent have been a popular research topic. The synthetically produced absorbents are used to absorb the SO_2, produced in the bed via sulfonation and then reduced in the presence of CO, and H_2, in a separate bed at a temperature of 1100 ^oC. The number of cycles that an absorbent may be used is a determining factor for the economics of synthetic absorbers. Synthetic absorbers also may open up means to produce sulphuric acid economically from the captured SO_2 since it is possible to concentrate this gas to above 12 %.

Fluidised bed systems are now accepted as up to date coal burning system which can comply with present environmental restrictions for SO_2, NO_x, and CO emissions. However, as the awareness to public health issues increase, the performance of fluidised bed combustors will be scrutinized for PAH emissions and heavy metal concentration in the fly ash.

Furthermore, the steady increase in CO_2 concentration in atmosphere puts pressure on the industrial states to ensure maximum return from fossil fuel combustion. Therefore, beyond a certain capacity FB will be forced to operate in combined heat and power mode.

5. CONCLUSION

Fluidised bed combustion is an industrially accomplished clean coal utilisation combustion technique which has tremendous potential in the control of SO_x and NO emission coupled with economy and versatitily. The technology is currently available at 200 MW and 1000 MW capacity to atmospheric and pressurized systems.

Even though it has emerged as an enviromentally friendly technology, problems related to relatively higher emissions of N_2O and PAH compared to conventional combustion methods is receiving attention.

Segregation is a relevant phenomena in lignite fired fluidised beds which is enhanced by the changing nature of lignite ash. Differential temperature measurements in the bed are succesfully used to control segregation in both ash flotsam and jetsam rich systems. This approach is also helpful to control agglomeration in lignite fired fluidised beds. In order to control agglomeration the ash sintering temperature of lignite is also needed to be known. The sintering temperature may either be measured using a dilatometer or conducting defluidisation experiments and extrapolation.

Volatile mater constitutes an important part of lignite combustion in fluidised beds. The mechanism generally portraited in literature is for ash flotsam rich systems which are quite different from the oil jetsam rich systems.

REFERENCES

1 Kunii, D., Levenspiel, O. (1977) Fluidization Engineering, Krieger and Reinhold.
2 Nach H., Kiang, K.D., Lui, K.T., Murthy, K.S., Smithson, G.R., Oxyley, S.H. (1975) Fluidised Bed Combustion Review, Inst. Fluid. Tech., Vol.2, Ed. Keairns, D.L., Hemisphere Pub. Co. , McGraw Hill.
3 Yeager, K.T. (1990) Coal in the 21st Century, Energy World, June I-XIV.
4 Harrison, J.S. (1988) Summing Up Speech, 4th International Fluidised Bed Combustion Conference, The Inst. of Energy, London.
5 Fear, R.C., Bense, P., Fischer, M.J., Jackson, P.E.H. (1988) 'In Bed Multi-Fuel Firing in Fluidised Bed Combustors', IV. Int. Fluidised Bed Combustion Conference, Paper III/1/, Inst. of Energy.
6 Kita, J.C., Puff, R. (1988) 'Coal Residues from Fluidised Bed Combustion and Their Environmental Effect', 4th Int. Fluidised Bed Combustion Conference, Paper II/5/, The Inst. of Energy.
7 Citiroglu, M., Yardim, M.F., Ekinci, E. (1991) 'Determination of Polycyclic Aromatic Hydrocarbons (PAH) in Cyclone and Bottom Ashes in a Fluidised Bed Combustor', accepted for 5th Int. Conference on FBC Technology to Meet the Environmental Challenge.
8 Eiceman, V. (1983) 'Absorption of PAH on fly ash from a municipal incinerator and a coal fired power plant', At. Env., 17:3.
9 Shure, N. (1981) 'Vapour Particle Association of PAH in Power Plant Emissions-Theoretical Predictions', 6th Int. Symp. on PAH.
10 Braly, H., Jacobs, J.P., Milhau, A., Verhoeff, F., Gulcat, O., Bernergard, L., Sobozynski, S.F. (1990) 'The Recent State of Ort in the Commercialisation of Fluidised Bed Combustion (FBC)', EEC Working Perry.
11 Hamdullahpur, F. (1991) 'Fluidised bed Combustion Modelling', Seminar at ITU.
12 Stubington, J.F. (1980) 'The Role of Coal Volatile in Fluidised Bed Combustion', J. Inst. Energy, 53, 191-195.
13 Pillai, K. K. (1981) 'The Influence of Coal Type on the Devolatilization and Combustion in Fluidised Beds', J. Inst. Energy, 54, 142-150.
14 Ekinci, E., Yalkin, G., Atakul, H., Senatalar, A. (1988) 'The Combustion of Volatiles from Some Turkish Coals in a Fluidised Bed', J. Inst. Energy, 61:449, 189-191.

15 Ekinci, E., Atakul, H., Tolay, M. (1990) 'Detection of Segregation Tendencies in a FB Using Temperature Profiles', Powder Tech., 61, 185-192.

16 Ekinci, E., Yardim, M. F., Atakul, H. (1988) 'Temperature Profiles in a Lignite Floatsome-Rich Fluidised Bed', Fuel, 67, 191-197.

17 Rowe, P.N., Nienow, A.W. (1976) 'Particle Mixing and Segregation in Gas Fluidised Beds.', A Review, Powder Tech., 15,141-147.

18 Nienow, A.W., Rowe, P.N., Cheung, L.Y.L. (1978) 'A Quantitative Analysis of the Mixing of Two Segregating Powders of Different Density in a Gas Fluidised Bed', Powder Tech. 20, 89-97.

19 Jackson, R. (1971) Fluid Mechanical Theory, Chp.3 in Fluidization, Ed. Davidson, D. and Richardson, J.F., Academic Press.

20 Abbi, Y.B., Banerji, M., Ghosh, M.K., Malliah, K.T.U., Sharan, H.N. (1978) 'Development of Fluidized Combustion Boiler for Indian Coals', AIChE Symp. Ser. 176, Vol. 74, 162.

21 Basu, P. (1982) 'Study of Agglomeration of Coal Ash in Fluidised Beds', The Canadian J. Chem. Eng., 60, 791-795.

22 Ercikan, D., Madrali, E. S., Ekinci, E. (1989) 'Determination of Sintering Temperatures Using Defluidization Method', Proc. II. Combustion Symp., ITU, 83-93.

23 Atakul, H., Ekinci, E. (1990) 'Agglomeration in Lignite Ash Floatsome and Jetsome Rich Systems', Powder Tech., 60, 77-82.

24 Atakul, H., Ekinci, E. (1989) 'Agglomeration of Turkish Lignites in Fluidised Bed Combustion', J. Inst. Energy, 62, 57-61.

25 Borghi, G., Sarofim, A.D., Beer, Y.M. (1985) 'A Model of Devolatilization and Combustion in Fluidised Beds', Combustion and Flame, 6, 1-16.

26 Atimtay, A. (1982) Combustion of Volatile Matter in Fluidised Beds, in Fluidization, ed. Grace, J.R. and Mathsen, J.M., Plenum Press.

27 Saxena, S.C. (1990) 'Devolatilization and Combustion Characteristics of Coal Particles', Prog. Energy and Comb. Sci., 16, 55-94.

28 Howard, J. B. (1981) Chemistry of Coal Utilization, 2nd suppl. vol. (Ed. Elliot, M.A.), John Wiley and Sons, New York, 665-784.

29 Anthony, D.B., Howard, J.B., Hottel, H.C., Meissner, H.P. (1975) 5th Symp. (Int.) on Combustion, The Combustion Institute, Pittsburgh, p.1303.

30 Pillai, K.K. (1982) 'A Schematic of Devolatilization in Fluidised Bed Combustors', J. Inst. Energy, 55, 132-133.

31 Agarwall, P.K., Genetti, W.E., Lee, Y.Y. (1984) 'Devolatilization of Large Coal Particles in Fluidised Beds', Fuel, 63, 1748-1752.

32 Stubington, J.F., Linjewille, T.M. (1989) 'The Effect of Fragmantation on Devolatilization of Large Coal Particles', Fuel, 68, 155-160.

33 Salam, T.F., Shen, X.L., Gibbs, M.B. (1988) 'A Technique for Determining Devolatilization Rates of Large Coal Particles in a Fluidised Bed Combustor', Fuel, 67, 414-419.

34 Saito, N., Sadokata, M., Sato, M., Sakai, T. (1987) 'Devolatilization Characteristics of Single Coal Particles for Combustion in Air and Prolysis in Nitrogen', Fuel, 66, 717-720.

35 Harris, J.P., Keairns, D.L., (1979) 'Coal Devolatilization Studies in Support of the Westinghouse Fluidised Bed Coal Gasification Process', Fuel, 58 (7), 465-476.

ADVANCES IN THE DEVELOPMENT OF COKE FREE IRON OXIDE REDUCTION PROCESSES

M. WEEDA, F. KAPTEIJN and J.A. MOULIJN*
*Department of Chemical Engineering, University of Amsterdam,
Nieuwe Achtergracht 166, 1018 WV Amsterdam, The Netherlands*
* *Department of Chemical Engineering, Delft University of Technology,
Julianalaan 136, 2628 BL Delft, The Netherlands.*

ABSTRACT. The blast furnace process is the dominant process for the reduction of iron oxide to iron. Although, blast furnace operation has reached near perfection, disadvantages such as the use of coke and the necessity to operate at a large scale (2-3 Mt/y) to be economic have initiated new developments. Two lines of research can be identified. One is concerned with the injection of pulverized coal into the blast furnace in order to replace as much coke as possible, thereby reducing the need for expensive coke and improving the process economy. The other is concerned with the development of new processes which aim to be more flexible with respect to the use of raw materials, produce liquid iron economically also at a small scale (e.g. 0.5 Mt/y), and cause less pollution than a blast furnace. In general, the new processes which are termed smelting reduction processes, make use of two separate reactors to replace the blast furnace. Coal is used directly, although there are processes that still require a small amount of coke. Some processes also use fine ore directly.

1. Introduction

The most important nonfuel use of coal is in the reduction of iron ore to produce steel. Today about 800 Mt/y of steel is produced. Approximately 500 Mt of the annual steel production originates from iron ore. The main process for the production of iron is the blast furnace. Direct reduction processes account for only about 2.5% of the iron production. Due to mainly economical factors the rapid development of direct reduction as forecasted up to the beginning of the 1970's has not taken place. In fact, ironmakers are no longer interested in these processes, and direct reduction is no longer seen as a promising route. The remaining 300 Mt of the annual steel production originates from scrap. About two thirds of the scrap is processed in electric arc furnaces, and one third is used in basic oxygen steelmaking.

Since its introduction the blast furnace technology has seen enormous advancement. In 1850 an efficient blast furnace produced 20 ton hot metal per day. By the turn of the century, 400 ton per day were already produced. In 1950, 2,000 ton per day could be produced. Today, more than 10,000 ton of hot metal per blast furnace can be tapped. Typically, production capacities of modern blast furnaces are in the order of 2-3 Mt/y. Together with the high production capacities, the main advantages of the blast furnace are a high energy efficiency and the production of liquid iron.

Although the blast furnace has always been the ironmaking process, and in spite of the fact that through gradual evolution blast furnace operation has reached near perfection, the process also has some disadvantages;

- It has a low flexibility with respect to the use of raw materials, it requires coke and iron ore of specified quality.
- It has a low flexibility with respect to production capacity.

Y. Yürüm (ed.), Clean Utilization of Coal, 261–275.

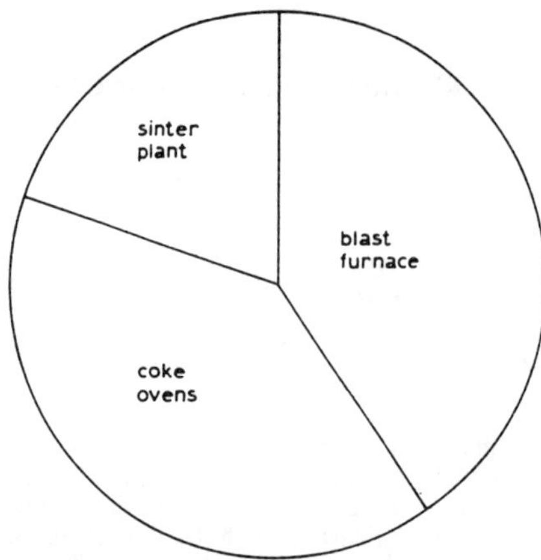

FIGURE 1. Capital cost breakdown of a conventional ironworks.

The major disadvantage is that the blast furnace requires the use of coke instead of using coal directly although a growing portion of the coke can be replaced by coal injected through tuyeres. In all-coke operation, today, about 500 kg/tHM is needed. Coke is prepared in coke ovens which significantly raise the investment costs of the blast furnace process (figure 1). They have high costs of maintenance, and its gas and dust emissions when charging and discharging the ovens are undesirable because of environmental reasons. Moreover, coke can only be prepared from coals which exhibit a plastic behaviour during heating, and as the coke must meet rather strict specifications high quality coking coals have to be used. Because these coals are relatively scarce they are more expensive than other coals. As a consequence, there is a growing concern in existing plants about rebuilding antiquated coke ovens both for financial and environmental reasons.

Furthermore, optimization of coke usage and productivity (production per unit inner volume) requires intensive pretreatment of the iron ore. The use of fine ore is undesirable. Usually, fine ore is pelletized, or sintered and screened for fines before being processed to liquid pig iron. Sometimes, high quality lump ore is used. As with the coke ovens, the pretreatment facilities raise the investment costs, operating costs and environmental problems of the blast furnace process.

Finally, because of high investment costs of coke ovens, pretreatment facilities, and the blast furnace itself, large scale units are necessary to achieve economic operation. As a consequence, there is only a small flexibility in production capacity, and problems can be anticipated when addition of small increments in production capacity is required to fulfill the predicted slow growth in market needs [1].

The disadvantages of the blast furnace process and the demise of direct reduction processes have created the opportunity for the development of a new generation of ironmaking processes which can produce liquid iron economically at a small scale. These processes are comprehensively termed smelting reduction processes. The ability to produce economically at a small scale mainly is the result of avoiding, or at least greatly reducing the need for the cost-intensive route of coking coal to coke by using coal directly. Also for some processes agglomeration facilities are avoided by direct use of fine ore.

A schematic overview of the different process routes for ironmaking and steelmaking is given in figures 2 and 3. Besides the new smelting reduction processes, the present developments in blast furnace ironmaking will also be highlighted as it appears that the blast furnace still has a large development potential. The electric arc furnace based process routes will not be treated any further.

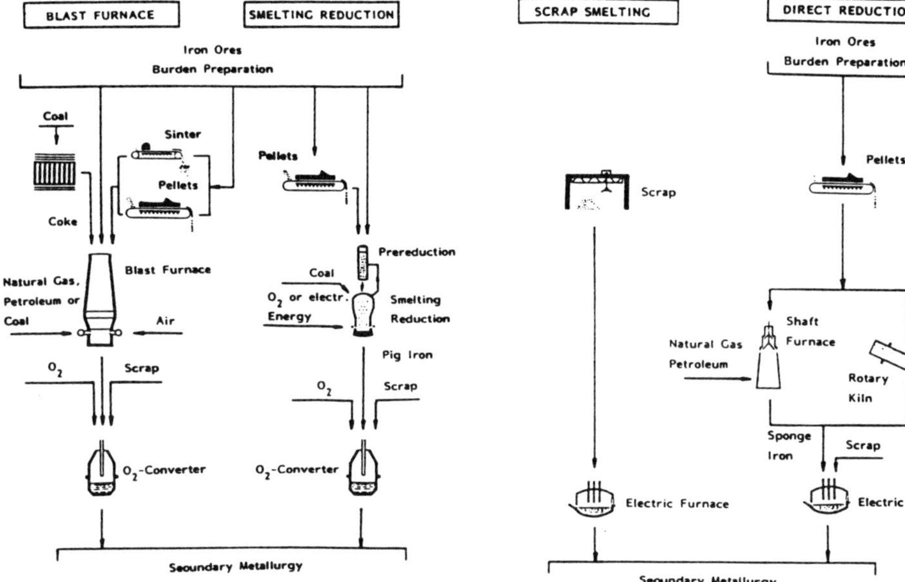

FIGURE 2. Oxygen converter based process route for the production of steel.

FIGURE 3. Electric arc furnace based process route for the production of steel.

2. The Blast Furnace

A blast furnace incorporates three process units:

- A reactor for reduction of iron oxide with reducing gas,
- A reactor for gasification of coke,
- A reactor for smelting of iron ore and slag.

Briefly, preheated air of about 1400-1600 K (the hot blast) is injected through tuyeres in the lower part of the blast furnace, creating a raceway in front of each tuyere. Coke moves down into the raceway where it is oxidized to carbon dioxide which is subsequently reduced to carbon monoxide in the area near the coke bed. In the raceway the temperature of the gas is approximately 2400-2500 K. When the high temperature gas ascends from the raceway it is cooled rapidly to about 1750 K by heat exchange with the permeable coke bed which constitutes the so-called dead man, by heat exchange with the molten materials trickling down through the coke bed countercurrently, and by the endothermic reduction of the molten iron oxide according to the overall reaction (3).

$$FeO + CO \rightarrow Fe + CO_2 \quad (1)$$

$$CO_2 + C \rightarrow 2\,CO \quad (2)$$

$$FeO + C \rightarrow Fe + CO \quad (3)$$

Subsequently, the gas ascends through coke slits in the cohesive zone. In this zone all burden materials except coke soften and are finally melted down. When the temperature of the ascending gas and the burden decrease below 1300 K gasification of coke does not proceed any further. The iron oxides are exclusively reduced by carbon monoxide according to reaction (1). The higher iron oxides hematite and magnetite are reduced in the upper part of the stack according to reactions (4) and (5), respectively.

$$3\ Fe_2O_3 + CO \rightarrow 2\ Fe_3O_4 + CO_2 \quad (4)$$
$$Fe_3O_4 + CO \rightarrow 3\ FeO + CO_2 \quad (5)$$

The molten iron and slag coming down from the cohesive zone accumulate in the hearth. Due to its lower density, slag floats on top of the hot metal. Hot metal and slag are tapped intermittently at intervals of a few hours. In figure 4 a schematic diagram is given of the internal situation in the blast furnace. Figure 5 shows the profiles of temperature and reduction degree in the blast furnace.

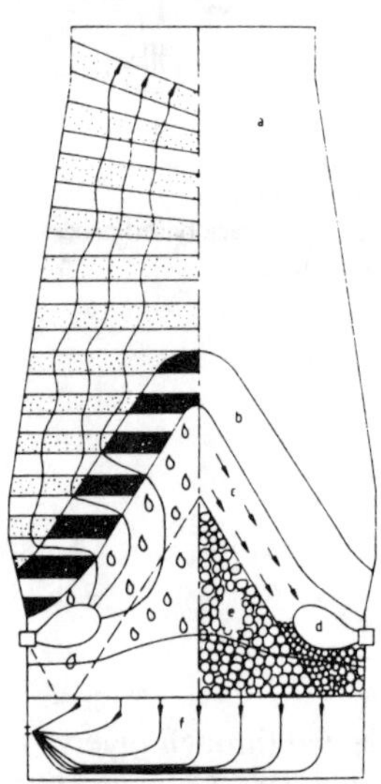

FIGURE 4. Schematic of the blast furnace internal situation: a) Stack; b) Cohesive zone; c) Dripping zone; d) Raceway; e) Dead man; f) Hearth

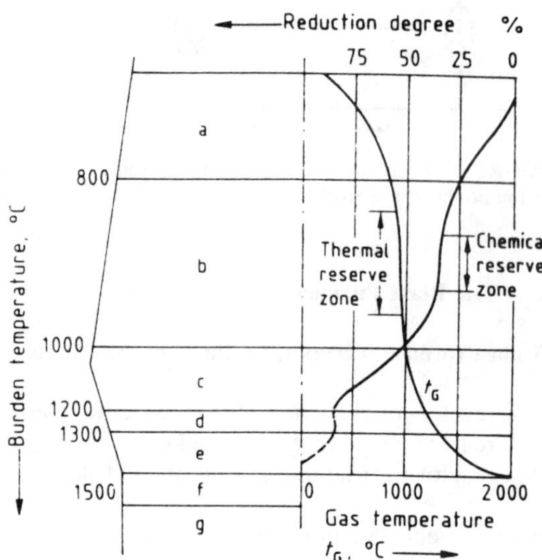

FIGURE 5. Reaction zones and profiles of temperature and reduction degree in the blast furnace: a) Indirect reduction of hematite (Fe_2O_3) and magnetite (Fe_3O_4); b) Indirect reduction of wustite (Fe_xO); c) Direct reduction of wustite and Boudouard reaction; d) Melting of iron and slag; e) Reduction of FeO in the slag; f) Combustion in front of the tuyeres; g) Hearth

In the blast furnace coke has three separate functions:

- It serves as a fuel,
- It acts as a reducing agent,
- It gives support to the burden and ensures good permeability especially in the cohesive zone where all other components are softened or molten.

The first and part of the second function can readily be performed by other fuels. Starting in the 1960's many auxiliary fuels such as natural gas, coke oven gas, tar and coal have been used, but the injection of oil has demonstrated the most favourable advantages. However, after the oil crises in the 1970's the injection of oil was more or less given up and blast furnace research and engineering activities focussed on injection of coal as coal appeared to be the cheapest and best alternative to oil. Worldwide, more and more blast furnaces are now being adjusted for injection of pulverized or granular coal directly into the raceway together with the hot blast. Current technology uses pulverized coal injection (PCI) rates in the range of 50-100 kg/tHM, and the technology for stable operation of the blast furnace with PCI rates of up to 100-150 kg/tHM is within reach [2].

The aim of coal injection is to replace as much coke as possible. The ratio of the amount of coke saved and the amount of coal used is often called the replacement ratio. In general, the replacement ratio is larger than one. This is mainly due to the higher reducing power of hydrogen as compared to carbon monoxide. Moreover, per mol of oxygen the combustion of coal yields more reducing gas than the combustion of coke. However, per mol of oxygen the combustion of coal generates less heat than the combustion of coke which results in a decrease of gas temperature and reduction rates. Consequently, with increasing coal rate the advantageous effects become gradually less, and the replacement ratio gets worse.

The blast furnace seems to be rather flexible with respect to the quality of the coal that is used for injection [2]. However, as with increasing coal rate the coal is fed in a gradually more concentrated form and the excess oxygen falls, and as the residence time in the raceway is very short (10-20 ms), special properties with respect to combustibility and calorific value are required to achieve complete combustion and efficient operation. To arrive at the desired properties blends of coals are used. Blending of coals appears to be effective if,

- one of the coals has a high content of volatile matter through which the ignition delay is reduced [3],
- and the other coal has a high content of solid carbon to meet the enthalpy requirement and to replace as much coke as possible.

With respect to combustibility it may be important not only to consider the volatile matter content, but also the tar yield [4] under PCI reaction conditions as at high temperatures tar readily forms soot. Soot is less reactive than the residual coal char and complete combustion of soot requires a high air/fuel ratio and residence times exceeding that of the raceway. As a consequence, excessive soot formation can cause serious operating problems such as blockage of raceways, decreased burden permeability and an undesirable temperature distribution within the furnace, and can thus lead to an increased fuel rate.

Upon increasing the coal rate, combustion of coal eventually becomes incomplete. As a result, unburnt coal is ejected at the top, gas permeability is reduced, and blast furnace operation becomes unstable. In conventional blast furnace operation a rate somewhere between 100 to 150 kg/tHM seems to be the upper limit of efficient PCI. To enable even higher PCI rates, process conditions have to be changed;

- The temperature of the blast must be increased.
- An effective blend of coals should be used and/or a catalyst should be added to reduce the ignition delay and to increase combustion rates. With respect to this, addition of a lignite

may be useful as a lignite already incorporates an effective gasification catalyst. Besides it has a high volatile matter content and a low tar yield upon pyrolysis.

- The contact between coal and oxygen should be enhanced. This can be achieved by:
    - oxygen enrichment of the blast,
    - intensive mixing of coal and air/oxygen.
- The coal residence time in the raceway should be increased by increasing the infurnace pressure.

Investigations on ways to significantly reduce the coke rate in the blast furnace are primarily being performed in Europe. Since 1983, British Steel, recently in cooperation with Hoogovens (The Netherlands) and Ilva (Italy), and supported by the European Coal and Steel Community, has been developing an "oxy-coal" blast furnace process. After small scale tests on the melting zone, trials have been performed first using a single tuyere and then 12 out of 26 tuyeres of a production blast furnace. PCI rates equivalent to 200-300 kg/tHM have been practised with good coke replacement ratios. A production trial on a 1,000 t/day blast furnace is now underway. The aim of this demonstration project is to push coal rates up to 50% of the total carbon input by means of oxygen enrichment of the blast of up to 40%, whilst retaining hot metal quality. A similar "oxy-coal" project is also under development by NKK (Japan).

Another development based on the conventional blast furnace is the Pirogas project. The process uses plasma heating to superheat the air blast to permit high levels of coal injection. Studies on an experimental blast furnace of CRM (Belgium) showed that for various hydrocarbons a suitable reducing gas could be produced by using plasma burners. With plasma-reformed natural gas the furnace was operated with coke rates as low as 100 kg/tHM without any operational problems. Trials to achieve the injection of coal by means of plasma burners in an existing blast furnace have been carried out in cooperation with IRSID (France) for a ferromanganese blast furnace. First results were so promising that plasma burners are also to be tried out for a blast furnace for the production of pig iron. The target injection rate is 200 kg/tHM.

Despite all these developments, in a blast furnace coke will always have to be used to support the burden and to provide gas permeability, especially in the cohesive zone. At present it is not yet possible to predict the minimum amount of coke required. The critical parameter may be coke strength as with decreasing coke rate, the coke must support an increasing amount of ore. The limit to coke reduction may also be determined by the ore pretreatment step or the attainable degree of reduction in the upper zone of the blast furnace. These factors are important with respect to the temperature where softening and melting of the ore takes place and thus determine the extent of the cohesive zone. As coke provides the gas permeability in the cohesive zone, with decreasing coke rate the cohesive zone should be as narrow as possible.

The use of pulverized coal injection is expected to become increasingly significant and this will have its impact on coking coal requirements (approximately 12% of the world coal reserves are coking coals). Less coke will be required, so less coking coal will be required. However, as the ore burden load per unit coke will increase high quality coke will be required. Thus the demand of prime coking coal will be at least sustained.

3. Smelting reduction processes

Smelting reduction processes is the generic term used for all new process concepts that aim to compete with the blast furnace route [5 - 8]. In general these processes make use of two separate reactors to replace the blast furnace. In the first reactor iron ore is prereduced and preheated with process gas. Prereduction is performed at temperatures suitable for direct reduction, i.e. 1073-1173 K. In the second reactor the iron ore is melted and final reduction takes place. The energy required for melting and final reduction, and the gas for prereduction are generated by partial oxidation of coal. Some processes also use electrical energy in the melting stage. A range of process technologies is being used for the separate stages.

3.1 PREREDUCTION

Prereduction is performed in packed bed shaft furnaces and fluidized bed reactors.

3.1.1 *Shaft furnace.* The advantages of the shaft furnace are its inherent efficiency due to countercurrent operation, and the availibility of proven designs.

The main drawback is its inability to use fine ore directly. As with the blast furnace, a prerequisite for smooth and efficient operation of a shaft furnace is a good gas permeability of the burden. This requires that the burden constituents have a particle size that is sufficiently large. However, the particles should not be too large as the rate of heating and reduction decreases with increasing size. The burden must meet specifications with respect to abrasion strength, crushing strength and swelling behaviour (swelling through lattice rearrangement during reduction may lead to particle disintegration) to minimize the formation of fines. Because of preferential collection in regions where the resistance is high, the presence of fines may lead to channeling. This is a situation in which gas flows at a high rate through individual channels and is not optimally utilized. Analogously to the blast furnace, pellets and sinter can best be used.

3.1.2 *Fluidized bed.* The ability to use fine ore is the principal advantage of a fluidized bed reactor. However, a fluidized bed has a series of disadvantages.

The gas utilization is low. In the bed partially reduced material is continuously mixed with unreduced material. This opposes further reduction of the partially reduced material as with increasing degree of reduction the reaction rates decrease and gas with a high reducing power is required for further reduction. To achieve a more efficient utilization of the gas a beds in series configuration can be used to simulate countercurrent operation.

Furthermore, the tendency of reduced iron ore particles to sinter is extremely troublesome [9]. As soon as the sintering process has started in any given location, it quickly spreads through the entire fluidized bed, eventually resulting in a large porous body that attaches itself to the bottom of the reactor. Sticking through sintering can be suppressed by incorporating inert material in the bed, such as char and limestone. Sticking can also be suppressed by not fully reducing the ore as with decreasing degree of reduction sticking properties become less unfavourable. Another possibility to avoid sticking is to reduce the ore at a lower temperature.

Finally, the technology for hot discharge from the bed and injection into the melting unit is a relatively unexplored area and requires intensive novel engineering development.

3.2 MELTING AND FINAL REDUCTION

Melting and final reduction is being performed in coal bed melter gasifiers, in coke bed melter gasifiers, and in iron bath melter gasifiers.

3.2.1. *Coal bed melter gasifier.* Examples of processes which use this type of unit are the Corex process [10] of Voest Alpine (Germany) and the Kawasaki XR process of the Kawasaki Steel Corp. (Japan). Lump coal is fed into the top of the reactor where it is dried and devolatilized by contact with off-gases from the melting process. The resulting char forms a bed which is partially fluidized by oxygen injection via sidewall tuyeres. Because the coal is devolatilized in the upper zone the bed contains char with a high solid carbon content. Consequently, a fairly wide range of coals is applicable.

Prereduced iron ore is gravity fed into the bed and melted by the heat liberated there. A pool of slag and metal forms below the bed which is periodically tapped. There is a degree of countercurrent operation in this reactor. The ascending melter off-gas is cooled by the coal injected at the top of the reactor through heat exchange and the endothermic gasification reactions:

$$CO_2 + C \rightarrow 2\,CO \qquad (2)$$

$$H_2O + C \rightarrow CO + H_2 \qquad (6)$$

In this way part of the excess sensible heat of the gas is converted to chemical energy available for reduction. The temperature of the gas is reduced from about 1773 K in the lower melting zone to around 1273-1473 K in the upper gasification zone. The off-gases contain only a small amount of oxidized species.

3.2.2 *Coke bed melter gasifier.* Processes which use a coke bed melter gasifier aim to reduce significantly the consumption of coke while maintaining as much as possible the use of blast furnace technology. Examples are the Sumitomo SC process [11] of Sumitomo Metal Industries (Japan) and the Plasmasmelt process [12] of SKF Steel (Sweden). Coal and oxygen or coal and plasma are injected through sidewall tuyeres in the lower part of the bed. Fine ore is also injected through tuyeres in the lower part (Plasmasmelt), or prereduced ore is fed to the top of the bed (Sumitomo). The coke bed reactor is a countercurrent reactor with efficient heat and mass transfer, including melting in the lower zone and gasification in the upper zone. The off-gases from this unit contain little oxidized species and particulate matter. The disadvantage of the reactor is that it still requires coke although less coke is needed and low grade coke can be used.

3.2.3 *Iron bath reactor.* A large number of smelting reduction concepts are based on liquid iron bath technology. Coal and oxygen are injected via top lances or via tuyeres in the bottom of the vessel. This induces mixing which facilitate rapid assimilation of the coal and the prereduced iron ore. Final reduction of the ore takes place in the molten bath. The reduction proceeds at high rates as a result of high heat and mass transfer rates in the bath.

When the coal particles are blown into the liquid iron they are rapidly heated and degassed, and subsequently dissolved. A part of the carbon is combusted to generate the heat required for melting and final reduction. In the melting zone the gas generated by combustion and reduction will be at or near equilibrium with the molten metal which in practice will have a temperature of about 1773 K. Consequently, the gases will be present as carbon monoxide and hydrogen which means that only a small portion of the energy is released. To minimize the coal consumption high calorific value coals should be used.

Several process concepts include post-combustion of carbon monoxide and hydrogen above the bath. It is the intent to transfer the heat generated back to the slag and liquid iron thereby reducing the amount of coal needed. Examples of processes that use post-combustion are the HIsmelt process [13] of CRA (Australia) and the NSC process [14] of the Nippon Steel Corp. (Japan). Examples of processes that do not use post-combustion are the Converted Blast Furnace (CBF) process [15] of British Steel and Hoogovens (UK/ The Netherlands) and the Coin process [16] of Krupp (Germany)

Post-combustion can significantly reduce the coal consumption [17]. However, the actual effectiveness of post-combustion depends on the heat transfer coefficient which is the ratio of the heat transferred to the bath and the heat generated by post combustion. With increasing degree of post-combustion ever more energy is released and if the heat transfer to slag and metal is too low, the gas temperature gets too high and the danger exists that the top of the reactor burns out. In addition, post-combustion reduces the reduction power of the gas and as a result lowers the rates in the prereduction reactor. The degree of post-combustion should therefore be carefully tuned.

An important factor with respect to heat transfer from the post-combusted gas to the bath is the slag as it serves as the main heat transfer medium. Because of the large gas volumes that are generated slag foaming is significant and may be a limiting factor in the rate of production. Control of slag foaming [18] and optimal use of the slag for heat transfer still require a lot of research. There are several other critical elements in the operation of an iron bath reactor. Refractory wear, especially at the slag line, is seen as a major problem. As there is no coal or coke bed, control of the carbon content of the iron is more difficult. Due to the high gas flows significant amounts of dust are entrained which may cause problems elsewhere in the process.

3.3 NEW AND EMERGING PROCESSES

There are probably around two dozen of smelting reduction processes. Up to this day, the Corex process is the only one that has reached the stage of commercial ironmaking [19]. The Plasmasmelt technology has been well developed but is used mainly for non-ferrous metals and ferro-alloys. A process that has reached the stage of full scale demonstration is the HIsmelt process. Other important smelting reduction developments are:

- the proposal of Hoogovens and British Steel to develop a 'Converted Blast Furnace',
- the development efforts of the American Iron and Steel Institute (AISI) with cofunding from the Department of Energy (DOE) to develop a direct steelmaking route [20],
- German developments (e.g., Coin process, MIP process).
- and Japanese developments with several processes (e.g., Kawasaki, Sumitomo, NSC, Kobe)

Processes that use an electric arc furnace for melting are not considered here. As many of the processes have similar characteristics only a few fairly different processes are treated in somewhat more detail.

3.3.1. *The Corex process.* The process is developed by Deutsche Voest-Alpine Industrieanlagebau GmbH (DVAI), formerly Korf Engineering, in cooperation with Voest-Alpina of Austria. A schematic diagram of the process is shown in figure 6. The process consists of a shaft furnace direct reduction unit and of a coal bed melter gasifier in which the coal bed is partially fluidized. The melter gasifier is a refractory lined vessel with a hemispherical top and cylindrical melting zone and hearth.The process is designed to operate under elevated pressure, up to 5 atm.

Reducing gas is generated in the fluidized bed melter gasifier by partial oxidation of coal. The gas mainly consists of carbon monoxide and hydrogen with only a small amount of carbon dioxide. After leaving the melter gasifier the gas is mixed with cooling gas to obtain a temperature suitable for direct reduction, i.e. approximately 1123-1173 K. The gas is then cleaned in a hot cyclone and fed into the shaft furnace. The fines captured in the hot cyclone are recirculated to the gasifier. In the shaft furnace a highly metallised iron ore (prereduction degree > 90%) is produced which, having a temperature of 1073-1173 K, is continuously charged into the melter gasifier. In the melter gasifier melting occurs and reduction is completed. Hot metal and slag are discharged analogous to blast furnace practice.

Originally, the process has been designed with surplus gas production. In this configuration coal use will be around 1,000 kg/tHM. If export gas cannot economically be utilized a modified flow sheet incorporating top gas recycling can be considered. In this configuration the coal use will be in the order of 700-750 kg/tHM. However, the necessary modifications heavily influence process economics.

In 1981, a 60,000 t/y pilot plant at Kehl-am-Rhine in Germany was started up. The plant, which was operated until 1987, has been the main vehicle for the process development through a series of extended trials. In 1985 a Corex plant designed for a production capacity of 300,000 t/y was sold to Iscor (South Africa). After some major start-up problems, the process has been in operation commercially since the end of 1989. A plant with a production capacity of up to 800,000 t/y is now being designed.

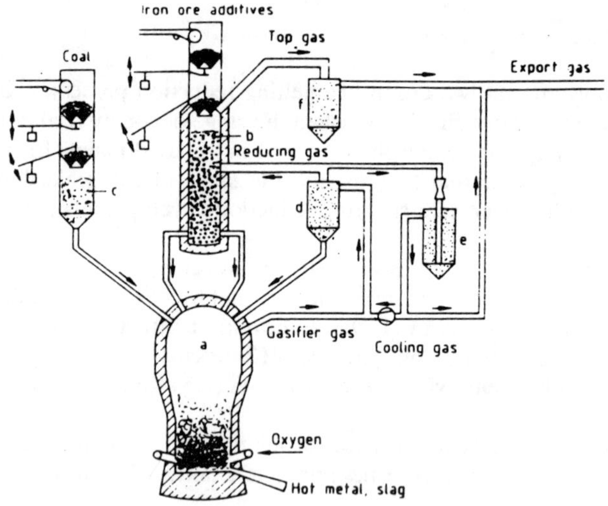

FIGURE 6. The Corex process with export gas: a) Coal bed melter gasifier; b) Reduction shaft furnace; c) Coal feed system; d) Hot dust cyclone; e) Cooling-gas scrubber; f) Top-gas scrubber.

As a result of former cooperation with Klöckner-Werke (Germany), CRA has been further developing the horizontal cylindrical iron bath reactor which was already used by KHD Humboldt Wedag AG (Germany) in their coal gasification process. Compared with the more conventional vertical design, the horizontal design provides a larger gas - melt interface. The result is a reduction of the exit velocity of the gas and, accordingly, of the amount of dust entrained by the off-gas. The reactor is operated at very high post combustion levels, typically 40-50%, and is therefore called the High Intensity smelting process (HIsmelt). The high post-combustion level inhibits the prereduction, but the high temperature in the melter allows the iron ore to be introduced into the iron bath in the FeO state. This is a favourable situation for fine ore reduction as sticking is largely avoided by a low degree of reduction.

Prereduction to the FeO state is performed in a single stage fluidized bed prereduction unit at a temperature of about 1073 K. After prereduction the ore is collected in a hot cyclone and fed into the melter gasifier.

Significant small scale and pilot plant testing and modelling work has been undertaken on aspects of the process and post combustion levels of up to 60% have been achieved, still resulting in stable operating conditions. In combination with its new development partner Midrex Corp. (USA), CRA have just started building a 150,000 t/y demonstration plant in Australia.

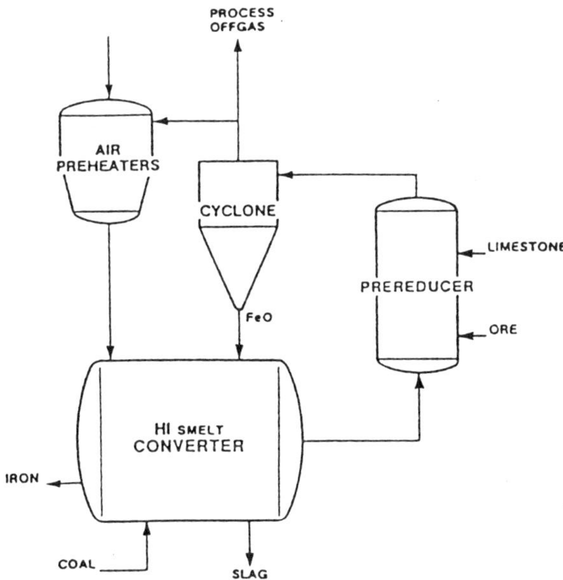

FIGURE 8. The HIsmelt process: a) Air preheater; b) Hot cyclone; c) Fluidized bed type prereducer; e) Horizontal cylindrical iron bath melter gasifier.

3.3.4. *The Converted Blast Furnace process.* The CBF process is a demonstration project in which Hoogovens (the Netherlands) and British Steel have combined their development in the field of coal based ironmaking. The process consists of an iron bath melter gasifier which is coupled with an iron ore prereduction shaft via a gas conditioning link. The process is represented schematically in figure 9.

3.3.2. *The Plasmasmelt process*. The Plasmasmelt process (figure 7) has been under development by SKF Steel Corp. of Sweden since the early 1970's. In this process fine ore is prereduced in a multi-stage fluidized bed so as to establish a quasi countercurrent reactor. Together with pulverized coal and oxygen, the iron ore with a prereduction degree of 50-60% is injected through tuyeres into a shaft furnace using a raceway similar to the blast furnace. Final reduction occurs in the raceway, especially in front of the tuyeres. There the temperature is so high that reduction takes place in the molten state.

The shaft furnace is filled with coke which among other things, ensures a strong reducing atmosphere, protects the refractory, and acts as a filter to remove the dust from the off-gas. Coke and coal usage in the process is about 50-100 kg/tHM and 200 kg/tHM, respectively, and the hot metal produced is similar to blast furnace hot metal.

As originally developed, the Plasmasmelt process uses electricity to supply thermal energy via plasma generators (1100 kWh/tHM). This version can be used where electricity is abundant and cheap. However, there is provision for oxygen injection at the same point, so that the heat source may also be a combination of electricity and combustion or, in principle, exclusively combustion. In both cases the coke usage will be considerably higher. A range of 140-270 kg/tHM is calculated for various process conditions.

In the near future SKF sees the best prospects for application of their technology in areas other than ironmaking. A melter with a capacity of 8 t/h is integrated in a modified process (Plasma Dust) where waste oxides from a ferrochrome plant are processed at Malmö (Sweden).

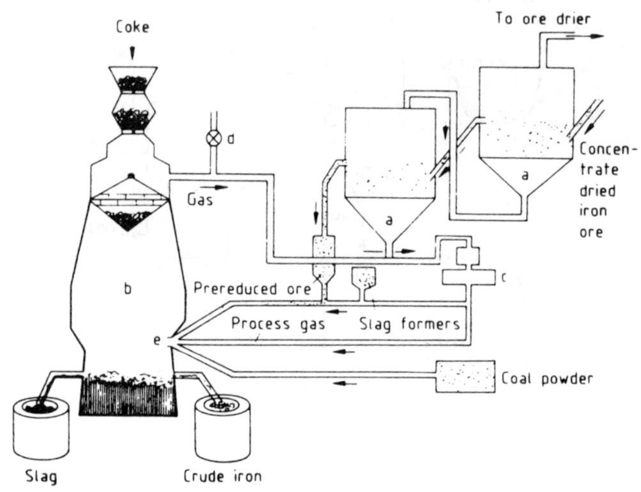

FIGURE 7. The Plasmasmelt process: a) Multi-stage fluidized bed prereduction; b) Shaft furnace/coke bed melter gasifier; c) Compressor; d) Pressure control; e) Plasma generator.

3.3.3. *The HIsmelt process*. The aim of CRA Ltd (Australia) has been to develop an ironmaking process that makes direct use of ore fines and low grade, low priced coal. The process thus comprises an iron bath melter gasifier and a fluidized bed prereduction unit. A simplified flow sheet of the process is given in figure 8.

The prereduced ore is charged hot into the iron bath. Coal and oxygen are fed by means of top lances. A special feature of the process is the twin chambered melter gasifier. In the first chamber melting and final reduction takes place. Metal and slag formed flow into a second chamber from which they are tapped periodically. This arrangement ensures a constant bath depth in the first chamber while also permitting addition of metallurgical reagents, e.g. desulphurents, in the second chamber. The reducing gas generated in the melter-gasifier contains about 10-15% of oxidized species. After cooling to approximately 1150 K and cleaning, the gas is used for prereduction in the shaft furnace. Although the concentration of oxidized species in the gas is significant, a high degree of prereduction (> 90%) is aimed at. With respect to this, the feasibility of the interposition of a free carbon containing reaction zone between the melting and reduction zone is examined [21]. Such a zone, which, in a sense, also is incorporated in e.g., the Corex process, provides a means of reducing the concentration of oxidized species in the off-gas through the endothermic gasification reactions of carbon dioxide and steam with the coal (reactions (2) and (6)), thereby also reducing the temperature of the off-gas.

The developed process flowsheet has been engineered to fit within the frame of an existing blast furnace in order to make use of the available infrastructure, especially with respect to its charging and tapping capabilities.

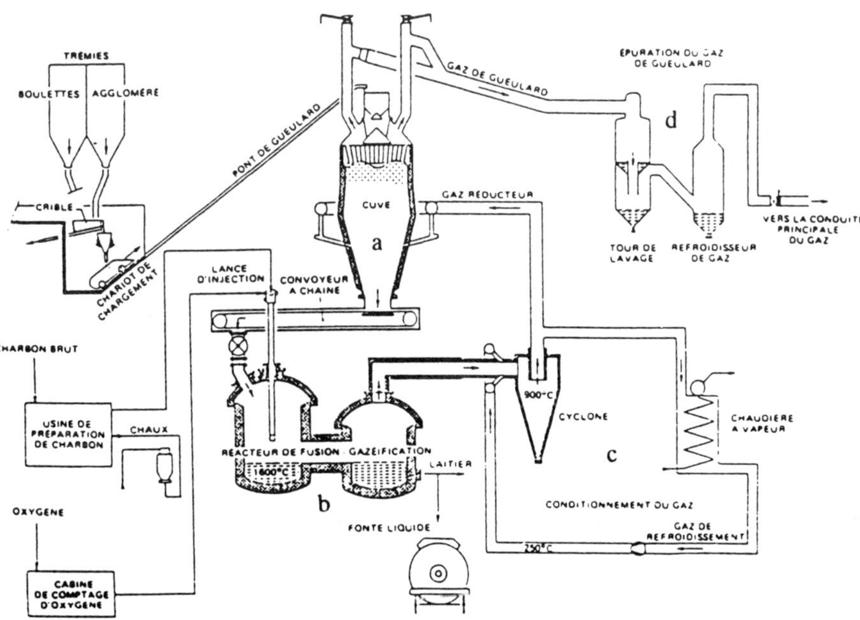

FIGURE 9. The Converted Blast Furnace process with direct cooling of the off-gas: a) Shaft furnace prereduction unit; b) Twin chambered iron bath melter gasifier with top lance injection of coal and oxygen; c) Gas conditioning; d) Top-gas Cleaning.

4. Outlook in the future

Although having significant economic and environmental advantages, large scale introduction of smelting reduction processes is not to be expected in the next 10 to 20 years;

- Many of the processes are still in an early stage of development and much research remains to be done before smooth and efficient operation can be achieved.
- At the moment, there is an overcapacity in the production of iron. With the present development in steel consumption no significant additional capacity will be required until at least the year 2000.
- Coal injection leading to a reduction of the use of coke is economically feasible. Thus, the blast furnace itself also has a significant development potential with respect to reduction of coke consumption and as the large integrated steel works are more or less tied to the use of the blast furnace because of previous investments, there is a preference for this option.

In the near future application of smelting reduction can be expected particularly when small increases in the production capacity of existing plants are required, or as the basis of new mini-steelworks. Incentives for such works are the desire to use local resources or to process the raw materials near their source. New mini-steelworks may also be required to enhance scrap recycling and to improve the quality of scrap based products.

Despite the expected slow start, in the long term, smelting reduction has good prospects. Because of significant advantages with respect to economics and environment, smelting reduction processes will be attractive when new investments are planned to replace or completely rebuild an aged ironworks. Thus it appears that there is a bright future for the development of coke free iron oxide reduction processes.

5. Acknowledgement

The authors wish to acknowledge G.A. Flierman, J.M.M. Droog, M. te Lindert, J.M. Driessen and A.G.S. Steeghs (Hoogovens Groep BV, IJmuiden, The Netherlands) for stimulating discussions.

6. References

1 Holschuh, L.J. (1989) 'The situation in the iron and steel industry', Metallurgical Plant and Technology 4, 8-16.

2 Peters, K.H., Peters, M., Korthas, B., Mülheims, K. and Kreibich, K. (1990) 'Limits of coal injection', Metallurgical Plant and Technology Int. 6, 32-43.

3 Jüntgen, H. (1987) 'Coal Characterization in relation to coal combustion', in Lahaye, J. and Prado, G. (eds.), Fundamentals of the Physical Chemistry of pulverized coal combustion', Martinus Nijhoff Publishers, Dordrecht, pp 4-51.

4 Hunty, W.P. and Price, J.T. (1989) 'Importance of tar contents in selecting coals for blast furnace injection', Ironmak. Steelmak. 16, 165-167.

5 Smith, R.B. and Corbett, M.J. (1987) 'Coal-based ironmaking', Ironmak. Steelmak. 14,

49-75.
6 Steffen, R. (1989) 'Direct reduction and smelting reduction - an overview', Steel Research 60, 96-103.
7 Swanson, A.R., Callcott, T.G., Callcott, R., Brown, K. and Atkinson, B.W. (1989) 'The role for coal in future metallurgical processes', Coal Research Conf. [Proc.], Wellington, New Zealand, 321-332.
8 Astier, J. (1990) 'Smelting reduction processes carve out an ironmaking nice', Steel Times 2, 84-86.
9 Gudenau, H.W., Fang, J., Hirata, T. and Gebel, U. (1989) 'Fluidized bed reduction as the prestep of smelting reduction', Steel Research 60, 138-144.
10 Hauk, R. and Flickenschild, J. (1986) 'KR-process; hot metal production on the basis of coals', PTD Proceedings, ISS-AIME 6, 1031-1039.
11 Hatano, M., Miyazaki, T., Yamaoka, H. and Kamei, Y. (1986) 'New ironmaking process by use of pulverized coal and oxygen', PTD Proceedings, ISS-AIME 6, 1049-1055.
12 Edström, J.O. and Ma, J. (1989) 'Influence of the type of coal on energy consumption in shaft furnace smelting reduction process', Scand. J. Metallurgy 18, 105-112.
13 Brotzmann, K. (1989) 'Smelting reduction in iron baths', Steel Research 60, 110-112.
14 Hayashi, Y., Nakamura, M. and Tokumitsu, N. (1986) 'Development of smelting reduction process using an iron bath', PTD Proceedings, ISS-AIME 6, 1057-1064.
15 van Langen, J.M. and Smith, R.B. (1989) 'The Converted Blast Furnace process', La Revue de Metallurgie-CIT, 759-764.
16 Neuschütz, D. and Hoster, T. (1989) 'Concept and present state of a coal-based smelting reduction process for iron ore fines', Steel Research 60, 113-119.
17 Fruehan, R.J., Ito, K. and Ozturk, B. (1989) 'Analysis of bath smelting processes for producing iron', Steel Research 60, 129-137.
18 Ito, K. and Fruehan, R.J. (1989) 'Slag foaming in smelting reduction processes', Steel Research 60, 151-156.
19 Steffen, R. (1990) 'The Corex-process: First operating results in hot metal making', Metallurgical Plant and Technology 2, 20-23.
20 Aukrust, E. (1990) 'AISI direct steelmaking program', Iron Steel Eng. 5, 125-128.
21 Weeda, M., Tromp, P.J.J. and Moulijn, J.A. (1990) 'The potential of coal gasification in a novel iron oxide reduction process', Chemical Engineering Science 45, 2721-2728.

HYDROGEN ENERGY SYSTEM AND HYDROGEN PRODUCTION METHODS

F. Barbir and T.N. Veziroğlu
Clean Energy Research Institute
University of Miami
Coral Gables, FL 33124
U.S.A.

ABSTRACT. Hydrogen is being considered as a synthetic fuel or energy carrier for the post-fossil fuel era. It has many properties to commend itself: it complements the renewable energy sources and presents them to the consumer in a convenient form at the desired location and time; it can be converted to various energy forms at the consumer end with higher efficiencies than other fuels; it is renewable; and it is the lightest and cleanest fuel. It can be used in every application where fossil fuels are used today. It can be produced from fossil fuels or from water using renewable energy sources. This paper contains an overview of the hydrogen production methods, those being commercially available today as well as those which are being developed.

1. INTRODUCTION

At the present time, about 80 percent of the world energy demand is met by fossil fuels - coal, petroleum and natural gas [1]. They are convenient to use and humankind has learned to exploit them relatively efficiently to produce the energy services it needs.

However, consumption of fossil fuels has become a destructive force, locally because of emissions, spills and leaks, and strip mining, regionally because of pollutants dispersion and acid rains, and even globally because of carbon dioxide accumulation and threatening consequences - global warming, climate changes and sea level rise [2].

Reserves of fossil fuels are finite, and sooner or latter we will run out of them. Known reserves of oil and natural gas are about 8,000 EJ (1 EJ = 10^{18} J). Coal reserves are much larger; known reserves are about 20,000 EJ, but estimated ultimately recoverable resources add up to 150,000 EJ [3]. Present rate of fossil fuel consumption is about 300 EJ/yr, and is increasing steadily.

Another problem with fossil fuels is that they are not distributed evenly among the countries of the world. Unequal distribution, together with the addiction to petroleum, and in a lesser extent to natural gas, is causing international problems between the suppliers of the fluid

Y. Yürüm (ed.), Clean Utilization of Coal, 277–293.

fossil fuels (i.e., petroleum and natural gas) and the consuming nations, and an increasingly intense rivalry among the super powers in their attempts to safeguard their energy supplies.

In order to make up for the shortcomings of fossil fuels and meet the energy demand, several alternative and nonconventional primary energy sources are being considered and researched by scientists, such as direct solar radiation, wind energy, currents, waves, tides, ocean thermal energy, geothermal energy, nuclear breeders and fusion reactors. It may be necessary to make use of several of these primary energy sources, depending on the economics and availability in different regions and through different time periods.

The nonconventional energy sources being considered do not possess all the advantages of fossil fuels, although some of them, such as solar, wind, currents, tides, waves and ocean thermal, are almost unlimited and are environmentally compatible. Some are only intermittently available. For example, solar energy is only available in the daytime when the skies are clear. Even then, the intensity of solar radiation is subject to diurnal and seasonal changes. Hence, solar energy needs to be stored to meet the demand when solar radiation is not available. Some of the new energy sources are continuously available, but they are too far away from the consumption centers. None of the new energy sources mentioned can be used as fuel for transportation because they are not transportable or storable by themselves.

The shortcomings of the non-conventional energy sources point to the need for an intermediary energy system to form the link between the new primary energy sources and the energy consuming sectors or the user.

Hydrogen -- with its vast and ready availability from water, its nearly universal utility, and its inherently benign environmental characteristics -- has all the properties that might be required for such an energy system.

2. HYDROGEN ENERGY SYSTEM

In the hydrogen energy system (see Fig. 1), hydrogen will be produced from water using any and all primary energy sources, and will be used in every application where fossil fuels are used today. In such a system hydrogen is not a primary source of energy. It is an intermediary or secondary form of energy or an energy carrier. Hydrogen complements the renewable energy sources, and presents them to the consumer in a convenient form at the desired location and time. It is relatively inexpensive to produce; it can be converted to other forms of energy more efficiently than other fuels; and it is the cleanest fuel because its combustion product is just water vapor [4].

The hydrogen energy system is a permanent energy system, because it is independent of the primary energy sources, it is inexhaustible and recyclable, and it is versatile. Even if the primary energy sources and the consumption patterns change due to technology development and due to changes in lifestyle, the hydrogen energy system as an intermediary link remains basically intact.

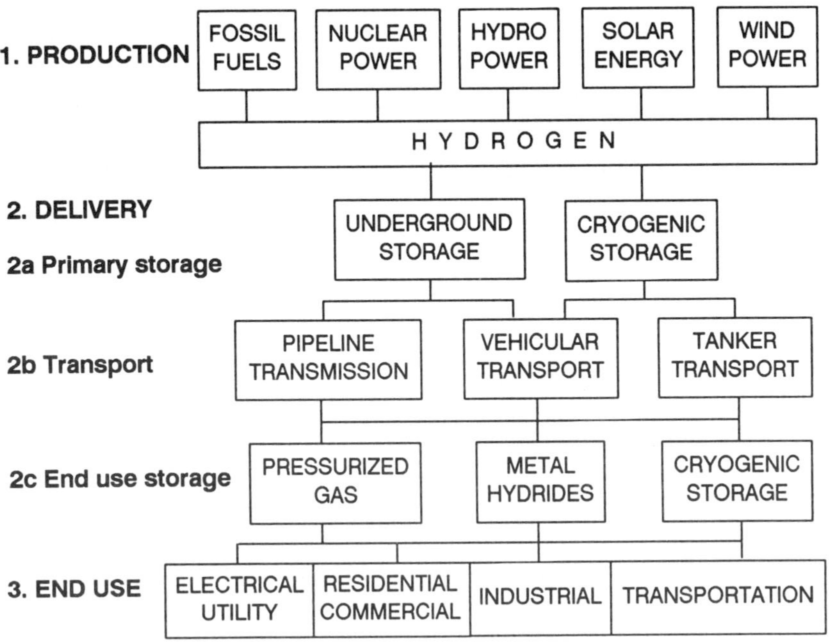

Figure 1. Hydrogen Energy System

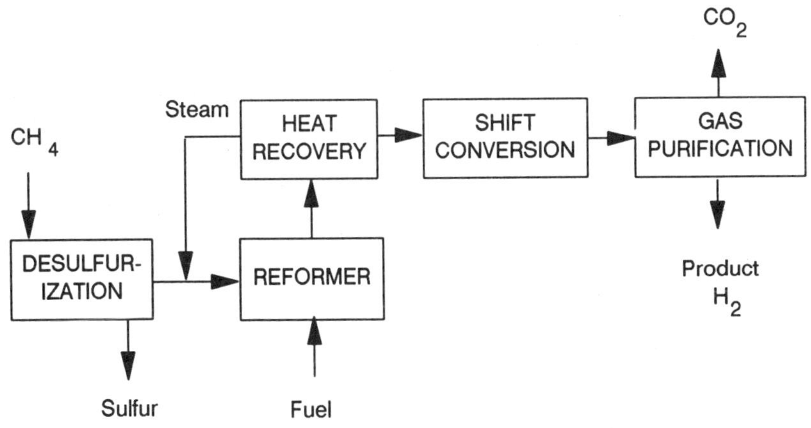

Figure 2. Natural Gas Steam Reforming Process

Since the energy crisis of 1973, the research on the hydrogen energy system has increased and the results have been presented in the proceedings of The Hydrogen Economy Miami Energy (THEME) Conference [5], in those of the eight World Hydrogen Energy Conferences [6-13] held to date, and in the International Journal of Hydrogen Energy [14]. Books and reports by Bockris [15], Veziroglu [16], Ohta [17] and Winter and Nitsch [18] cover hydrogen production methods and utilization in some depth.

Hydrogen is a synthetic fuel that must first be produced. Water is the most abundant available source of hydrogen. A variety of hydrogen production methods using solar energy (and its derivatives) have been or are currently being developed. They are: direct thermal decomposition (thermolysis), thermochemical cycles, electrolysis, and photolysis which includes photo-electrolysis, catalytic photolysis and bio-photolysis. In the transition period hydrogen can be produced from available fossil fuels by methane reforming and/or by coal gasification.

In order to accommodate daily and seasonal discrepancies between energy source availability and demand, hydrogen must be stored. It can be stored as a pressurized gas in underground caverns (aquifers, salt caverns, depleted oil wells, etc.), in above-ground large pressure vessels, or as a liquid in cryogenic reservoirs. From the production plants hydrogen is transported (by means of pipelines or supertankers) to energy consumption centers, where it is used in every application in which fossil fuels are being used today, except where carbon is specifically needed. It can be used as a fuel for air and surface transportation, to produce heat through flame or catalytic combustion, or it can be converted directly to electricity in fuel cells.

3. HYDROGEN PRODUCTION METHODS

Presently, hydrogen is mostly being produced from fossil fuels (i.e., fossil fuels provide the source of hydrogen and energy required to produce hydrogen), but it is rather used as a feedstock not as an energy carrier (except for the space program). However, in the future energy system where hydrogen will be an energy carrier, it would make no sense to produce hydrogen from fossil fuels, except during the transition period until a solar based hydrogen system is established. In the solar hydrogen system, hydrogen will be produced from water, using renewable energy sources (i.e., direct and indirect solar energy forms). Following is an overview of hydrogen production methods from fossil fuels and from renewable energy sources.

3.1. Hydrogen Production from Fossil Fuels

3.1.1. Steam Reforming of Natural Gas

Steam reforming of hydrocarbons (mainly natural gas) has been the most efficient, economical and widely used process for hydrogen production. A simplified basic flow diagram of the conventional steam reforming process is shown in Fig. 2 [19]. The process basically consists of three main steps: (1) synthesis gas generation, (2) water-gas shift, and (3) gas purification. To protect the catalysts, natural gas has to be desulphurized before being fed to the reformer. The desulphurized feedstock is then mixed with process steam and reacted over a nickel based catalyst contained inside of a system of alloyed steel tubes. The following reactions take place in the reformer:

$$CH_4 + H_2O \longrightarrow CO + 3H_2$$

$$CO + H_2O \longrightarrow CO_2 + H_2$$

$$CO + 3H_2 \longrightarrow CH_4 + H_2O$$

The reforming reaction is strongly endothermic, with energy supplied by combustion of natural gas or fuel oil. The metallurgy of the tubes usually limits the reaction temperature to 700-925 ^oC. A synthesis gas leaving a catalytic reformer is typically a mixture of H_2, CO, CO_2 and CH_4. After the reformer, the gas mixture passes through a heat recovery step cooling down to about 350 ^oC and is fed into water-gas shift reactor to produce additional H_2. The cold gas next passes through gas purification units to remove CO_2, the remaining CO and other impurities to deliver purified hydrogen. Several commercial processes can be used for removing CO_2, such as wet scrubbing and pressure swing adsorption.

Efficiency of the steam reforming process (expressed as the ratio of the heating value of produced hydrogen and energy input (feedstock, fuel and small amount of electricity) is about 65%-75%. The cost of produced hydrogen is about $6/GJ, but is dependent upon the cost and availability of natural gas or other fuel feedstock [19].

3.1.2. Partial Oxidation of Heavy Oil

Partial oxidation is used for converting hydrocarbons heavier than naphtha (for which steam reforming is not applicable) to hydrogen. A simplified basic flow diagram is shown in Fig. 3 [19]. There are three main steps: (1) synthesis gas generation, (2) water-gas shift reaction, and (3) gas purification. The shift reduction and gas purification (either by the conventional wet scrubbing or by the pressure swing adsorption process) are similar to those of steam reforming.

In the synthesis gas generation step, the hydrocarbon feedstock is partially oxidized with oxygen, and the carbon monoxide is shifted with steam to produce CO_2 and H_2. Because of the difficulties of separating nitrogen to produce pure hydrogen, pure oxygen is used in this process. The partial oxidation reactions are typically as follows:

$$C_nH_m + n/2\ O_2 \longrightarrow nCO + m/2\ H_2 + heat$$

$$C_nH_m + nH_2O + heat \longrightarrow nCO + (n + m/2)H_2$$

$$CO + H_2O \longrightarrow CO_2 + H_2 + heat$$

where n=1 and m=1.3 for residual fuel oils.

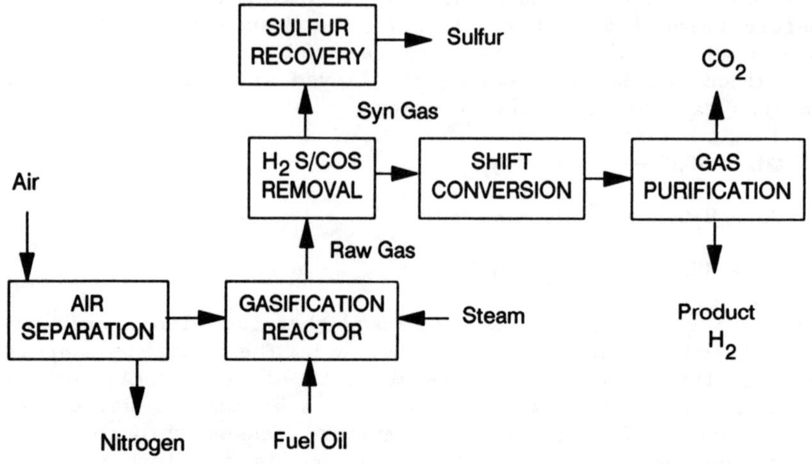

Figure 3. Heavy Oil Partial Oxidation Process

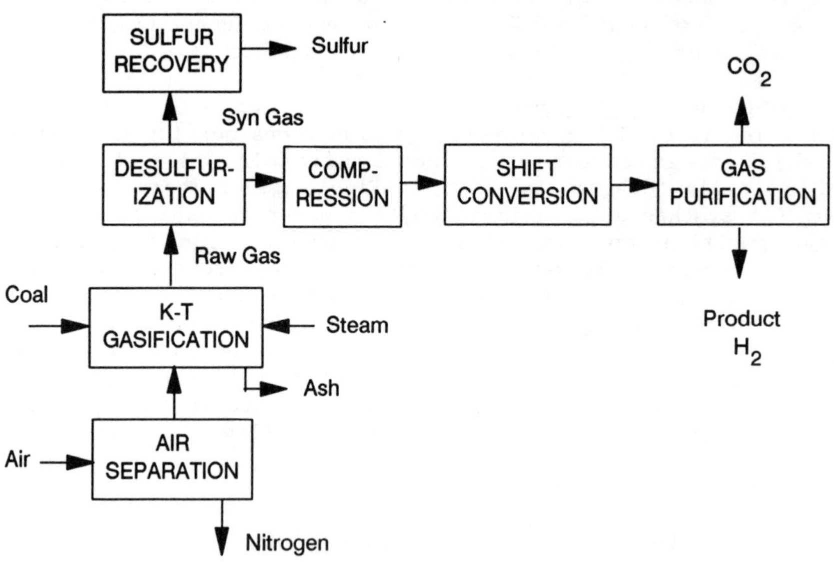

Figure 4. Koppers-Totzek Coal Gasification Process

While the hydrocarbon-oxygen reaction is exothermic, the additional energy required for the endothermic hydrocarbon-steam reaction, which is similar to steam reforming, is usually supplied by burning additional fuel. Operating temperatures are rather high, 1150° to 1315°C. The product stream (a very hot mixture of hydrogen, oxygen, CO, CO_2, steam, and a small amount of methane with entrained molten ash or slag) is typically quenched with water as it exits the reactor to solidify the slag for downstream removal with cyclones, gas filtration or in a slurry. The sulphur contained in the feedstock is converted mainly into H_2S and only a small portion to COS. After sulphur removal, the gas is treated in the same way as the product gas of the steam reforming process.

The efficiency of partial oxidation is generally lower than for steam reforming, usually about 50%. Hydrogen can be produced for approximately $10/GJ if cheap oil is available (less than $4/GJ) [19].

3.1.3. Coal Gasification

In the coal gasification process (so called Koppers-Totzek process), the pulverized coal is rapidly partially oxidized with oxygen and steam at essentially atmospheric pressure under slagging conditions. The raw gas is then cooled to recover waste heat, followed by quenching with water to remove entrained ash particles before going through the steps of compression, shift and gas purification (again by conventional wet scrubbing or pressure swing adsorption processes). The product is hydrogen at about 2.8 MPa with purity higher than 97.5%. The simplified block flow diagram for the process is shown in Fig. 4 [19].

Since hydrogen compression is highly energy consuming, and most of the hydrogen users need hydrogen at higher pressures, it is important to gasify coal at elevated pressures. In that case, compression of the synthesis gas is not necessary. The Texaco has developed a coal gasification process which operates at high pressures, around 5.5 MPa. By operating in the direct quench mode at this pressure, a high steam content in the synthesis gas is desirable for use in the shift reaction for hydrogen production. The raw gas is then desulphurized, shifted, and purified. The product hydrogen is at about 4 MPa with a purity higher than 97%.

The coal gasification process is complicated because of the necessity to handle solid fuels and to remove large amounts of ash. The solids handling problem has a significant impact on costs. Generally, the lower cost of coal does not compensate for the higher capital cost of coal gasification systems, compared to steam reforming, and the costs of produced hydrogen are $12-$14/GJ [19].

3.1.4. Steam-Iron Process

Although it is a coal-based process, hydrogen is derived from decomposition of steam by reacting with iron oxide, rather than synthesis gas generated from coal, i.e., the synthesis gas is mainly used for the reduction of iron oxide to iron. Since hydrogen is derived from the synthesis gas, air can be used in the gasifier. A basic block flow diagram for the process is presented in Fig. 5 [19]. It consists of four main steps: (1) coal gasification, (2) iron regeneration, (3) hydrogen gener-

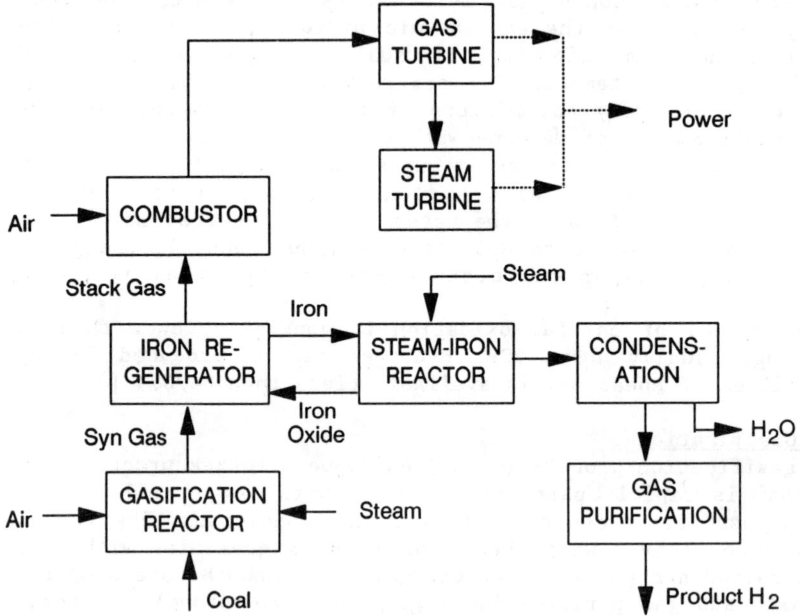

Figure 5. Steam-Iron Process

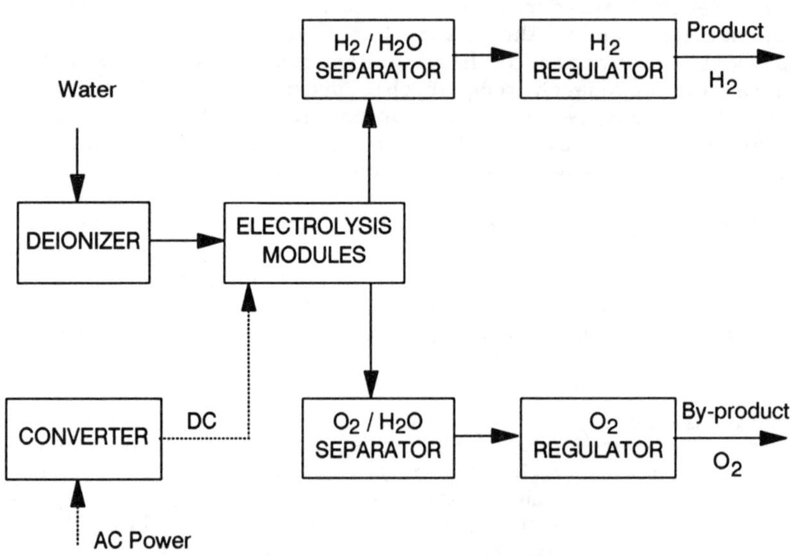

Figure 6. Water Electrolysis Process

ation, and (4) purification. Coal is gasified with steam and air to produce a synthesis gas. In the iron regenerator, the synthesis gas is mixed with iron oxide and the following reduction reactions occur:

$$Fe_3O_4 + H_2 \longrightarrow 3FeO + H_2O$$

$$Fe_3O_4 + CO \longrightarrow 3FeO + CO_2$$

$$FeO + H_2 \longrightarrow Fe + H_2O$$

$$FeO + CO \longrightarrow Fe + CO_2$$

The synthesis gas, actually CO and H_2, are not completely converted in reducing iron oxide. Most of the remaining heating value plus sensible heat at 825°C in the gas exiting from the iron regenerator can be used to generate power with a gas-steam combined cycle turbine generator. In a sense, this can be considered as a cogeneration plant producing hydrogen and electrical power.

The regenerated iron then enters the steam-iron reactor where it is oxidized by steam to produce Fe_3O_4 and a hydrogen-rich gas. At temperatures of of about 815°-870°C, the following reaction occurs:

$$3FeO + H_2O \longrightarrow Fe_3O_4 + H_2$$

The effluent from the steam-iron reactor contains 37% hydrogen and 61% steam (plus small amounts of nitrogen and carbon oxides). By condensing out the steam, the hydrogen concentration increases to about 96% with 1.6% carbon oxides and 2.4% nitrogen. A methanation reaction, which follows, reduces the carbon oxides to 0.2%. There is no need for CO shift or acid gas scrubbing.

The costs of produced hydrogen are similar to those of coal gasification process, i.e., about $12/GJ. However, it should be noted that the economics of this process relies heavily on by-product credit for steam and/or electric power, which amounts to 25% of the gross production costs [19].

3.2. Water Electrolysis

Production of hydrogen by water electrolysis is a fundamentally simple process that involves no moving parts and is a technology that is more than 50 years old. It is the only industrial hydrogen process which does not necessarily rely on fossil energy. The electrolytic process uses electrical energy to split water into its basic elements - hydrogen and oxygen. Required electrical energy can be derived either from fossil or from nuclear or from any renewable energy source (hydro, solar-thermal, solar-photovoltaic, OTEC, wind, etc.). Although presently not economically competitive with hydrogen production from hydrocarbons, water electrolysis has several advantages, such as high product purity and flexibility of operation. An electrolysis plant can inherently operate over a wide range of capacity factors and is convenient for wide range of operating capacities, which makes this method particularly interesting for coupling with renewable energy sources for the post-fossil fuel era.

Water electrolysis involves basically splitting water into hydrogen and oxygen by passing a direct electrical current through water that has been made electrically conducting by addition of hydroxyl ions (e.g., alkaline potassium hydroxide, KOH). The electrochemical reaction at the cathode, where hydrogen is produced, is [15,17]:

$$2H_2O + 2e^- \longrightarrow 2OH^- + H_2$$

At the anode oxygen is generated by following reaction:

$$2OH^- \longrightarrow H_2O + 0.5O_2 + 2e^-$$

The minimum amount of electrical energy required is determined by the free enthalpy of the overall reaction ($H_2O \longrightarrow H_2 + 0.5O_2$), which at 25°C and 1 bar is ΔG^o=237 kJ/mol, which corresponds to 1.23 V. Thermoneutral potential, at which heat is neither being taken from the surroundings nor being rejected, is 1.47 V. However, the cell voltages that are achievable in practice with the present commercial electrolyzer technology are considerably higher, about 1.7 to 1.8 V.

The simplified block flow diagram of the water electrolysis plant is shown in Fig. 6 [19]. The plant consists of a water purification unit, electrolyzer section, gas separation, power transformer, regulator and rectifiers, compressor system, cooling water system and inert gas system. The most of commercially available electrolyzers are "bipolar" type, in which the individual cells are linked electrically and geometrically (in the filter press configuration) in series. Thus each metal wall separating one cell from another, functions in bipolar fashion - it supports a cathode on one side and an anode on the other.

The current state of the art offers very reliable electrolyzers which have an energy efficiency (based on hydrogen enthalpy) of up to 80%. However, today's commercial electrolyzers were not designed or built to supply hydrogen in large quantities to be used as an energy vector. They are simple multiples of units stacked in parallel to match the target production rates. No serious engineering analysis and development effort has been practiced for proper scale-up and design of these electrolyzers [20]. Several advanced electrolyzer technologies are already available today such as:

(a) Advanced alkaline electrolysis, which employs new materials for membranes and electrodes that allow further improvement in efficiency - up to 90% [21,22].

(b) Solid polymer electrolytic (SPE) process, which employs a proton-conducting ion exchange membrane as electrolyte and as membrane that separates the electrolysis cell. The water to be dissociated does not require dissolved electrolytes to increase its conductivity, and is added solely to the anode side [22,23].

(c) High temperature steam electrolysis which, operates between 700° and 1000°C and which employs oxygen ion-conducting ceramics as electrolyte. The water to be dissociated is entered on the cathode side as steam which forms a steam-hydrogen mixture during electrolytic dissociation. The O^{2-} ions are transported through the ceramic material to the anode where they are discharged as oxygen [24].

At present, steam electrolysis is still relatively far from technical maturity whereas advanced alkaline and membrane electrolyzers are offered commercially by a considerable and rising number of manufacturers.

The cost of electricity amounts to 80% of the total production cost of hydrogen. For a quick estimate of electrolytic hydrogen cost, following formula can be used [25]:

$$Ch = 350\ Ce + 3.5$$

where Ch is hydrogen cost in $/GJ, and Ce is electricity cost in $/kWh. Unless low-cost electricity is available (<$0.02/kWh), for example from off peak hydropower, electrolytic hydrogen cannot compete with fossil-based technologies.

3.4. Hydrogen Production from Renewable Energy Sources

There are several hydrogen productions technologies based on solar energy, which are being developed. They include thermolysis, thermochemical cycles, photolysis and biological hydrogen production. They are all candidates for hydrogen generation in the post-fossil fuel era. However, they are not commercially available today, and few or no data exist on their economics.

3.4.1. Thermolysis

Water can be split thermally with very high temperatures (T>2,000 K). The overall reaction of thermal dissociation of water can be represented as [17,26]:

$$H_2O \longrightarrow x_1H_2O + x_2OH + x_3H + x_4O + x_5H_2 + x_6O_2$$

The product is therefore a mixture of different gases. The equilibrium composition depends on pressure and temperature applied. For example at 2500 K the percentage of water decomposition is only 8.5% and at 3000 K the percentage of water decomposition rises to 34.3% [27].

The main problems in connection with the direct thermal decomposition of water are related to materials required for extremely high temperatures and separation of hydrogen and oxygen at high temperatures.

3.4.2. Thermochemical cycles

Thermochemical production of hydrogen involves the chemical splitting of water at temperatures lower than those needed for thermolysis, through a series of cyclical chemical reaction which ultimately release hydrogen.

Temperatures up to 1,100-1,200 K are regarded as the upper limit for executing thermochemical processes, in part because of the material engineering problems, but also because process heat is available at these temperature ranges from solar power plants' receivers and also from the high-temperature nuclear reactors [18]. Under these circumstances at least three process steps (or three chemical reactions) are required. However, model computations show that the theoretical efficiency of a thermochemical cycle decreases with the increasing number of reaction steps. Since the mid-1960s, research has been performed to investigate a number of potential thermochemical cycles for the production of hydrogen, and some 2,000 to 3,000 cycles have been invented. These research efforts have ranged from pure theoretical studies of potential reaction sequences, to the investigation of specific chemical reactions, and to the fabrication and testing of laboratory bench-scale pilot plants. After examining their practicability in terms of reaction and process technology, only 20 to 30 remained applicable for the large scale hydrogen production. Some of the more thoroughly investigated thermochemical process cycles are listed hereinbelow [18,28,29]:

- Sulfuric acid - iodine cycle,
- Hybrid sulfuric acid cycle,
- Hybrid sulfuric acid - hydrogen bromide cycle,
- Calcium bromide - iron oxide cycle (UT-3),
- Iron chlorine cycle (Mark 9).

Solar energy conversion technologies capable of producing high temperature heat are heliostats with solar tower, parabolic troughs, and parabolic dishes. Thermochemical reactions require a continuous heat supply. Intermittent operation would result in tremendous heat losses associated with need to heat the whole system including the reactants to the working temperature at the beginning of each working cycle. Unfortunately, solar energy is intermittently available and to make it applicable as a constant heat source for the thermochemical processes, heat storage and appropriate interface are required.

Efficiencies of thermochemical hydrogen production are different for different processes, but the average efficiency is about 40%-50%. When the efficiencies of solar to thermal energy conversion and of thermal energy storage are taken into account, the overall efficiency is still higher than that for PV-electrolysis system [30]. Although more efficient, thermochemical methods are not likely to become feasible because of the following reasons [20,30]:

- Movement of large mass of materials is required in the chemical reactions to produce relatively small amounts of hydrogen.
- Numerous chemicals are involved in the thermochemical processes which may not be 100% recyclable, and some of them can be toxic and environmentally unacceptable for large scale applications.
- High temperature thermal energy storage and appropriate interface installation are required, which would be very difficult to design and to operate with current technologies.
- Special construction materials are required because of the high temperatures and corrosive chemicals used in the processes.

3.4.3. Photolysis

Photolytic production of hydrogen involves direct splitting of water by solar energy. The energy required to split the hydrogen and oxygen bond in water is 1.229 eV [31]. Thus a single photon would require at least 2.458 eV to cause an oxidation reduction to occur to split the water molecule. A two-photon process would require at least 1.229 eV per each photon to accomplish the same purpose. The ideal limit for the efficiency of a one photon system is only 5.3%, but for a two photon system is 30.7%. The difference is principally the portion of the solar spectrum these systems can use.

Photolysis, or photon conversion, may be carried out through photobiological, photochemical or photoelectrochemical processes.

Photobiological systems are photosynthetic organisms that use sunlight to convert water or carbon compounds into hydrogen. The organisms are of three types: the green algae, cyanobacteria, and photosynthetic bacteria. Algae and cyanobacteria function like plants, have two photosystems, and can split water to hydrogen, although efficiencies are rather low (about 1-2%). The photosynthetic bacteria have only one photosystem and cannot split water, but they produce hydrogen by acting on carbon substrates such as organic wastes, and much higher efficiencies are possible [31].

Photochemical systems contain synthesized molecules in a homogeneous solution. The molecules are typically porphyrin-quinones that are designed to mimic photosynthesis. Such photochemical systems suffer losses from back reactions and at present have low conversion efficiencies [31].

Photoelectrochemical systems employ semiconductors immersed in aqueous solutions. Solar photons are absorbed in semiconductors and produce electron-hole pairs. These pairs separate at the conductor-liquid junction and drive chemical reactions at the surface. Hydrogen is reduced on one surface of a semiconductor plate and oxygen oxidized at the other surface of the plate in the presence of a catalyst. Because the semiconductors have been well studied for photovoltaic and other applications, considerable progress has been made in improving conversion efficiencies and in the basic understanding of such systems. The major challenges in producing hydrogen economically from photoelectrochemical systems are to reduce photocorrosion of the semiconductor in the electrolyte, increase the photon conversion efficiency and reduce the cost of fabrication of the devices. Current photon conversion efficiencies have reached 13%, and the theoretical limit is about 32%, so there remains substantial opportunity for improvement [31].

3.4.4. Conversion of Biomass

Biomass may offer one of the least expensive near term renewable paths to hydrogen production. Hydrogen production from biomass would involve collecting, transporting and preparing the biomass for pyrolysis/gasification. The process involves preparation of biomass/water slurry which is heated to high temperature under pressure in a reactor. This decomposes and partially oxidizes the biomass, producing a gas stream rich in CO_2, CO, nitrogen, methane and hydrogen. Ash would be removed as a slag from the bottom of the reaction vessel. The gas stream would then

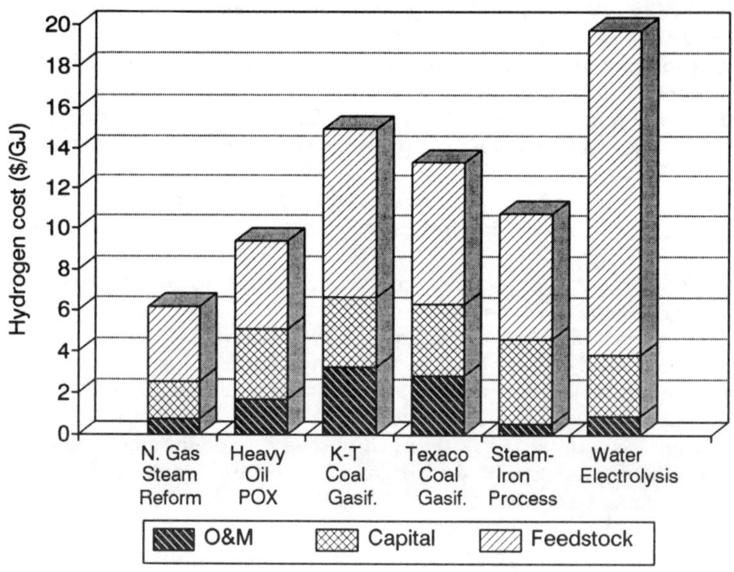

Figure 7. Hydrogen Production Cost for Commercially Available Processes

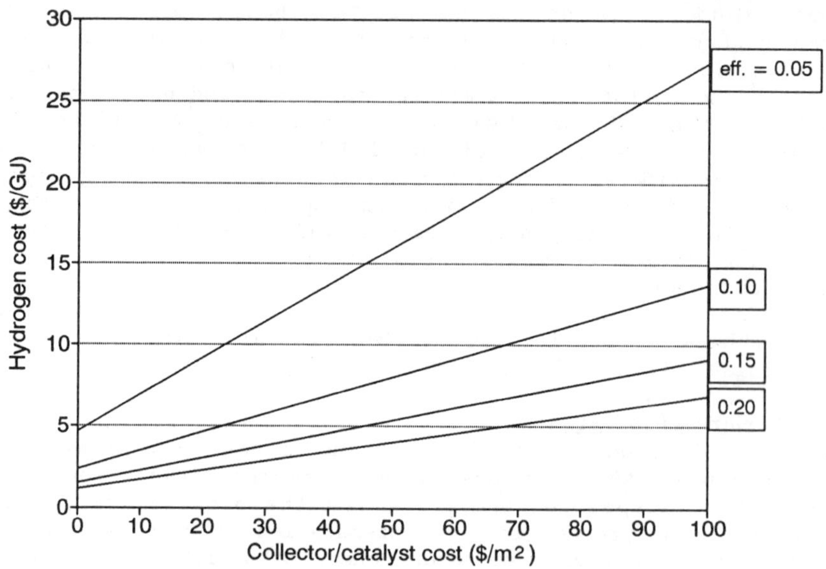

Figure 8. Hydrogen Production Cost as Function of Collector Cost and Efficiency for Processes Using Solar Energy

go to a high temperature shift reactor and a pressure swing adsorption (PSA) unit to produce high purity hydrogen. Except for pretreatment of the biomass, and the design of the reaction vessel, the system is similar to a coal gasification plant. The processing facility would have to be larger than that of comparably sized coal gasification plant because of the lower heat content per unit mass of biomass as compared to coal, which might result in higher capital costs. This system is still in the research phase [20].

4. CONCLUSIONS

Two groups of hydrogen production methods have been described above: (i) already commercially developed methods, which are mainly based on fossil fuels, and can produce hydrogen at \$6-\$20/GJ, and (ii) methods which utilize solar energy and are still being developed. The latter group of methods can be used for large scale hydrogen production in post fossil fuel era. Only water electrolysis is already developed and mature technology, which can be applied in conjunction with renewable energy sources. Costs of commercially available processes are shown in Fig. 7 [19]. It can be noted that water electrolysis has the highest costs. However, the most of the costs are due to very expensive fuel being used in the process - electricity. Therefore, development of the processes which can produce cheap electricity from renewable energy sources will significantly reduce the cost of hydrogen. Electricity can be produced through photovoltaic conversion, solar-thermal conversion, wind energy, OTEC, geothermal, etc. Some of these technologies are already cost competitive with fossil fuels (geothermal and wind at some locations), but for hydrogen production the costs have to be reduced even further. The most promising technology is photovoltaic conversion. It is expected that the cost of photovoltaic cells will continue to decrease, and by the end of the century, it will be possible to produce hydrogen economically through photovoltaic-electrolysis systems.

Electrolysis can be coupled directly to photovoltaic cells since they produce DC which is required in the electrolysis process. Several experimental facilities have already been built and tested. The results indicate a low specific energy consumption (<4 kWh_e/Nm^3), good gas purities, a fast dynamic response, and a wide range of operation [32]. The PV-electrolysis system is therefore the prime candidate for large scale hydrogen production.

Processes which can convert sunlight directly into hydrogen (without intermediary step of conversion in electricity), offer potential for even further cost reductions. However, their efficiency must be improved to reduce areas (and consequently the costs) required to collect sunlight. In addition they must use relatively cheap materials. Fig. 8 shows the relationships between the costs of hydrogen and costs per unit area for different efficiencies.

Solar hydrogen energy system offers a clean, inexhaustible and far more sensible alternative in the long run than the energy system on which we now rely.

5. REFERENCES

1. Davis, G.R., (1990) Energy for Planet Earth, *Scientific American*, Vol. 263, No.3, pp. 55-62
2. Barbir, F., T.N. Veziroglu and H.J. Plass, Jr., (1990) Environmental Damage Caused by Fossil Fuel Consumption, *Int. J. Hydrogen Energy*, Vol. 15, No. 10, pp. 739-749.
3. Fulkerson, W., R.J. Judkins and M.K. Sanghvi, (1990) Energy from Fossil Fuels, *Scientific American*, Vol. 263, No.3, pp. 129-135
4. Bockris, J. O'M., and T.N. Veziroglu, (1985) A Solar-Hydrogen Energy System for Environmental Compatibility, *Environmental Conservation*, Vol. 12, No. 2, pp. 105-118
5. Veziroglu, T.N., (ed.), *(1975) Hydrogen Energy*, Parts A and B, Proc. Hydrogen Economy Miami Energy Conf., Plenum Press, New York
6. Veziroglu, T.N., (ed.), (1976) *Proc. 1st WHEC* (3 vols.), Clean Energy Research Inst., University of Miami, Coral Gables, FL
7. Veziroglu, T.N., and W. Seifritz, (eds.), (1979) *Hydrogen Energy System* (5 vols.), Proc. 2nd WHEC, Pergamon Press, Oxford
8. Veziroglu, T.N., K. Fueki and T. Ohta (eds.), (1981) *Hydrogen Energy Progress* (4 vols.), Proc. 3rd WHEC, Pergamon Press, Oxford
9. Veziroglu, T.N., W.D. Van Vorst and J.H. Kelley (eds.), (1982) *Hydrogen Energy Progress IV* (4 vols.), Proc. 4th WHEC, Pergamon Press, Oxford
10. Veziroglu, T.N., and J.B. Taylor (eds.), (1984) *Hydrogen Energy Progress V* (4 vols.), Proc. 5th WHEC, Pergamon Press, Oxford
11. Veziroglu, T.N., N. Getoff and P. Weinzierl, (eds.), (1986) *Hydrogen Energy Progress VI* (3 vols.), Proc. 6th WHEC, Pergamon Press, Oxford
12. Veziroglu, T.N., and A.N. Protsenko, (eds.), (1988) *Hydrogen Energy Progress* VII (3 vols.), Proc. 7th WHEC, Pergamon Press, Oxford
13. Veziroglu, T.N., and P.K. Takahashi, (eds.), (1990) *Hydrogen Energy Progress* VIII (3 vols.), Proc. 8th WHEC, Pergamon Press, Oxford
14. Veziroglu, T.N., (ed.), (1976-1991) *International Journal of Hydrogen Energy* (16 vols.), Pergamon Press, Oxford
15. Bockris, J. O'M., (1975) *Energy: The Solar-Hydrogen Alternative*, Australia and New Zealand Book Co., Sydney
16. Veziroglu, T.N., (1975) Solar hydrogen energy system and hydrogen production. In *Heliotechnique and Development*, Development Analysis Associates, Cambridge, Massachusetts
17. Ohta, T., (ed.), (1978) *Solar-Hydrogen Energy System*, Pergamon Press
18. Winter, C.-J., and J. Nitsch, (1988) *Hydrogen as an Energy Carrier*, Springer-Verlag, Berlin, Heidelberg
19. Steinberg, M., and H. C. Cheng, (1988) Modern and Prospective Technologies for Hydrogen Production from Fossil Fuels, in T. N. Veziroglu and A. N. Protsenko (eds.) *Hydrogen Energy Progress VII*, Vol. 2, pp. 699-740, Pergamon Press, Oxford
20. National Hydrogen Association, (1991) THe Hydrogen Technology Assessment, Phase I, A Report for the NASA, Washington, D.C.

21. Bonner, M., et al., (1984) Status of Advanced Electrolytic Hydrogen Production in the United States and Abroad, *Int. J. Hydrogen Energy*, Vol. 9, No. 4, pp. 269-275
22. Dutta, S., (1990) Technology Assessment of Advanced Electrolytic Hydrogen Production, *Int. J. Hydrogen Energy*, Vol. 15, No. 6, pp. 379-386
23. Wendt, H., (1988) Water Splitting Methods, in C.-J. Winter and J. Nitsch, (eds.) *Hydrogen as an Energy Carrier*, Springer- Verlag, Berlin Heidelberg, pp. 166-238
24. Liepa, M.A., and A. Borhan, (1986) High-Temperature Steam Electrolysis: Technical and Economic Evaluation of Alternative Process Designs, *Int. J. Hydrogen Energy*, Vol. 11, No. 7, pp. 435-442
25. Plass, Jr., H.J., H. Miller and F. Barbir, (1989) Hydrogen Systems Application Analysis, Task A, in T.N. Veziroglu (P.I.) Solar Hydrogen Energy System, Report to DoE/SERI #XL-9-19168-1, Clean Energy Research Institute, Coral Gables, FL
26. Baykara, S.Z., and E. Bilgen, (1989) An Overall Assessment of Hydrogen Production by Solar Water Thermolysis, *Int. J. Hydrogen Energy*, Vol. 14, No. 12, pp. 881-889
27. Baykara, S.Z., and E. Bilgen, (1988) Solar Hydrogen by Hybrid Process of Water Thermolysis and Electrolysis, in T. N. Veziroglu and A. N. Protsenko (eds.) *Hydrogen Energy Progress VII*, Vol. 2, pp. 809-818
28. Engels, H., et al., (1987) Thermochemical Hydrogen Production, *Int. J. Hydrogen Energy*, Vol. 12, No. 5, pp. 291-295
29. Yalcin, S.,(1989) A Review of Nuclear Hydrogen Production, *Int. J. Hydrogen Energy*, Vol. 14, No. 8, pp. 551-561
30. Plass, Jr., H.J., F. Barbir and T.N. Veziroglu, (1991) Hydrogen Systems Application Analysis, Proc. DoE/SERI Hydrogen Program Review Meeting, Washington, D.C.
31. Bull, S.R., (1988) Hydrogen Production by Photoprocesses, Proc. Int. Renewable Energy Conf., Honolulu, Hawaii, pp. 413-426
32. Hug, W., et al., (1990) High Efficient Advanced Alkaline Electrolyzer for Solar Operation, in T. N. Veziroglu and P. K. Takahashi (eds.), *Hydrogen Energy Progress VIII*, Vol. 2, pp. 681-690, Pergamon Press, Oxford

ECONOMIC COMPARISON OF HYDROGEN AND FOSSIL FUEL SYSTEMS

T.N. Veziroğlu and F. Barbir
Clean Energy Research Institute
University of Miami
Coral Gables, FL 33124
U.S.A.

ABSTRACT. The fuels most considered for the post petroleum and natural gas era, hydrogen (gaseous and liquid), and coal and coal derived synthetic fluid fossil fuels, have been compared by taking into account production costs, external costs and utilization efficiencies. The results show that hydrogen is a much more cost effective energy carrier than coal and synthetic fossil fuels, as well as being the environmentally most compatible fuel.

1. INTRODUCTION

The economic growth of modern industrialized society has been based on utilization of energy locked in fossil fuels. At the present time, about 80 percent of the world energy demand is met by fossil fuels - coal, petroleum and natural gas [1]. They are readily available in one form or another, although they are concentrated in certain regions of the world. Humankind has learned to exploit them relatively efficiently to produce the energy services it needs. They are storable and portable, which makes them excellent fuel for transportation.

A simplified version of today's energy system is shown in Fig. 1. Fossil fuels are used for transportation (mostly petroleum products), for heat generation in residential, commercial and industrial sectors, and for electric power generation. For transportation, mostly petroleum products are used (gasoline, diesel fuel, jet fuel, etc.). Heat generation includes space heating, domestic water heating, cooking, steam generation and direct heating and/or drying in various industrial processes. All three forms of fossil fuels are used for these purposes. In electric power generation, coal is used mainly for the base load generation, and natural gas and heating oil are used for peak load. Part of the electric power is produced by hydro and nuclear power.

However, consumption of fossil fuels has become a destructive force, locally because of emissions, spills and leaks, and strip mining, regionally because of pollutants dispersion, and even globally because of carbon dioxide accumulation and threatening consequences - global warming, climate changes and sea level rise.

Y. Yürüm (ed.), Clean Utilization of Coal, 295–313.

If environmental stress does not force a shift to other energy sources and other fuels, the finite supply of fossil fuels will mandate such a change eventually. Reserves of fossil fuels are finite, particularly those of oil and natural gas. Known reserves of oil and natural gas are about 8,000 EJ (1 EJ = 10^{18} J), which would be enough for the next 40 years at the current consumption rate [2]. If the exponential population growth and the demand growth are taken into account they would only last about 25 years. Even if the estimated additional undiscovered resources were added, that would satisfy energy needs for fluid fuels an additional 20 years or so. Coal reserves are much larger, known reserves are about 20,000 EJ, but estimated ultimately recoverable resources add up to 150,000 EJ [2]. These large amounts of coal could eventually be used to produce synthetic liquid fuels, allowing the society to continue employing the present energy system, but also to continue with detrimental environmental effects, and at higher rates, which would ultimately lead to the environmental calamity with unforeseeable consequences.

Another alternative is to use renewable energy sources, mainly solar energy in its direct and indirect forms. These sources are not as convenient as fossil fuels. They are only intermittently available, they are not storable or transportable by themselves, and they cannot be used as a fuel for transportation. These shortcomings of the renewable energy sources point to the need for an energy carrier or an intermediary energy system to form the link between the new primary energy sources and the energy consuming sectors.

Hydrogen - with its vast and ready availability from water, its nearly universal utility, and its inherently benign environmental characteristics - has all the properties that might be required for such an energy system. In the hydrogen energy system, it is envisaged that hydrogen will be produced from water using any and all primary energy sources, and will be used in every application where fossil fuels are used today [3]. In this system hydrogen is not a primary source of energy. It is an intermediary or secondary form of energy or an energy carrier. Hydrogen particularly complements the renewable energy sources, and presents them to the consumer in a convenient form at the desired location and time. It is relatively inexpensive to produce; it can be converted to other forms of energy more efficiently than other fuels; and it is the cleanest fuel because its combustion product is just water vapor. The hydrogen energy system is a permanent energy system, because it is independent of the primary energy sources, it is inexhaustible and recyclable, and is versatile. So, even if the primary energy sources and the consumption patterns change due to technology development, due to changes in lifestyle and/or due to other reasons, the hydrogen energy system as an intermediary link remains basically intact.

2. TECHNO-ECONOMIC ANALYSIS OF FUELS

Hydrogen energy technology is already technically feasible, but the economics of hydrogen produced from renewable energy sources remains practically the major obstacle for its implementation. Comparing strictly the prices of the fuels, hydrogen produced using clean and renewable energy sources is more expensive than existing (fossil) gaseous and liquid fuels.

The comparison between competing fuels should be based on the effective costs of the services these fuels provide. The effective costs include the utilization efficiency, the cost of the fuel, and the costs associated with fuel consumption but which are not included in its price (so called external costs). External costs should include the costs of the physical damage done to humans and to the environment due to harmful emissions, oil spills and leaks, coal strip mining, as well as governments' expenditures for pollution abatement and expenditures for military protection of oil supplies, costs of international conflicts and other economic effects (such as employment, balance of trade, depletion of non-renewable resources, etc.).

In economic considerations, it is also important to compare the future costs of hydrogen (which will be considerably lower than they are today because of assumed market and technology development) with the future costs, both internal and external, of fossil fuels (which will unavoidably be higher than today's prices due to depletion, international conflicts and environmental impact).

2.1. Fossil Fuel Prices

Table I shows prices of fossil fuels today and projected values for the year 2000 in the United States [4]. Prices are given for different fuels and for different end-use sectors. It should be mentioned here that the prices given in Table I are the end-use prices, which means that they include the production costs, as well as the costs of transportation and distribution. As expected, the projected prices of the fossil fuels for the year 2000 are much higher than today, particularly those for petroleum products.

2.2. Production Costs of Synthetic Fuels

The production costs of interest here are for the large-scale production of the synthetic fuels, including hydrogen - also a synthetic fuel. Hydrogen can be produced by several methods, using various primary energy sources. Today hydrogen is mainly produced by methane steam reforming and coal gasification. As the post-fluid fossil fuel era is under consideration, the main fossil fuel resource of interest would, of course, be coal. Coal could also be used to produce synthetic natural gas, synthetic gasoline and synthetic jet fuel. Processes for converting coal into liquid and gaseous fuels already exist, but are not used on a large scale because synthetic fuels would be more expensive than available and relatively cheap oil and natural gas.

Hydrogen can also be produced by water splitting employing solar energy in its direct and indirect forms. The most widely used method of water splitting is electrolysis, which is a mature and very efficient technology. The cost of electrolytically produced hydrogen is directly proportional to the cost of electricity. Therefore, cheap electricity is required to produce hydrogen at competitive prices. One of the possible sources of cheap electricity is hydro-power. Electricity generated by solar energy, at the present time is not competitive with coal, hydro or nuclear electricity, except in the case of wind generated electricity, which is already competitive at some locations. However, expected development, particularly of the thin film photovoltaics, could in the near future provide electricity at $0.20-$0.35/kWh, which will make possible to generate hydrogen at large scale and at competitive prices [5].

Table I Present and Projected Prices of Fossil Fuels (in 1990 US$/GJ)

Energy Sector/Fuel	1990	2000
Residential Sector		
Natural Gas	5.61	7.04
Fuel Oil	6.65	9.14
Industrial Sector		
Natural Gas	3.26	4.32
Petroleum Fuels	4.45	7.83
Coal	1.81	2.20
Electric Utilities		
Natural Gas	2.63	3.83
Petroleum Fuels	3.11	5.66
Coal	1.47	1.86
Transportation		
Gasoline	8.98	11.23
Jet Fuel	5.93	7.42

Table II presents the averages of large scale production costs taken from recent literature [5-9]. In the case of hydrogen, the costs are classified by the primary energy source used in the production, namely coal, hydropower and solar energy. All the costs have been brought up to 1990 US$ by taking into consideration the applicable inflation.

Although gaseous hydrogen can be used in most of the applications where gaseous or liquid fossil fuels are currently being used, there are some applications where liquid hydrogen must be used, e.g., in rocket engines for space travel and in jet engines for air transportation. Consequently, the prices of liquid hydrogen have also been considered. Hydrogen liquefaction requires some energy (about 35% of the hydrogen's heating value), and it is associated with some costs. Preliminary studies show that the magnetic liquefaction process, which is currently being developed, will need less capital investment and less maintenance than conventional liquefaction systems [10,11]. It then becomes reasonable to assume a 25% add-on for the liquid hydrogen production costs.

From Tables I and II it is clear that synthetic fuels will be more expensive than fossil fuels. Among the synthetic fuels, hydrogen from renewable sources (hydro and solar) is more expensive fuel, as far as the production costs are concerned. It should be mentioned here that the costs of transportation, storage and distribution of synthetic fuels must be added in order to compare costs in Tables I and II. For synthetic fossil fuels these costs can be assumed to be $2.60/GJ and for hydrogen fuels $3.50/GJ [5].

Table II Costs of Synthetic Fuels - Estimates for 2000 (in 1990 US$/GJ)

Fuel	Production Cost	Storage & Distribution	End-Use Cost
Gaseous fuel			
Coal-Hydrogen	8.58	3.50	12.08
Hydro-Hydrogen	11.51	3.50	15.01
Solar-Hydrogen	15.44	3.50	18.94
Synthetic N. Gas	8.85	2.60	11.45
Liquid Fuels			
Coal-Hydrogen	10.73	3.50	14.23
Hydro-Hydrogen	14.39	3.50	17.89
Solar-Hydrogen	19.30	3.50	22.80
Synthetic Gasoline	17.25	2.60	19.85
Synthetic Jet Fuel	13.05	2.60	15.65

2.3. External Costs of Fossil and Synthetic Fuels

Throughout the process of fossil fuel consumption (extraction, transportation, processing, and particularly their end use - combustion) there are harmful impacts on the environment, which cause direct and indirect negative effects on the economy.

Excavation of coal devastates the land, which has to be reclaimed, and is out of use for several years. During the extraction, transportation and storage of oil and gas, spills and leakages occur, which cause water and air pollution. Refining processes also have an environmental impact.

Most of the fossil fuel environmental impact occurs during the end use. The end use for all fuels is combustion, irrespective of the final purpose (i.e., heating, electricity production or motive power for transportation). The main constituents of fossil fuels are carbon and hydrogen, but also some other ingredients, which are originally in the fuel (e.g., sulfur), or are added during refining (e.g., lead, alcohols). Combustion of the fossil fuels produces various gases (CO_x, SO_x, NO_x, CH), soot and ash, droplets of tar, and other organic compounds, which are all released into the atmosphere and produce the air pollution. Air pollution causes damage to human health, animals, crops and structures, and reduces visibility.

Once in the atmosphere, triggered by sunlight or by mixing with water and other atmospheric compounds, the primary pollutants may undergo chemical reactions, change their form and become secondary pollutants, like ozone, aerosols, peroxyacyl nitrates, various acids, etc.

Precipitation of sulfur and nitrogen oxides, which have dissolved in clouds and in rain droplets to form sulfuric and nitric acids is called acid rain; but also acid dew, acid fog, and acid snow have been recorded. Acid deposition (wet or dry) causes soil and water acidification, resulting in damages to the aquatic and terrestrial ecosystems, affecting humans, animals, vegetation and structures.

The remaining products of combustion in the atmosphere, mainly carbon dioxide, together with other so called greenhouse gases (methane, nitrogen oxides and chlorofluorcarbons), result in thermal changes by absorbing the infrared energy the Earth radiates back into the atmosphere and reradiating it back to the Earth partially, causing global temperature increases. The effects of the temperature increases are melting of the ice caps, sea level rise and climate changes, which include heat waves, droughts, floods, stronger storms, more and bigger wildfires, etc.

The cost of the above described negative effects are not included in the market price of fossil fuels, and can be considered as external costs. These costs are paid by the society, and/or eventually will be paid by the society, since in the long term any disturbed ecosystem will affect the human society, its environment and its economy.

Environmental damage is not the only external cost of fossil fuels. In a complete economic study, one must take into account the military costs for protecting oil and gas supplies, full costs of international conflicts, and other external economic effects (such as employment, balance of trade, depletion of non-renewable resources, etc.).

Taking into account various reported damages, Table III has been prepared [12]. It summarizes the damages attributed to each of fossil fuels. Environmental damage due to coal utilization is the highest $9.82/GJ, then comes petroleum with $8.47/GJ, and natural gas is the cleanest (or the least harmful) amongst fossil fuels with environmental damage estimated at $5.60/GJ.

Since some of the synthetic fuels will be produced from coal, more than 1 GJ of coal will be used for each GJ of a synthetic fuel manufactured and consumed. Consequently, the environmental damage caused by 1 GJ of a synthetic fuel manufactured from coal is greater than the estimated damage caused by coal consumption. This can be expressed as follows:

$$E_{syn} = E_{coal} \cdot p_s \tag{1}$$

where E_{syn} is the environmental damage due to 1 GJ of a synthetic fuel (including hydrogen) produced from coal, E_{coal} the environmental damage due to coal and p_s the synthetic fuel pollution factor. Because of the dearth of data, total CO_2 generated per unit energy has been taken as a measure of the pollution factor. Table IV presents the pollution factors for various synthetic fuels compared with coal.

Table III Environmental Damage Caused by Fossil Fuel Combustion

Type of Damage	Environmental Damage (1990 $/GJ)		
	Coal (E_{coal})	Petroleum (E_{pet})	N. Gas (E_{gas})
Effect on humans	3.48	2.83	2.09
Effect on animals	0.51	0.42	0.30
Effect on plants and forests	1.35	1.09	0.81
Effect on aquatic ecosystems	0.18	1.05	0.11
Effect on man-made structures	1.12	0.90	0.67
Other air pollution costs	0.98	0.79	0.59
Effect of strip mining	0.49	-	-
Effect of climatic changes	1.39	1.13	0.84
Effect of sea level rise	0.32	0.26	0.19
Totals	9.82	8.47	5.60

Table IV CO_2 Emissions and Pollution Factor

Fuel	CO_2 emissions (kg/GJ)	Pollution factor
Coal	85.5	1
Coal GH_2	116.3	1.36
Hydro GH_2	0	0
Solar GH_2	0	0
SNG	116.3	1.36
Coal LH_2	145.4	1.70
Hydro LH_2	0	0
Solar LH_2	0	0
Syn-Gasoline	131.6	1.54
Syn-Jet	131.6	1.54

2.4. Utilization Efficiency

In comparing the fuels, it is important to take into account the utilization efficiencies at the user end. For utilization by the user, fuels are converted to various energy forms, such as mechanical, electrical and thermal. Studies show that in almost every instance of utilization, hydrogen can be converted to the desired energy form more efficiently than the fossil fuels (or the synthetic fuels) [13]. In other words, conversion to hydrogen would result in energy conservation due to its higher utilization efficiencies.

Thermal energy generation: In many industrial, commercial and residential applications, fuels are converted to thermal energy (space heating, water heating, cooking, steam generation and direct heat in industrial processes). Hydrogen combustion in industrial applications is state-of-the-art. Hydrogen is burned in large amounts in the chemical industry to produce heat, especially when it is not economical to purify it [14]. However, some design changes in the existing burners are required to match hydrogen burning properties, which are somewhat different than those of methane and other gaseous fuels, but the efficiency is about the same as for fossil fuel combustion.

However, catalytic combustion of hydrogen (which is a unique property of hydrogen to combine with oxygen in the presence of a catalyst such as platinum or palladium in an exothermic reaction without a flame), is more efficient than flame combustion. In some applications, e.g., for space heating catalytic combustion can be up to 99% efficient, since all the heat of the catalytic reaction remains inside the heated space; there are no exhaust gases [15]. Catalytic burners can also be used in kitchen ranges [16].

Hydrogen burns with oxygen producing pure steam at quite high temperatures. Then an appropriate amount of water is added in order to bring the steam temperature down to the required temperature. Such a device, called the H2/O2 steam generator, has been developed and tested at German Aerospace Research Establishment (DLR) [17]. This steam generator is very simple, compact and extremely efficient (up to 99%). Conventional steam generators are usually about 75% to 80% efficient.

Electrical power generation: Hydrogen can be converted to electricity in fuel cells with much greater efficiencies than those possible in thermal power plants using fossil fuels. While conversion efficiencies for the latter are in the range of 35%-38%, practical efficiencies in hydrogen fuel cells are in the range 50%-70% [18]. In the advanced hydrogen fuel cells which are now being developed, even higher efficiencies (up to 80%) can be expected. This is another important, unique property of hydrogen, which can also increase the conversion efficencies in transportation vehicles, and consequently reduce transportation fuel consumption. Hydrogen fuel cell/electric motor combinations would yield more than two times greater conversion efficiencies than an internal combustion engine running on gasoline or diesel engine [19].

<u>Transportation:</u> Hydrogen powered vehicles with internal combustion engines have been proven to be more efficient than gasoline vehicles. Hydrogen can be considered more thermally efficient than gasoline, primarily because it burns better in excess air and permits the use of a higher compression ratio. Data from engine tests indicate that hydrogen combustion is 15%-50% more thermally efficient [20]. The overall fuel efficiency of a hydrogen vehicle, which takes into account the thermal efficiency as well as the weight of the vehicle is also better when compared with a gasoline vehicle (in average, hydrogen vehicles are 22% more efficient [20]). As hydrogen can burn in lean fuel/air mixtures as well as in rich mixtures, it can cause large improvements in fuel-use efficiencies in the stop-start type city driving.

Investigations show that for a given number of passengers and a given payload, a subsonic jet passenger airplane would use 19% less energy, if it were to use liquid hydrogen instead of fossil-based jet fuel [21]. In the case of a supersonic jet plane, the efficiency advantage of hydrogen is even greater [21]; it is 38% better than jet fuel.

The above discussed advantageus hydrogen utilization efficiencies are summarized and shown in Table V.

Table V Utilization Efficiency Advantage of Hydrogen

Application	Utilization Efficiency Ratio (η_F/η_H)
Thermal energy - flame combustion	1.00
Thermal energy - catalytic combustion	0.80
Thermal energy - steam generation	0.80
Electric power - fuel cells	0.54
Surface transportation - I.C. engines	0.82
Surface transportation - fuel cells/EM	0.40
Subsonic Jet transportation	0.84
Supersonic Jet transportation	0.72

2.5. Effective costs

As already mentioned, the comparison between competing fuels should be based on the effective costs of the services these fuels provide. The effective costs include the utilization efficiency and both the internal and external costs of the fuel. Internal costs are the costs included in the end-user price of fuel (production, storage, transportation and distribution costs). External costs, as discussed above, are the costs associated with fuel consumption but which are not included in its price (e.g., environmental damage).

Effective costs of energy services can be expressed as $/GJ$_{th}$ for heat generation, $/kWh$_{el}$ for electric power generation and $/passenger-km or $/freight-km for transportation. In order to bring these costs to a common basis and to be able to add them together, they will be expressed as $/GJ of fossil fuel equivalent, or in other words $/GJ of fossil fuel required to provide the same service.

Effective cost of fossil fuels by definition is:

$$C_{eff} = (C_{in} + C_{ex}) \tag{2}$$

where C_{in} is internal cost (from Table I) and C_{ex} is external cost (from Table III).

Effective cost of synthetic fuels (including hydrogen) must take into account the utilization efficiency, and is given as follows:

$$C_{eff} = (C_{in} + C_{ex}) \frac{\eta_F}{\eta_S} \tag{3}$$

Internal costs (C_{in}) for synthetic fuels are given in Table II, and external costs ($C_{ex} = E_{syn}$) can be calculated by using equation (1). For coal derived synthetic fossil fuels, the utilization efficiency factor is assumed to be $\eta_F/\eta_S = 1$, i.e., the utilization efficiency of synthetic fossil fuels is equal to the utilization efficiency of fossil fuels. For hydrogen, this factor is $\eta_F/\eta_S = \eta_F/\eta_H$ and it can be found in Table V.

Using the above equations with the data developed earlier, Tables VI, VII, VIII and IX have been prepared to give the effective costs for various types of applications for thermal energy generation, electric power generation and transportation respectively.

Table VI shows that the catalytic combustion of gaseous hydrogen would produce the most economical thermal energy with the costs comparable to those of using fossil fuels as they are being used today. Heat produced from synthetic fossil fuels would be the most expensive. When electric power generation is considered (see Table VII), hydrogen powered fuel cells would be the most cost effective method of generating electricity because of a high conversion efficiency. The fuel cells appear to be the most cost effective option for surface transportation also. Air transportation requires the most expensive fuels. Liquid hydrogen from renewable sources (hydro and solar) would be much cheaper than synthetic jet fuel when effective cost is considered. It should be noticed that hydrogen from coal has the lowest production cost amongst hydrogen fuels, but its effective cost would be the highest. This is of course due to the addition of the environmental damage.

Table VI Effective Cost of Thermal Energy Generation (in 1990 US$/GJ)

Application	Fuel	Effective Cost
Flame Combustion	Natural Gas	11.82
	Petroleum Fuels	16.96
	Coal	12.02
	SNG	24.81
	Syn-Fuel Oil	30.77
	Coal-GH_2	25.44
	Hydro-GH_2	15.01
	Solar-GH_2	18.94
Catalytic Combustion	Coal-GH_2	20.35
	Hydro-GH_2	12.00
	Solar-GH_2	15.15

Table VII Effective Cost of Electric Power Generation (in 1990 US$/GJ)

Application	Fuel	Effective Cost
Thermal Plant	Natural Gas	9.43
	Petroleum Fuels	14.13
	Coal	11.68
	SNG	24.81
	Syn-Fuel Oil	30.77
Thermal Plant with H_2-O_2 Steam Generator	Coal-GH_2	20.35
	Hydro-GH_2	12.00
	Solar-GH_2	15.15
Fuel Cells	Coal-GH_2	13.74
	Hydro-GH_2	8.10
	Solar-GH_2	10.23

Table VIII Effective Cost of Surface Transportation (in 1990 US$/GJ)

Application	Fuel	Effective Cost
I.C. Engines	Gasoline	19.70
	SNG	24.81
	Syn-Gasoline	34.97
	Coal-GH_2	20.86
	Hydro-GH_2	12.31
	Solar-GH_2	15.53
	Coal-LH_2	25.36
	Hydro-LH_2	14.67
	Solar-LH_2	18.70
Fuel Cells	Coal-GH_2	10.17
	Hydro-GH_2	6.00
	Solar-GH_2	7.58

Table IX Effective Cost of Air Transportation (in 1990 US$/GJ)

Application	Fuel	Effective Cost
Subsonic Jets	Jet Fuel	15.89
	Syn-Jet Fuel	30.77
	Coal-LH_2	25.98
	Hydro-LH_2	15.03
	Solar-LH_2	19.15
Supersonic Jets	Jet Fuel	15.89
	Syn-Jet Fuel	30.77
	Coal-LH_2	22.27
	Hydro-LH_2	12.88
	Solar-LH_2	16.42

3. COMPARISON OF THREE SCENARIOS FOR THE FUTURE ENERGY SYSTEM

The following analysis will compare the effective costs of three possible scenarios for the future energy system: (1) fossil fuel energy system as it is today, (2) coal and coal based synthetic fuels, and (3) hydrogen based on renewable energy sources.

<u>(1) Fossil fuel system:</u> In this scenario, it is assumed that the present fossil fuel system will be continued as it is today. Therefore, it is assumed that 40% of primary energy (in fossil fuel equivalent units) will be used for thermal energy generation, 30% for electric power generation and 30% for transportation (2/3 for surface transportation and 1/3 for air transportation). Energy supplied by hydro and nuclear power plants (mostly in the form of electric power) and by other non-fossil fuel sources has not been taken into account, since it is assumed that it will be the same for all three considered systems. Actually, it is reasonable to expect that in the future even more energy will be supplied by these sources.

In the fossil fuel system (see Fig. 1), it is assumed that one half of the thermal energy will be supplied by natural gas, and the rest by petroleum fuels (fuel oil and residual oil) and coal. Coal is assumed to be the main energy source for electricity generation, gasoline for surface transportation and jet-fuel for air transportation. This is of course a simplified version of the fossil fuel energy system, but it is close enough to the present patterns of energy consumption, and can be used as the basis for further analyses and comparisons.

The prices of fossil fuels are expected to rise, as already shown in Table I. The projected costs for the year 2000 are used in calculations of the effective costs, which are shown in Table X.

Table X Effective Cost of the Fossil Fuel System (inn 1990 US$/GJ)

Application	Energy Consumption Fraction	Fuel	Effective Cost	Fraction X Cost
Thermal Energy				
	0.20	Natural Gas	11.82*	2.36
	0.10	Petroleum Fuels	16.96*	1.70
	0.10	Coal	12.02	1.20
Electric Power	0.30	Coal	11.68	3.50
Surface Transport.	0.20	Gasoline	19.70	3.94
Air Transportation	0.10	Jet Fuel	15.89	1.59
Totals	**1.00**	**Overall Effective Cost**		**$14.29/GJ**

* Average for residential and industrial sector.

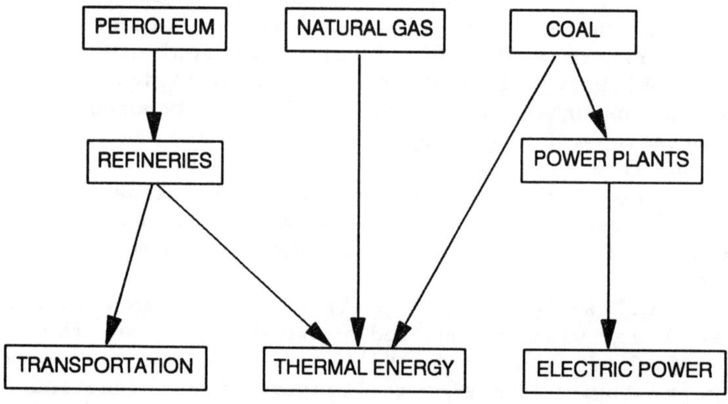

Fig. 1 Fossil Fuel Energy System

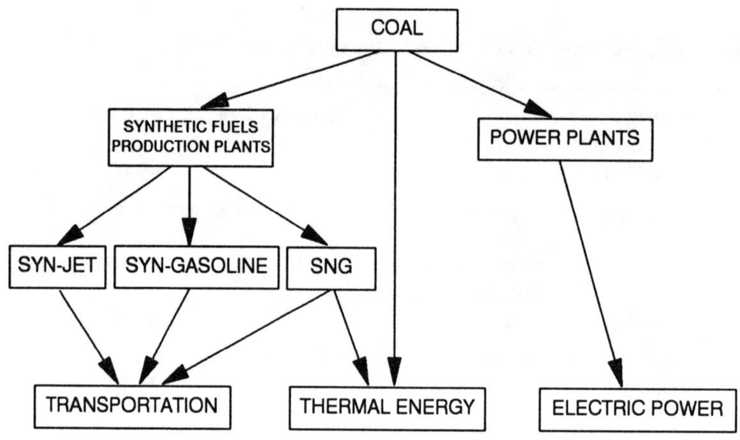

Fig. 2 Coal/Synthetic Fuel Energy System

(2) Coal/Synthetic Fossil Fuel System: In this case, it is assumed that the present fossil fuel system will be continued by the substitution with synthetic fuels derived from coal wherever convenient and/or necessary. Patterns of energy consumption are also assumed to be unchanged. Coal will be used extensively for thermal power generation and for electric power generation, because it is much cheaper than synthetic fuels. However, some end-uses require fluid fuels; therefore it has been assumed that synthetic natural gas (SNG) will be used for thermal energy generation (primarily in the residential sector) and also as a fuel for surface transportation, where it will share the market with synthetic gasoline. Synthetic jet fuel will be used in air transportation.

The energy system based on coal and synthetic fossil fuels is shown in Fig. 2. Table XI presents the effective costs.

Table XI Effective Cost of the Coal/Synthetic Fuel System (in 1990 US$/GJ)

Application	**Energy Consumption Fraction**	**Fuel**	**Effective Cost**	**Fraction X Cost**
Thermal Energy				
	0.30	Coal	12.02	3.61
	0.10	SNG	24.81	2.48
Electric Power	0.30	Coal	11.68	3.50
Surface Transport.				
	0.10	SNG	24.81	2.48
	0.10	Syn-Gasoline	34.97	3.50
Air Transportation	0.10	Syn-Jet	30.77	3.08
Totals	**1.00**	**Overall Effective Cost**		**$18.65/GJ**

(3) Solar Hydrogen Energy System: In this case it is assumed that the conversion to the hydrogen energy will take place, and one-third of hydrogen needed will be produced from hydropower and two-thirds by direct and indirect (other than hydropower) solar energy forms. The same percentage of energy demand sectors as the above scenarios will be assumed. It will further be assumed that one half of the thermal energy will be achieved by flame combustion, one quarter by steam generation with hydrogen/oxygen steam generators and the last quarter by catalytic combustion; electric power will be generated by fuel cells; one-half of the surface transportation will use gaseous hydrogen burning internal

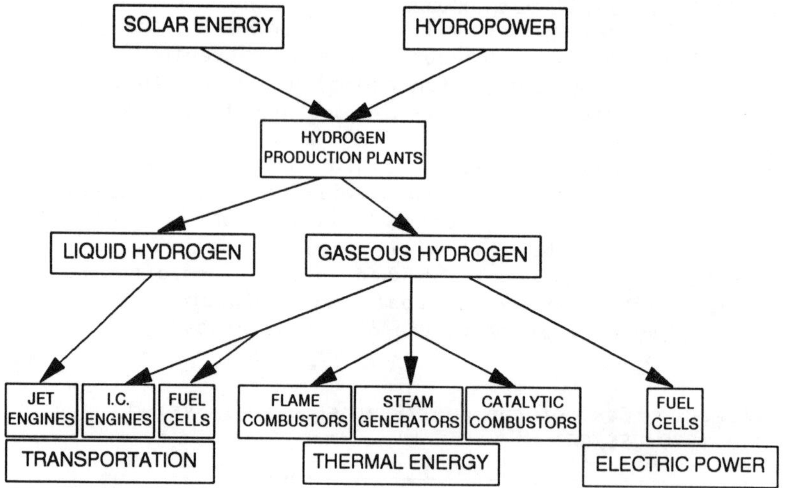

Fig. 3 Solar Hydrogen Energy System

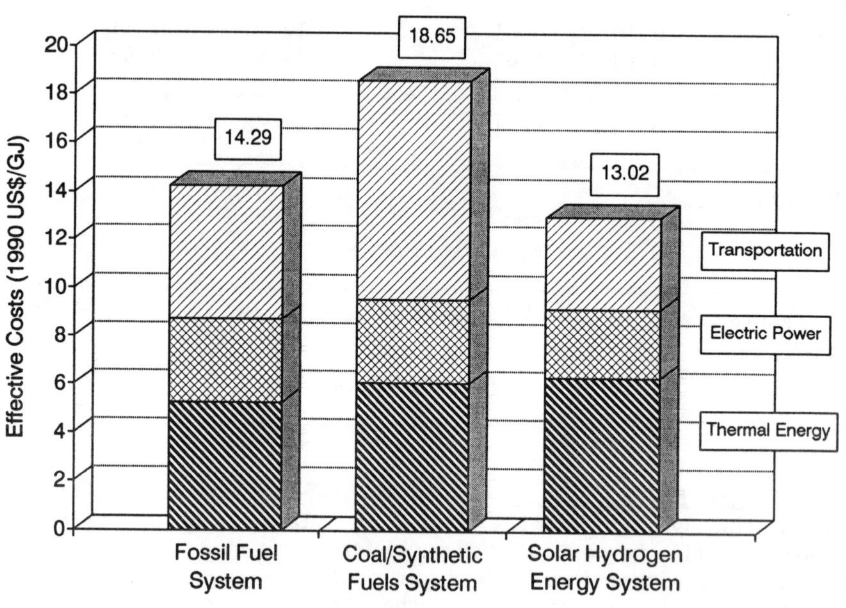

Fig. 4. Effective Costs of Analyzed Energy Systems

combustion engines and the other half will use fuel cells. In air transportation, both subsonic and supersonic, liquid hydrogen will be used. Fig. 3 shows a proposed solar hydrogen energy system. Its effective costs have been calculated and are given in Table XII.

Table XII Effective Cost of the Solar Hydrogen Energy System (in 1990 US$/GJ)

Application	**Energy Consumption Fraction**	**Fuel***	**Effective Cost**	**Fraction X Cost**
Thermal Energy				
Flame Combustion	0.20	GH_2	17.63	3.53
Steam Generation	0.10	GH_2	14.10	1.41
Catalytic Combstn.	0.10	GH_2	14.10	1.41
Electric Power				
Fuel Cells	0.30	GH_2	9.52	2.86
Surface Transport.				
I.C. Engines	0.10	GH_2	14.46	1.45
Fuel Cells	0.10	GH_2	7.05	0.71
Air Transportation				
Subsonic	0.05	LH_2	17.78	0.89
Supersonic	0.05	LH_2	15.24	0.76
Totals	1.00	**Overall Effective Cost**		$13.02/GJ

* It has been assumed that 1/3 of hydrogen will be produced from hydropower and 2/3 from solar.

Comparing the results from Tables X to XII, it becomes clear that hydrogen energy system has lower effective cost than today's fossil fuel system (with the costs projected for the year 2000 for both systems). Energy system based on coal and coal derived synthetic fuels would be the most costly alternative (see Figure 4). Off course, conversion to a new energy system is not possible in such a short time period. These calculations show that hydrogen energy system can become cost competitive at the beginning of the next century, but the transition should start as soon as possible, because only with adequate technology and market developments the projected costs of hydrogen can be achieved.

4. CONCLUSIONS

As a result of the foregoing study, the following conclusions are reached:

(a) The solar hydrogen energy system is environmentally more compatible than the fossil fuel system; it is actually the only solution for carbon dioxide induced greenhouse effect and its threatening consequences.

(b) The utilization efficiencies of hydrogen are greater than those of fossil and synthetic fossil fuels; therefore, in the hydrogen energy system, less energy will be required to perform the same services.

(c) The solar hydrogen energy system is the most cost effective system if effective costs (the costs which society pays for the energy services) are taken into account. It is not only more cost effective than the synthetic fossil fuel system -- it is even more cost effective than the present fossil fuel system.

The transition to the solar hydrogen energy system can help saving our economy and our planet.

5. REFERENCES

1. Davis, G.R., (1990) Energy for Planet Earth, *Scientific American*, Vol. 263, No.3, pp. 55-62
2. Fulkerson, W., R.J. Judkins and M.K. Sanghvi, (1990) Energy from Fossil Fuels, *Scientific American*, Vol. 263, No.3, pp. 129-135
3. Bockris, J. O'M., and T.N. Veziroglu, (1985) A Solar-Hydrogen Energy System for Environmental Compatibility, *Environmental Conservation*, Vol. 12, No. 2, pp. 105-118
4. *Annual Energy Outlook, Long Term Projections*, (1990), Energy Information Administration, Washington, D.C.
5. Ogden, J.M., and R.H. Williams, (1989) *Solar Hydrogen: Moving Beyond Fossil Fuels*, World Resources Institute, Washington, D.C.
6. Moore, R.B., and D. Dahmias, (1991), Gaseous Hydrogen Markets and Technologies, *Proceedings: Transition Strategies to Hydrogen as an Energy Carrier - First Annual Meeting of the U.S. National Hydrogen Association*, prepared by Technology Transition Corporation, Washington, D.C., pp. 11.1-17
7. Williams, R.H., (1985) Potential Roles for Bioenergy in an Energy-Efficient World, *Ambio*, Vol. 14, No. 405, pp. 201-209
8. Steinberg, M., and H.C. Cheng, (1988) Modern and Prospective Technologies for Hydrogen Production From Fossil Fuels, in T.N. Veziroglu and A.N. Protsenko (eds.), *Hydrogen Energy Progress VII*, Vol. 2, pp. 699-739
9. Bockris, J.O'M., and J.C. Was, (1988) About the Real Economics of Massive Hydrogen Production at 2010 A.D., in T.N. Veziroglu and A.N. Protsenko (eds.), *Hydrogen Energy Progress VII*, Vol. 2, pp. 101-151

10. Barklay, J., (1991) Magnetic Liquefaction of Hydrogen, presented at the DoE/SERI Hydrogen Program Review Meeting, Washington, D.C.
11. Sherif, S.A., et al., (1991) Analysis and Optimization of Hydrogen Liquefaction and Storage Systems, Proc. DoE/SERI Hydrogen Program Review Meeting, Washington, D.C.
12. Barbir, F., and T.N. Veziroglu, (1991) Environmental Damage Caused by Fossil Fuel Consumption, *Int. J. Energy·Environment·Economics*, Vol. 1, No. 4.
13. Chen, D.Z., I. Gurkan, T.N. Veziroglu and J.W. Sheffield, (1982) Effective Cost of Fuels: Comparison of Hydrogen with Fossil Fuels, in T.N. Veziroglu, W.D. Van Vorst and J.H, Kelley (eds.), *Hydrogen Energy Progress IV*, Vol. 4, pp. 1523-1537, Pergamon Press, Oxford
14. Peschka, W., (1988) Technologies for the Energetic Use of Hydrogen, in C.-J. Winter and J. Nitsch (eds.), *Hydrogen as an Energy Carrier*, pp. 30-55, Springer-Verlag, Berlin
15. Haruta, M., Y. Souma and H. Sano, (1982) Catalytic Combustion of Hydrogen, *Int. J. Hydrogen Energy*, Vol. 7, No. 9, pp. 729-740
16. Ledjeff, K., (1990) New hydrogen appliances, in T. N. Veziroglu and P. K. Takahashi (eds.), *Hydrogen Energy Progress VIII*, Vol. 3, pp. 1429-1444, Pergamon Press, Oxford
17. Sternfeld, H.J., and P. Heinrich, (1989) A demonstration plant for the hydrogen/ oxygen spinning reserve, *Int. J. Hydrogen Energy*, Vol. 14, No. 10, pp. 703-716
18. Kordesch, K., and J.C.T. Oliveira, (1988) Fuel Cells: The Present State of the Technology and Future Applications, With Special Consideration of the Alkaline Hydrogen/Oxygen (Air) Systems, *Int. J. Hydrogen Energy*, Vol. 13, No. 7, pp. 703-716
19. Billings, R.E., and M. Saunders, (1990) Automotive Hydrogen Fuel Cell, presented at the 8th World Hydrogen Energy Conference, Honolulu, Hawaii
20. DeLuchi, M., (1989) Hydrogen Vehicles: An Evaluation of Fuel Storage, Performance, Safety, Environmental Impacts, and Cost, *Int. J. Hydrogen Energy*, Vol. 14, No. 2, pp. 81-130
21. Brewer, G.D., (1982) The Prospects for Liquid Hydrogen-Fueled Aircraft, *Int. J. Hydrogen Energy*, Vol. 7, No. 1, pp. 21-41

SUBJECT INDEX

W

X

Z